1st International Conference on Maritime Infrastructure and Development (ICMID 2018)

IOP Conference Series: Materials Science and Engineering
Volume 797

Kendari, Indonesia
19 - 21 November 2018

ISBN: 978-1-7138-2672-9
ISSN: 1757-8981

Printed from e-media with permission by:

Curran Associates, Inc.
57 Morehouse Lane
Red Hook, NY 12571

proceedings.com

Some format issues inherent in the e-media version may also appear in this print version.

Institute of Physics
Dirac House, Temple Back
Bristol BS1 6BE UK

Phone: 44 1 17 929 7481
Fax: 44 1 17 920 0979

techtracking@iop.org

Additional copies of this publication are available from:

Curran Associates, Inc.
57 Morehouse Lane
Red Hook, NY 12571 USA
Phone: 845-758-0400
Fax: 845-758-2633
Email: curran@proceedings.com
Web: www.proceedings.com

TABLE OF CONTENTS

IOP Conf. Series: Materials Science and Engineering **797** (2020) 011001 doi:10.1088/1757-899X/797/1/011001

IOP Conf. Series: Materials Science and Engineering **797** (2020) 011001 doi:10.1088/1757-899X/797/1/011001

Preface

Welcome Remarks,
Chair of Steering Committee

It is a great pleasure to welcome all of you to Kendari and to The 1st International Conference on Maritime Infrastructure and Development 2018 (ICMID 2018) held by Faculty of Engineering, Universitas Halu Oleo (UHO). I also conveyed welcome to all delegations from various countries and universities. I hope you all enjoy the natural beauty of Kendari as part of the Indonesian archipelago, beautiful and charming maritime continent.

Halu Oleo University (UHO) Kendari as part of the nation's community has a strong vision of maritime development. This is reflected in UHO's vision, namely in 2045 to become the 250 leading universities in the world in science and engineering for the continent of maritime, harmonization and sustainable prosperity.

To realize the vision of UHO 2045 specifically related to maritime affairs, UHO needs to develop maritime research in the form of basic research, applied research and development. UHO researchers need to collaborate with and collaborate with other researchers from both domestic institutions and universities.

The UHO Faculty of Engineering as the core of applied research and development in UHO in the field of technology needs to work together with various parties, especially with the government, business world and society in supporting the acceleration of maritime infrastructure development. The UHO Faculty of Engineering has a great responsibility to integrate and collect various expertise resources that exist both internally and externally to jointly develop maritime research. The acceleration of collaborative research between researchers and the government as policy makers and the business world and the community as users of research results is needed to support Indonesia's maritime vision. Therefore, as a Dean of Faculty of Engineering UHO, I am very happy that our Faculty can hold an international seminar to accelerate the collaboration of researchers from home and abroad in all fields related to maritime affairs.

I hope this international seminar can continue in the future to build a positive contribution to the development of the Indonesian maritime continent. Please enjoy the seminar.

Kendari, 19 November 2018
Chair of ICMID2018 Steering Committee,

Dr. Edward Ngii

IOP Conf. Series: Materials Science and Engineering **797** (2020) 011001 doi:10.1088/1757-899X/797/1/011001

Preface

Welcome Remarks,
Head of Organizing Committee

Welcome to Kendari, Indonesia, to all delegates and presenters. It is my pleasure and privilege to welcome all of you to the 1st International Conference on Maritime Infrastructure and Development 2018 (ICMID 2018) held by Faculty of Engineering, Universitas Halu Oleo (UHO).

ICMID (International Conference on Maritime Infrastructure and Development) is a biennial event to provide a forum for researchers, academicians (lecturers and students), government agencies, consultants and contractors to exchange experiences, technological advancement and innovation related to maritime infrastructure and development.

The Organizing Committee of ICMID 2018 would like to express our gratitude for the support from many parties. The support from Faculty of Engineering UHO, keynote speakers, and support from our Steering and Organizing Committees are really acknowledged.

The last but not the least, I would also like to thank the reviewers for their valuable time and advice which helped in improving the quality of the papers selected for presentation at the conference and for publication. Finally, I want to thank the authors, the members of the organizing committee, the chairpersons, sponsors, and all other conference participants for their support of ICMID 2018.

Kendari, 19 November 2018
Head of ICMID 2018 Organizing Committee,

Dr. La Ode Muh. Golok Jaya

IOP Conf. Series: Materials Science and Engineering **797** (2020) 011002 doi:10.1088/1757-899X/797/1/011002

The 1st International Conference on Maritime Infrastructure and Development 2018 (ICMID 2018) List of Committees

STEERING COMMITTE

Chair:
Dr. Edward Ngii, ST, MT.

Honorary Members:
Prof. Dr. Muhammad Zamrun F, S.Si, M.Si, M.Sc.
Prof. Kiril Tenekedjiev
Dr. Iwan Hermawan
Prof. Dr. Ir. Riri Fitri Sari, MM., M.Sc.
Riyanti Djalante, Ph.D
Abu Hasan, M.Pd.

Members:
1. Mustarum Musaruddin, ST, MIT, Ph.D.
2. Dr. Abdul Kadir, ST, MT.
3. Dr. Eng. Sudarsono, ST, M.Eng.
4. Yuspian Gunawan, ST, MT.
5. Dr. Ishak Kadir, ST, MT.
6. Dr. Eng. Minson Simatupang, ST, MT.
7. Dr. Ir. La Ode Muh. Magribi, MT
8. Dr. Adris Ade Putra, ST, MT.
9. Dr. H. M. Tahir Azikin, ST, MT
10. Dr. Eng. Lukas Kano Mangala, ST, MT.
11. Dr. Ir. Muh. Ihsan Sarita, M.Kom.
12. Romy S. Edwin Tamburaka, ST, MT, Ph.D.
13. Ir. H. Ridway Balaka, M.Eng.
14. Ir. Baso Mursidi, M. Proc. Mgmnt.
15. Dr. Ir. Burhan Said, M.Pd.

ORGANIZING COMMITTEE

Head:
Dr. La Ode M. Golok Jaya, ST, MT.

Vice Head:
Masykur Kimsan, ST, MT.

Secretary:
Santi, ST, MT.

Treasures:
Kurniati Ornam, ST, MT.

Secretariat:
1. Hasmina Tari Mokui, ST, ME, Ph.D.
2. Al Azis Bachrun, ST, M.Sc.
3. Jayanti Yusmah Sari, ST, M.Kom.
4. Mutmainnah Muchtar, ST, M.Kom.

Program:

1. Belinda Amri, ST, MT.
2. Syafrianto Amsyar, ST, MT.
3. Rizal Adi Saputra, ST, M.Kom.
4. Adha Mashur Sajiah, ST, M.Eng.

Website:
1. Muh. Nadzirin Anshari Nur, S.Kom, M.T.
2. Muh. Yamin, ST, M.Eng.

Design:
1. Mr. Hartono
2. Ms. Nurul Fajriani

REVIEWER

1. Prof. DSc PhD Kiril Tenekedjiev (Australian Maritime College, University of Tasmania)
2. Prof. Dr. El Shafie (University Technology Malaya, Malaysia)
3. Dr. Iwan Hermawan (EOS, Singapore)
4. Prof. Dr. Ir. Riri Fitri Sari, MM., M.Sc. (UI, Indonesia)
5. Riyanti Djalante, Ph.D (UNU, Japan)
6. Prof. Ketut Wikantika (ITB, Indonesia)
7. Prof. S. Hendriatiningsih (ITB, Indonesia)
8. Prof. Ma'ruf Kasim (UHO, Indonesia)
9. Mustarum Musaruddin, Ph.D (UHO, Indonesia)
10. Dr. Agung Budi Harto (ITB, Indonesia)
11. Prof. Nyoman Nuarsa (Udayana University, Indonesia)
12. Prof. Satria Bidjaksana (ITB, Indonesia)

Conference Organizing Committee:

Fakultas Teknik, Universitas Halu Oleo (UHO),
Jl. HEA. Mokodompit No. 8, Kampus Hijau UHO Bumi Tridharma Anduonohu, Kendari, Sulawesi Tenggara, Indonesia.
Phone: +62 813-2196-6683
Email: laodemgj@uho.ac.id, icmid2018@uho.ac.id
Website: eng.uho.ac.id/icmid2018

IOP Conf. Series: Materials Science and Engineering **797** (2020) 011002 doi:10.1088/1757-899X/797/1/011002

CONFERENCE PHOTOGRAPHS

FLOATING MARKET IN
MUARA ANGKE
ICMID
"TOWARDS on RESEARCH, DEVELOPMENT, and APPLICATION on MARITIME
INFRASTRUCTURE and TECHNOLOGY for INDONESIA FUTURE
as WORLD MARITIME AXIS"

ICMID
Sustainability of Sulaa Coastal Areas as a
Heritage Assets in Baubau City
Southeast Sulawesi Province

ICMID 2018
Halu oleo University
In Kendari City
ICMID
"TOWARDS on RESEARCH, DEVELOPMENT, and APPLICATION on MARITIME
INFRASTRUCTURE and TECHNOLOGY for INDONESIA FUTURE
as WORLD MARITIME AXIS"

IOP Conf. Series: Materials Science and Engineering **797** (2020) 011004 doi:10.1088/1757-899X/797/1/011004

IOP Conf. Series: Materials Science and Engineering **797** (2020) 011004 doi:10.1088/1757-899X/797/1/011004

IOP Conf. Series: Materials Science and Engineering **797** (2020) 011004 doi:10.1088/1757-899X/797/1/011004

IOP Conf. Series: Materials Science and Engineering **797** (2020) 011004 doi:10.1088/1757-899X/797/1/011004

ICMID
THE FIRST INTERNATIONAL
CONFERENCE ON MARITIME
INFRASTRUCTURE AND
DEVELOPMENT (ICMID)

ICMID

IOP Conf. Series: Materials Science and Engineering **797** (2020) 011005 doi:10.1088/1757-899X/797/1/011005

Peer review statement

All papers published in this volume of *IOP Conference Series: Materials Science and Engineering* have been peer reviewed through processes administered by the proceedings Editors. Reviews were conducted by expert referees to the professional and scientific standards expected of a proceedings journal published by IOP Publishing.

IOP Conf. Series: Materials Science and Engineering **797** (2020) 012001 doi:10.1088/1757-899X/797/1/012001

Sustainability of Sulaa Coastal Areas as Heritage Assets in Baubau City Southeast Sulawesi Province

Ishak Kadir[1*], **Sachrul Ramadan**[1], **Muhammad Zakaria Umar**[1]
[1] Department of Architecture, Faculty of Engineering, Universitas Haluoleo, 93232 Kendari, Sulawesi Tenggara, INDONESIA

[*]Corresponding author's e-mail: ishak69kadir@gmail.com

Abstract. Sulaa is one of the coastal villages in Baubau, most people work as fishermen, therefore the sea is a very important part of their lives. Sulaa became Buton's traditional weaving center in Baubau with diverse cultural traditions which were maintained as the value of local wisdom. This paper aims to show of sustainability of Sulaa coastal areas in heritage area of Sulaa. This research uses rationality qualitative method. The data collecting done by measurement, interviews, observations, documents collecting and literature study. This research showed there are innovation of sustainable settlement and ecotourism development of Sulaa settlement, includes: 1) the diversity of natural heritage assets include the Nirwana beach, Lakeba beach, Lakusa's cave, etc; 2) sustainability of Butonese traditional house concepts shown in the sustainable culture, architecture, and nature; 3) sustainability of cultural traditions of society, and; 4) creation and use of space based on the values of society.

1. Introduction

The Indonesian state has the potential for cultural heritage. Indonesian culture is diverse and rich. This potential lies in the arts, customs, language, sites, architecture and historical areas. This wealth and diversity contribute to cities in Indonesia. This potential shapes the character, uniqueness and distinctive image of each city. This potential can also form the image of the city. These cultural assets have historical value and become a heritage series. These cultural assets need to be conserved, developed and sustainable. The era of globalization can lead to a shift in cultural values. The era of globalization has an impact on the uniformity of aspects of life, as the face of architecture in the city of Indonesia tends to be the same. The era of globalization ignores the local tropical culture and climate. Cultural shift causes symptoms of cultural backflow and returns to search for the identity of a city. This phenomenon makes architecture tends to return to the culture and values of local wisdom, so it is predicted that this period is based on culture and will dominate in the future. City is a manifestation of events. This event is different in each period, economic, socio-cultural and environmental situation. City management should recognize this diversity. The heritage city not only provides a unique city identity, but also provides a functional role. Urban planning is an instrument to maintain the shape and function of the city. Heritage areas are managed as conservation efforts and integrated with urban development [1].

Baubau City is one of the heritage cities in Indonesia. The city of Baubau has many cultural heritage, beautiful coastal areas, unique culture and traditional buildings. Buton's traditional building is called *banuatada*. *Banuatada* is still used as an alternative house in Baubau City. The development of the city of Baubau is based on morphology. Morphology is grouped by means of heritage assets. The grouping of heritages is adjusted to the plan of urban space patterns, land use, and strategic areas. This is due to be in accordance with current conditions and planned by the government city. Grouping The Baubau City heritage area is grouped, as follows: 1) The *Keraton* Fortress and the Old Port area; 2) underwater nature conservation area and coastal culture of Makassar Island; 3) *Tirta Rimba* heritage area; 4) *Alam Bungi* heritage area and Samparona; 5) *Sulaa* Cultural Reserve area.

Baubau City began in the *Sulaa* region. *Sulaa* area is also called *Wolio* Village. The village developed and joined other villages, so that four small countries were formed. This inter-country coalition formed the kingdom of Buton. Buton Kingdom developed rapidly in coastal areas. Buton Kingdom experienced a cultural change. This cultural change affected the formation of the city of Baubau. This cultural change greatly affected the Sultanate of Buton. Wolio Fortress is used as a defense and this is an important event in the Sultanate of Buton. After the fortress stood firm, the development of Baubau City was centered on the fort area. The fort area is used as the center of government. The development of Baubau city is concentrated in coastal areas. The development of Baubau City is mainly around sea shipping and trade lanes. The development of Baubau City is divided into several stages, as follows: the period before the kingdom *(Mia Patamiana),* the period of the kingdom, the sultanate, and the City of Baubau at this time. This paper is intended to demonstrate the potential of the *Sulaa* region as an heritage asset to the coastal area of Sulaa and the City of Baubau as a Heritage City so that it is sustainable. This study provides benefits, as follows: 1) this research can provide input to policy makers. Policy makers are devoted to the preservation of heritage assets; 2) this research can form the image of the city, so that the City of Baubau has a unique and character.

2. Literature Review

2.1. Sustainability Concepts in Heritage City

The heritage city can be sustainable by means of, as follows: 1) heritage city is understood in physical and non-physical; 2) heritage city is understood with essence; 3) concept of sense of place, conservation, and regional development; 4) area-based conservation approach; 5) areas divided by zones; 6) intensification of socialization; 6) the old city area was preserved; 7) tourist attractions, direct interaction and management of tour tours, as well; 8) old buildings and new buildings synergized [2, 3, 4, 5, 6, 7, 8, 9, 10, 11].

2.2. Butonese Traditional House Concepts

Butonese traditional house is in the form of stage known as *banuatada*. The house has an elbow on the pillar of the house which symbolizes of social status. This can still be seen in the old Butonese house. Butonese traditional house has existed since the days of the kings before Islam entered in Buton. When the time of Sultan Laelangi (4th Buton Sultan), the concept of the traditional house of Buton is adapted to the views in Islamic law [12]. There are three types of building Butonese traditional houses based on the social level in the community that is: 2) *banuatada kambero* (fan), house of sultanate and family; 2) *Banuatada*, house of *Walaka* (ordinary people); (3) *kamali* or *Malige*, house of *Kaomu* (nobility). Butonese traditional house is a symbol of the microcosm which is based on a spatial arrangement patterns consisting of three parts of space, namely *bamba, tanga, suo*. Vertically *kapeo, karona banua* and *pa* are markers that becomes a symbol of human nature legs, body and head which is the same as the cosmic macro, namely the greatness of God. Horizontally explains that the *bamba* space is a symbol of men and the *tanga* is a symbol of women. The *suo* space is an expression of the husband and wife symbol. *Pa* (attic) space for women who undergo traditional *posuo* activities [13, 14, 15].

The surface structure of the floor of the house consists of three levels of the floor as a physical space differentiator. The Floor steps are a symbol of the human breath up and down. Buton's traditional house is restricted in gender. This restriction is found in the configuration of the room and is taken from the principles of the teachings of Islam. In cosmology, there is a hierarchical concept. This hierarchical concept is found on the right and left of the house. The right side of the house is considered a sacred space. *Banuatada* consists of *bamba, tanga* and *suo* rooms. The left side of the house is considered impure. The left side of the house is *sasambiri, rapu,* and WC [12, 14, 15, 16]. Settlement consists of, as follows: 1) the content in the form of humans and activities; 2) the container is called the place where humans live. This place includes natural and artificial elements. The contents and place are divided into five, as follows: 1) nature is called a place. This place limits and creates settlements; 2) man is referred to as subject. Subjects create and fill settlements; 3) society is referred to as society and adapted to needs; 4) Shells are referred to as structures and adjusted to functions; 5) The network is used as a facility [17]. Settlements are referred to as a single system. Houses, settlements and landscapes are referred to as cultural systems. Cultural systems can influence each other [18]. Three-dimensional space surrounds humans. Space relations are included in spatial organization. The settlement is referred to as the result of human relations and the environment. Settlements function as centers and gathering places for humans. Houses are also referred to as part of a large environment [19].

The way of life and the system of activities determine the types and containers of activities. Containers are interconnected spaces. Spaces are in one spatial system. The spaces function as places of activity. Space systems and space activities are used to formulate cosmology, formulate values of life, describe life. All of this is contained in the system of activities and space. Aspects of human psychology and community culture determine the form of activities and containers [20]. Formation of space is called a process. This process is derived from the reflection of the relationship between human activities, social and environmental order. Relationships are emphasized on cultural backgrounds, such as life views, beliefs, values and norms. This relation can determine a person's behavior and is reflected in the way of life, as well as the role of the community [21, 22, 16].

3. Research Methode

This study uses a rationalistic qualitative approach to the potential of *Sulaa* Region. *Sulaa* area is known as a supporter of the Baubau Heritage City. The main instrument uses the researcher himself. This is based on the researcher being able to see objects in their context and conduct participatory observation. Systematics, data processing and discussion through theories are determined by the ability of researchers. Purposive sampling is used to capture information [23, 24]. The study begins with the initial survey and determines the case study and unit of analysis. Data retrieval is done by recording information, measurement, drawing, documentation and interviews.

4. Result and Discussion

4.1. Variety of Heritage Assets in the Sulaa Coastal Area

Sulaa area has various heritage assets. This variety of assets is used as the identity of the heritage city, as follows: 1) natural heritages are formed from nature, as follows: *Nirwana* Beach, *Lakeba* Beach, *Tobetobe* Beach, *Manjawari* Waterfall Baths, *Kasulana Tombi* area and *Lakasa* cave, and; 2) physical Heritage is produced from creativity, taste, intention, and work. Physical heritage, as follows: traditional Buton house, woven ward, *gode-gode* (gathering room), *Topa* Mosque, *Sapati Manjawari* Tomb and *Betoambari* tomb; 3) the intangible heritage is produced by creativity, taste, and intention. Non-physical heritages, such as: palantu activities (ritual of changing seasons), *maludhu haroa* (Prophet's birthday), *haroana rajabu* (early Friday in Rajab month), *posuo* (seclution girls), *peruo* (pregnancy), *pokurui* (haircut), and so on (Fig. 1).

IOP Conf. Series: Materials Science and Engineering **797** (2020) 012001 doi:10.1088/1757-899X/797/1/012001

4.2. Tangible Heritage in the Sulaa Coastal Area

The pattern of the Buton traditional house space consists of *bamba* (front room), *tanga* (living room) and *suo* (back room). Space patterns are referred to as symbols of human nature. The symbols are equalized on the feet, body and head of a human. The *suo* room is always made higher than other rooms. In general, the occupancy of the Baubau community is made in the form of a stilt house *(Banuatada). Banuatada* consists of *bamba, tanga* and *suo*. *Bamba, tanga,* and *suo* visualize the shape of the human body. *Banuatada* has undergone facade modification, structure, material and space. However, spatial planning is maintained. This is seen in the core space, such as *bamba* space, *tanga* space and *suo* space). This core space pattern is maintained. However, the space pattern experienced a shift in function, such as: the *bamba* room (living room) was used as a sleeping room and weaving cloth, the *suo* room (the head of the family's bedroom) was used as a kitchen and weaving cloth. Shifting of space functions occurs due to space requirements. The factor of space needs is caused by the increase in the number of family members and other activities. Shifting space functions also occurs vertically (bottom space/*kapeo*). *Kapeo* is used to store goods and equipment for fishing (past). Today, *kapeo* has become a residential space and space for daily activities. *Banuatada* has changed the function of space. However, when carrying out ritual activities, the function of the core space is maintained.

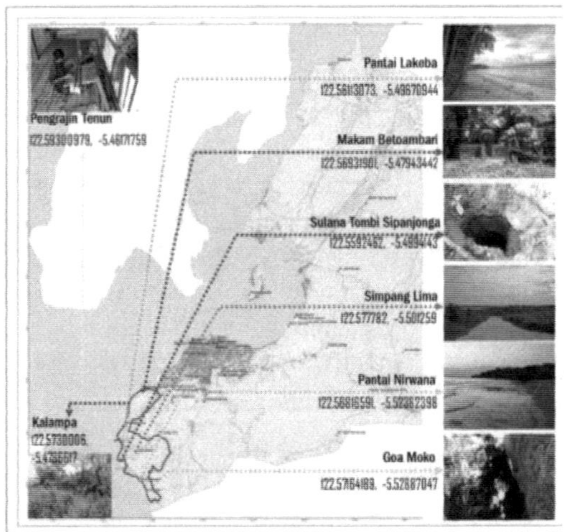

Figure 1. Sulaa Region Heritage Assets Distribution Map [Source: 26]

The application of the *banuatada* space concept is also found in non-stage buildings. Space patterns are equated with *banuatada*. Spaces consist of core space *(Bamba, tanga and suo)*. The spaces are placed to the right of the house (hierarchical). The characteristics of *banuatada* are in the location of the stairs. Stairs are placed on the left side of the house. Space functions are related to the hierarchy of space. This is applied to every ritual tradition activity and is equated with traditional activities in the *banuatada* building. The structure and construction of the banada have functions, as follows: 1) stiffener structure; 2) structure and construction indicate social status. This can be seen in *tada kambero* (fan-shaped house pillars). *Tada kambero* can only be used by nobility and officials who have a position in the Buton Sultanate period. *Banuatada* has a pole in one row. The number of poles in one row consists of *tare talu pale* and *tare pata pale. Tare talu pale* uses three poles. The pillar consists in a row of bonds. This pillar is used in people's homes that do not have a position in the Buton Sultanate period. Poles are made in round shape. Pale pata tare uses four poles. Poles are made in a row of bonds. This pole is used in the kaomu class house (nobility) which has a position in the Buton Sultanate period. The pole is made in a rectangular shape.

4.3. Tangible Heritage in the Sulaa Coastal Area

Sulaa community ritual activities are carried out with *haroa*. This is intended to pray for the soul to move to another world and to ask for salvation. *Haroa* rituals have important meaning. Therefore, *haroa* is categorized as local wisdom. *Banuatada* has a spiritual meaning for its inhabitants. Conception, *banuatada* knows the concept of space value categories. Space values are distinguished on the right and left side of the house. The right side is categorized as sacred space. The right side is the *bamba* room, *tanga* room and *suo* room. The left side of the house is categorized as unholy. The left side is *sasambiri* room, staircase entrance, rapu room, back stairs, and water closet. This concept is different from the *Banuatada* concept in *Sulaa* [26].

Changes and shifts in the function of space have no effect on ritual activities in *banuatada*. Spiritual awareness (*haroa* activity) in *banuatada* has implications for the existence of the core room of the house *(bamba, tanga* and *suo).* The core room is used as a ritual room. This pattern is also a reference in the construction of non-staged houses in *Sulaa*. The house continues to pay attention to the concept of cosmology. The concept of cosmology is based on the value and function of space in the house. In the scale of residential areas, the existence of spiritual space is awakened from the spiritual awareness of society. The *Sulaa* community uses springs, *Kasulaa Natombi* holes, the sea and the *Sulaa* mosque. These places are used as ritual spaces. These places are considered sacred and have high spiritual values. The ritual spaces are found in the *Topa* spring, *Kasulaa Natombi* pit, and the *kadatua* Strait sea. These places are believed to be a source of blessing, peace of mind, protection and source of strength. Springs, *Kasulaa Natombi* holes and the sea are used as the basis for the *Sulaa* community layout. This is because the *Sulaa* community has a functional and transcendental pattern. Functional patterns are obtained from the space pattern of the tire.

4.4. Sustainability of the Banuatada Development Concept

The process of building a house is carried out together. The development process starts from the preparation of building materials, site selection, and building orientation determination. The construction of a house begins with a hole in the house. *Ariy kabelai* pole (main pole) is perforated and continued with punching other poles. The house was established by making a row of main poles and *tare pale. Tare talu pale* (Pole is made with three rows). Three pillars are made for ordinary people's homes *(Walaka). Pata pale tare* (Pole made with four rows). Four pillars are made for the house of noble people *(Kaomu).* The establishment of the hood *(Tutumbu)* is the last stage.

5. Conclussion

The results of the analysis and discussion above concluded that heritages that can support the sustainability of the coastal area of *Sulaa*, so that the coastal area of *Sulaa* can support the City of Baubau, as follows: 1) natural heritages, such as: *Nirwana* beach, *Lakeba* beach, *Tobetobe* beach, *Manjawari* Water Bathing, *Kasulana Tombi Sipanjonga*, and *Lakasa* Cave; 2) physical heritage, such as: Buton traditional house, weaving ward, *gode-gode*, *Topa* Mosque, *Sapati Manjawari* Tomb, *Betoambari* Tomb, and; 3) non-physical cultural heritage such as: *haroa* ritual.

References

[1] Fadhil S 2013 Prosiding Temu Ilmiah IPLBI p D-25 – D-30
[2] Laretna T, Adishakti 2009 Disampaikan dalam Kongres I Jaringan Kota Pusaka Indonesia di Sawahlunto p 3-4
[3] Nuryanti W, Suwarno N 2008 J. *Manusia dan Lingkungan* **15** 104

[4] Martokusumo W 2014 Makalah disampaikan dalam diskusi Temu Pusaka Indonesia di Galeri Cemara Jakarta **1** 5
[5] Abdul N N, Arifin S S 2017 Prosiding Seminar Nasional Arsitektur Populis p B -42
[6] Bazher N M, Handayani K N, Iswati T R 2017 *J. Arsitektura* **15** 474
[7] Heldiansyah J C, Aufa N, Wastuty P W 2016 Prosiding Seminar Nasional Lahan Basah Jilid 3 p 992
[8] Adrisijanti I 2007 Disajikan dalam Diskusi Sejarah Kota dan Perubahan Sosial Dalam Perspektif Sejarah diselenggarakan oleh Balai Pelestarian Sejarah dan Nilai Tradisional p 8
[9] Purwaningsih L H, Kusuma H E 2016 Prosiding Temu Ilmiah IPLBI p I018
[10] Anurogo W 2017 *J. Ketahanan Nasional* **23** 258-259
[11] Kwanda T 2004 *J. Dimensi Teknik Arsitektur* **32** 104
[12] Ramadan S 2003 Tesis (Undergarduate) Magister Study Program Architecture of Engeneering Gadjah Mada University Yogyakarta p 34
[13] Kadir I 2000 Thesis (Undergraduate) Magister Study Program Architecture of Engeneering Gadjah Mada University Yogyakarta p 11-12
[14] Kadir I, Djunaedi A, Sudaryono, Wibisono B H 2013 Prosiding Seminar Nasional Reinterpretasi Identitas Arsitektur Nusantara Department of Architecture Faculty of Engeneering Udayana Universiy Denpasar p 247-254
[15] Kadir I, Djunaedi A, Sudaryono, Wibisono B H 2015 Prosiding Seminar Nasional finding the fifth element after water, earth, wind, and fire SCAN#6 Architecture Study Program Faculty of Engeneering Atmajaya University Yogyakarta p 133-141
[16] Kadir I, Ma'ruf A, Hasan L A 2016 Prosiding Seminar Nasional The Lost World SCAN#7 Architecture Study Program Faculty of Engeneering Atmajaya University Yogyakarta p 62-70
[17] Doxiadis A C 1968 Ekistics, An Introduction to The Science of Human Settlements (London 21: Hutchinson)
[18] Rapoport A 1969 House Form and Culture (Ngelwood Cliffs: Prentice – Hall Inc) N J 73
[19] Rapoport A 1977 Human Aspects of Urban Form (USA: Pergamon Press) p 36
[20] Haryadi, Setiawan B 2010 Arsitektur Lingkungan dan Perilaku, Pengantar ke Teori, Metodologi dan Aplikasi (Yogyakarta: Gadjah Mada University Press) p 24
[21] Kadir I, Djunaedi A, Sudaryono, Wibisono B H 2014 Prosiding Seminar Nasional SERAP, APRFDepartment of Architecture and Planning Gadjah Mada University Yogyakarta p 131-137
[22] Kadir I, Djunaedi, A Sudaryono, Wibisono B H 2015 Majalah Ilmiah Teknologi Forum Teknik Faculty of Engeneering Gadjah Mada University Yogyakarta 13-22
[23] Nasution S 1996 Metode Penelitian Naturalistik – Kualitatif (Bandung: Tarsito) p 4-64
[24] Moleong L J 2012 Metodologi Penelitian Kualitatif (Bandung: PT. Remaja Rosdakarya) p 111-119
[25] Anonim 2013 Direktorat Jenderal Penataan Ruang P3KP (Jakarta) p 158-159
[26] Kadir I 2015 Dissertation S3 Study Program architectural science Gadjah Mada University Yogyakarta p 190 244

IOP Conf. Series: Materials Science and Engineering **797** (2020) 012002 doi:10.1088/1757-899X/797/1/012002

Stability Analysis and Strengthening on Zona Karst of Seropan Cave Walls

Sulha Sulha[1*], Ahmad Rifa'i[2], Minson Simatupang[1], Umran Sarita[1]
[1]Department of Civil Engineering, Universitas Haluoleo, 93232 Kendari, Sulawesi Tenggara, INDONESIA
[2]Program Studi Teknik Sipil, Universitas Gadjah Mada, Jln. Grafika No. 2 Kampus UGM Yogyakarta, 55281, INDONESIA

[*]Corresponding author's e-mail: sulha@uho.ac.id

Abstract. The Indonesian state has many caves in several regions with unique legends and high cultural values, including Sang Hyang Sirah Cave in West Java, Seropan Cave and Semar Cave in Central Java, Liangkabori Cave on Muna Island and others. Seropan Cave has a large potential of water resources, especially in the karst area in the form of water flows in underground rivers. This study aims to analyze strain stress that occurs due to karst conditions on variations in groundwater level. Analysis method used finite element and plan strain model with the help of Plaxis v8.6 Software. The analytical approach is performed on the appearance of the cave with normal water level conditions, flood water levels and extreme use of shotcrete reinforcement, rock bolt and a combination of both. Material properties at the location are taken from the results of research and testing of the Seropan Cave IWRM laboratory and UGM Technical Team. The simulation results show that the existing conditions require wall reinforcement. Wall reinforcement uses 7 cm thick shotcrete and rock bolt with a diameter of 2 cm 4 m long with the distance between rock bolt is 2 m. Strengthening combination of rock bolt and shotcrete gives the highest safety factor of 2.01 while in earthquake conditions, it reaches 1.76. Combination wall reinforcement is quite stable against static and dynamic loads.

1. INTRODUCTION

Cai, M. (2011) examined the characteristics of rock mass by considering the diversity of rock properties for the design of large tunnels and caves. He conducted a quantitative, probabilistic approach using the Geological Strength Index (GSI) system to characterize rock masses. He said that this method provided a theory of a systematic analysis approach to the uncertainty of rock mass characteristics, and this can help us to know the part of knowing how uncertainty arises and how to decide the design of rock support systems that might be affected by it.

The structure of some caves cannot be avoided against unsafe and unstable conditions. This provides uncertainty, which results in the construction process becoming complicated and increasing risk. Therefore, the development of efficient scientific steps to ensure the quality and safety of construction projects are very important (Bao Jin at all, 2015).

Deng Hongling et all. (2015) examined the dynamic response of tunnels through karst caves to impact loads. Merela reported that the tunnel was only destroyed when it received a collision load of not less than 0.1 seconds, a vertical displacement of 2.88 m, a horizontal displacement of 0.80 m. The tunnel falls when the rock mass is destroyed at a maximum impact of 7E + 6 KN on the tunnel.

In this study, estimates of rock mass properties are based on GSI values using Rocklab software. The support system used based on rock mass classification RMR (Bieniawski, 1989). Based on the determined rock mass and support system, numerical simulations are carried out using Plaxis software to estimate the stability of the cave wall. The cave code that is located on karst land is important for stability analysis on static and dynamic loads.

2. STRENGTHENING OF THE CAVE WALL

2.1. Criteria for Rock Failure

The collapse criteria are based on Hoek-Brown 2002 as follows:

$$\sigma'_{cm} = \sigma_{ci}\frac{(m_b + 4s - a(m_b - 8s))\left(\frac{m_b}{4} + s\right)^{a-1}}{2(1+a)(2+a)} \tag{1}$$

$$m_b = m_i.\exp\left(\frac{GSI - 100}{28 - 14D}\right) \tag{2}$$

$$m_i = \frac{\sigma_{ci}}{\sigma\tau} \tag{3}$$

$$s = \exp\left(\frac{GSI - 100}{9 - 3D}\right) \tag{4}$$

$$a = \frac{1}{2} + \frac{1}{6}\left(e^{-GSI/15} - e^{-20/3}\right) \tag{5}$$

where σ_{ci} is rock compressive stress from the UCS test (MPa), $\sigma\tau$ is tensile stress of the point load test (MPa), m_i is compressive stress ratio and tensile stress ratio, m_b is reduction of mi value, s and a are rock mass constants, GSI is geological stress index based on fracture arrangement and surface conditions D is disturbance factor during excavation.

For estimating GSI, D values and determine rock shear strength based on Hoek-Brown 2002 criteria. Analysis of software approach to soil and rock is generally based on Mohr-Coulomb criteria, so an approach is needed from the Mohr-Coulomb equation (c, φ) with the Hoek-Brown criteria, namely:

$$\phi' = \sin^{-1}\left[\frac{6am_b\left(s + m_b\sigma'_{3n}\right)^{a-1}}{2(1+a)(2+a) + 6am_b\left(s + m_b\sigma'_{3n}\right)^{a-1}}\right] \tag{6}$$

$$c' = \frac{\sigma_{ci}\left[(1+2a)s + (1-a)m_b\sigma'_{3n}\right]\left(s + m_b\sigma'_{3n}\right)^{a-1}}{(1+a)(2+a)\sqrt{1 + \left(6am_b\left(s + m_b\sigma'_{3n}\right)^{a-1}\right)/((1+a)(2+a))}} \tag{7}$$

where $\sigma'_{3n} = \sigma'_{3max}/\sigma_{ci}$

2.2 Rock Mass Classification

The technical properties of rock mass can be obtained from the classification of rock mass. The classification aims to estimate the strength of the rock mass according to the type and class of rock and the permanent support system to be used. To estimate rock mass, rock classification can be used the RMR criteria. There are six basic parameters used in the RMR system (Bieniawski, 1989), namely: Intact rock strength test, which is based on index point load or uniaxial compressive strength, Rock Quality Designation (RQD), discontinuous field spacing and conditions, water condition's soil and disjointed plane orientation. The total RMR weight from the six parameters above is used to determine the rock mass classification and the support system to be used. The support system used is determined according to Table 1.

2.3. Safety Factor

According to Terzaghi and Peck, (1967) and the Canadian Geotechnical Society, (1992), shear failure for soil retaining structures was 1.5-2 as in table 2. Safe factor analysis of collapse in PLAXIS, using the Phi-c reduction method which principally refers to the Coulomb criteria with the following equation:

$$\frac{c}{c_{cr}} = \left(\frac{\tan.\varphi}{\tan.\varphi_{cr}}\right) = \sum Msf \tag{8}$$

Table 1. The procedure for determining supporting systems is based on value of the RMR (Bieniawski, 1989).

Rock mass class	Excavation	Rock bolts (20 mm diameter, fully grouted)	Shotcrete	Steel set
I-Very good rock RMR. 81 – 100	Full face, 3 m advance.	Genarally no support required except spot bolting		
II-Good rock RMR. 61 – 80	Full face, 1 – 1,5m advance, complete support 20 m from face	Locally, bolts in crown 3 m long, spaced 2,5m with occasional wire mesh	50 mm in crown where required	None
III-Fair rock RMR. 41 – 60	Top heading and bench 1,5 – 3 m advance in top heading. Commence support after each blast. Complete support 10 m from face.	Systematic bolts 4 m long, spaced 1,5-2 m in crown and walls with wire mesh.	50 – 100 mm in crown and 30 mm in sides.	None
IV-Poor rock RMR. 21 – 40	Top heading and bench 1,0 – 1,5 m advance in top heading. Install support concurrently with excavation, 10 m from face.	Systematic bolts 4-5 m long, spaced 1-1,5 m in crown and walls with wire mesh.	100-150 mm in crown and 100 mm in sides.	Light to medium ribs spaced 1,5 m where required.
V-Very poor rock RMR. < 20	Multiple drifts 0,5 – 1,5 m advance in top heading. Intall support concurrently with excavation. Shotcrete as soon as possible after balsting.	Systematic bolts 5-6 m long, spaced 1-1,5 m in crown and walls with wire mess, blot invert.	150-200 mm in crown, 150 mm in sides, and 50 mm on face	Medium to heavy ribs spaced 0,75 m with steel Lgging and forepoling if required. Close invert.

Table 2. The minimum SF for shear failure according to Terzaghi and Peck, (1967) *and the Canadian Geotechnical Society,* (1992) in Wyllie and Mah, (2005).

Failure type	Category	Safety factor
Shearing	Earth works	1,3 – 1,5
	Earth retaining structures, excavation	1,5 – 2,0
	Foundation	2 – 3

3. NUMERICAL SIMULATION

3.1. Finite element method in Plaxis

The modeling of the cave wall following procedure is the generation of initial stress by K_o procedure, determine of water level, activate surcharge load, installation of shotcrete, activate of rock bolts and earthquake load.

3.2. Research Procedures

The characteristics of the Seropan cave soil / rock structure are based on the results of the research report and the Seropan Cave IWRM laboratory test and UGM Technical Team and the results of Seropan cave lay out measurements. The loading and condition of the river water level to determine the behavior and displacement tendencies and stresses that occur due to static or dynamic loading. The simulation results obtained a safe and stable reinforcement method to be recommended at the repair stage in the field.

From the results of research and testing of Seropan Cave material in the laboratory of the Faculty of Engineering UGM, the rock mass was in the third class category (fair rock class) with an RMR = 53. The reinforcement method was in accordance to the guidelines of Bieniawski in 1989, in the form of rock bolt with a depth of 4 m and an average distance of 1.5-2 m. In addition, shotcrete needs to be added with a thickness of 50-100 mm on the crown and 30 mm in the cave wall. Taking into account the dynamic load, thick shotcrete is used 70 mm for the entire surface of the cave wall. RocLab software can be used as an alternative determination of shear strength. Based on the RocLab software approach, obtained output data in the form of cohesion (c) = 0.158 MPa, friction angle in rock (φ) = 39.72°, and rock elasticity (Em) = 210.32 MPa based on Hoek-Brown 2002 criteria.

The rock mass classification in this case is moderate (rock fair). The reinforcement method is in accordance to the guidelines of Bieniawski in 1989, in the form of rock bolt with a depth of 4 m and an average distance of 1.5-2 m. In addition, shotcrete needs to be added with a thickness of 50-100 mm on the crown and 30 mm on the cave wall. Taking into account the dynamic load, thick shotcrete is used 70 mm for the entire surface of the cave wall of the foundation construction area. Recapitulation of PLAXIS input data for each geometry model is presented in table 3.

Topographic measurement results in the form of layouts and cross sections of the initial design of the foundation to be analyzed are presented in Figure 1a. The Seropan underground river is located at a depth of about 70 meters below the soil surface. The dimensions of geometry modeling are (55,60) meters. The upper rock layer from the ground surface to a depth of 42.77 meters is considered as a uniform load. The amount it is taken from the weight of the rock unit multiplied by the height of the rock, amounting to 780 kNm^{-2}. The modeling used is plane strain, where Seropan cave is a non-irregular and oval-shaped non-tunnel. The existing condition of Seropan cave and observation point is modeled as in Figure 1b. To calculate the dynamic load, in the stability analysis of the cave wall uses a numerical constant which is usually called the earthquake coefficient (k_g). This coefficient is given in percent of gravity such as the 10% gravity coefficient (0.1g). According to the 2017 earthquake source map of Indonesia, Yogyakarta region has the earthquake acceleration coefficient of 0.3 - 0.4g.

3.3. Effect of water level fluctuations

There are three variations in the regulation of groundwater level, which are reviewed, namely normal water level was 0.92 meters. Flood water level was 1.92 meters and maximum (extreme) flood water level as high as 17.72 meters. Calculation of water pressure (generate water pressure) is done by defining the groundwater conditions (phreatic level) on the model geometry, so that the calculation of water pressure can be carried out. For flood water levels and extreme steps taken with the normal water level, the only difference is the water level. The relationship between the increase in total displacement and the safe number of construction stages in different river water conditions can be seen in Figure 2a.

3.4. Displacement and Safety Factors

To get the results of the analysis, the displacement value is observed at a certain point. This point is a part of the cave wall that experiences the largest displacement. Observation points are placed on the surface of the cave wall that is Point A, B, C and D according to Figure 1a. Based on the results of numerical simulations, a safe factor and displacement value is obtained for each construction stage in accordance with Figure 2b.

Tabel 3. PLAXIS input material data

Parameters	Symbols	Value	Unit
Material model	-	Mohr	-
Type of material behavior	-	Tak	-
Weight of rock content above the phreatic	γ_{unsat}	18,37	kNm^{-3}
Weight of rock content below the phreatic	γ_{sat}	19,28	kNm^{-3}
Horizontal direction permeability	k_x	0,000	ms^{-1}
Vertical direction permeability	k_y	0,000	ms^{-1}
Young's Modulus	E_{ref}	210.3	kNm^{-2}
Poisson's ratio	ν	0,25	-
Cohesion	c_{ref}	158	kNm^{-2}
Friction angle	φ	39,72	°
Dynamic friction angle (φ - 2) °	φ_{din}	37,72	°
Dilatancy angle	ψ	3°	
The interface shear strength reduction	R_{inter}	0,52	

IOP Conf. Series: Materials Science and Engineering **797** (2020) 012002 doi:10.1088/1757-899X/797/1/012002

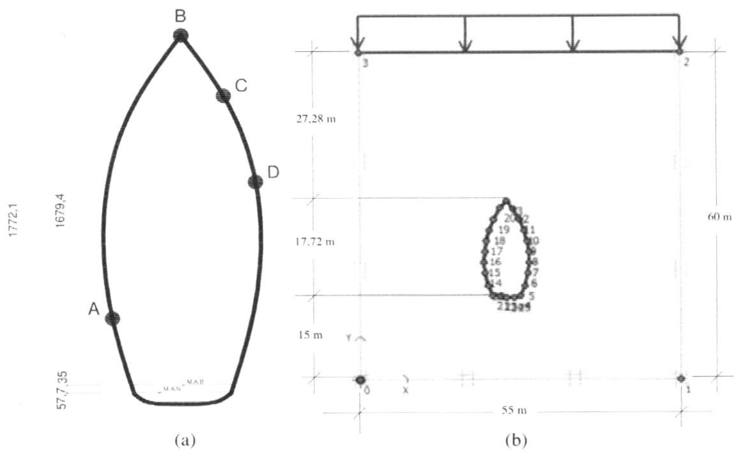

Figure 1. (a) Cross section, the dimensions and the observation points of the cave wall (b) Modeling the existing condition of the cave.

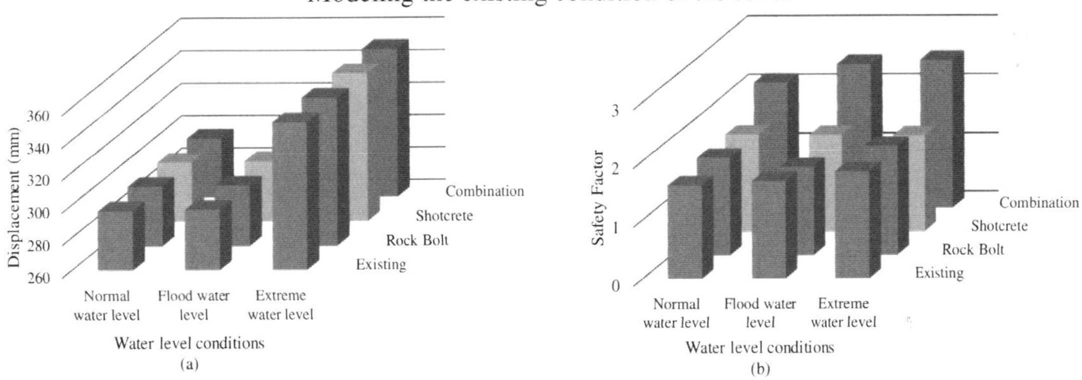

Figure 2. (a) Chart relationship of the total displacement to water level conditions, (b) Chart relationship of safety factor in each of strengthening with different water level conditions.

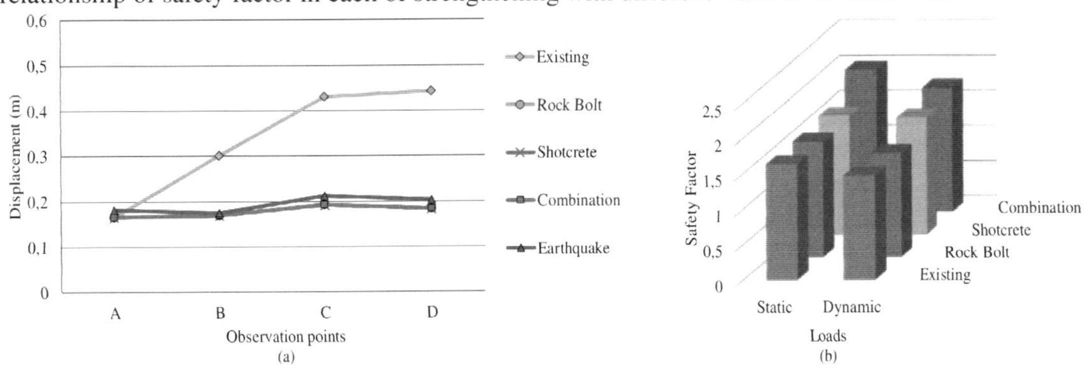

Figure 3. (a) Displacement chart on the cave wall at the observation points, (b) Chart relationship of safety factor to loading at each stage of construction.

From the picture above shows that the installation of a combination support system results in a decrease in displacement values of the cave wall. The safe factor after installing a support system against the collapse of the cave due to earthquake load with base rock acceleration of 350 cms^{-2} is reached 1.76. Rock bolt and shotcrete combination can reduce displacement and significantly increase the safe factor of the cave wall against static and dynamic loads. It can be applied in the maintenance and development stages of the cave in that location.

4. CONCLUSION

The location of the study included is in the category of fair rock. The reinforcement method used rock bolt with a length of 4 m. The distance between rock bolt is 2 m and thickness of shotcrete are 70 mm. Installation of rock bolt and shotcrete support system results in a decline displacement of the cave wall. The rock bolt and shotcrete support system used in this study are strong enough to support earthquake loads below 350 cms^{-2}.

REFERENCES

[1] Bieniawski, Z.T., 1989, *Engineering Rock Mass Classifications*, John Wiley and Sons. New York.
[2] Cai, M., 2011, Rock Mass Characterization and Rock Property Variability Considerations for Tunnel and Cavern Design, Rock Mechanics & Rock Engineering, **44** Issue 4, 379-399,doi 10.1007/s00603-011-0138-5.
[3] Deng Hongliang at all, 2015, The dynamic response of tunnel through the Karst cave under impact load, Applied Mechanics & Materials., **777,** 135 – 142, doi 0.4028 / www.scientific.net /AMM.777.135
[4] Geotechnical Engineering Office, 2006, *Foundation Design And Construction,* Geo Publication No. 1, 2006.
[5] Hoek, E., 2002, *Hoek-Brown Failure Criterion, Program"RocLab",* http://www.rockscience.com. 27 Maret 2013.
[6] Irsryam, M. at all, 2017, *Peta Sumber Bahaya Gempa Indonesia Tahun 2017*, Pusat Penelitian dan Pengembangan Perumahan dan Permukiman.
[7] Jin Bao, Zhang Hongjun dan Jianyang Zhang, 2015, A new Technique for construction of limestone cave, Key Engineering Materials, **648**, 9 – 16, doi 10.4028 / www.scientific.net /KEM.648.9
[8] National Earthquake Study Center, 2017, Indonesian Earthquake Source and Hazard Map of 21017, Housing and Settlement Research and Development Center, Research and Development Ministry of Public Works and Public Housing, Bandung.
[9] Rifa'i, A., 2011, Survey and Test of Soil Seropan Underground Mechanics Soil Laboratory Laboratory Phase I in Gunung Kidul Regency, Faculty of Engineering Department of Civil Engineering and Environmental Soil Mechanics Laboratory, Gadjah Mada University, Yogyakarta.
[10] Wyllie, D.C. and Mah, C.W., 2005, *Rock Slope Engineering*, Spon Press Taylor & Francis Group, London and New York.

IOP Conf. Series: Materials Science and Engineering **797** (2020) 012003 doi:10.1088/1757-899X/797/1/012003

Modern Analogue of Shaly Sand Reservoir on Coastal Plain in Selat Baru, Bengkalis Island, Indonesia

Yuniarti Yuskar[1*], Tiggi Choanji[1], Dewandra Bagus Eka Putra[1], Shanaz Noveta Bastira[1,2], Adi Suryadi[1], Dilla Permata Sari, Harisma[3]

[1]Department of Geological Engineering, Faculty of Engineering, Universitas Islam Riau, Jl. Kaharuddin Nasution No 113, Pekanbaru, Riau, INDONESIA

[2]School of Environmental Science, Graduate School, Universitas Padjadjaran, Jl. Dipati Ukur No. 35 40132 Bandung INDONESIA

[3]Department of Geological Engineering, Faculty of Earth Science and Technology, Universitas Haluoleo, 93232 Kendari, Sulawesi Tenggara, INDONESIA

[*]Corresponding author's e-mail: yuniarti_yuskar@eng.uir.ac.id

Abstract. Coastal environment is one of main reservoir depositional system in the oil industry and becomes the main challenges in defining its characteristic, especially in lateral correlation. Research conducted in Bengkalis Island which has a unique characteristic of coastal depositional environment that affected by a tidal system related to modern analogue for reservoir characterization. The method used for this research is analysing lithofacies from drilled core data, surface sampling, and lateral measurement from foreshore to the tidal limit. It is confirmed from core data that research area consists of sand, shale, and peat with lamination, cross laminations sedimentary structure and also filled with organism activity, which clustered into eight lithofacies. By integrating the core data with surface sampling and lateral measurement, it is interpreted that facies is changing only in near distance (about 30 – 40 m) with sloping 1 – 2 degrees in the surface, so by defining the same characteristic on lithofacies, correlation can clearly define on a shaly sand reservoir in the tidal coastal environment.

1. Introduction

Selat Baru is a small area that located in the northern coast of Bengkalis Island, Riau, Indonesia and contiguously to Malacca Strait (Figure 1). The Malacca Strait is located between the east coast of Sumatra Island in Indonesia and the west coast of Peninsular Malaysia, it is linked with the Strait of Singapore at the south-east end [1,2]. The study area was found to have brackish water in its river flow owing to a certain amount of salt water mixing [3,4]. The study was initiated from the result of water quality analysis in the western area of Bantan Tua, Jangkang and Deluk village which adjacent directly to Selat Baru Village.

The organic matter preserved in marine sediments contains varying contributions of terrestrial and marine source inputs, an issue that relevant to an area along the coastal margins where most carbon burial occurs [5]. Terrestrial organic matter deposited in marine sediments might then be expected to undergo less efficient re-mineralization and therefore be preferentially buried [6]. Large amounts of terrestrial organic carbon are annually transported from the continents to the oceans mainly by fluvial transport or, in lower amounts, by aeolian dust [7]. Understanding the fate of Terrestrial Organic Matter (TOM) in marine sediment is important for some reasons. In part, this interest stems from the observation that marine organic matter is broadly considered to be more reactive than terrestrial organic matter [6].

Stratigraphy of the study area composed of rocks that include surface deposits which are Young Superficial Deposit (Qh), and Older Superficial Deposit (Qp) [8]. Young Superficial Deposit consists of clays, silts, clean gravel, vegetation rafts, peat swamps and Older Superficial Deposits consist of clays, silts, clayey gravels, vegetation rafts [9].

The Northern Coast of Bengkalis Island, which is located on the estuary, have a thicker TOM compared to the beach that located further. The Selat Baru Beach is relatively cleaner because it is not in the estuary part, but the number of mangrove forests around it becomes a source of TOM that was carried to the coast. This study aims to determine the characteristics of coastal sediments and their changes vertically and laterally, it can be used as an analogue reservoir in the coastal environment which is influenced by the presence of high terrestrial organic matters.

Figure 1. Study Area in Selat Baru Area, Bengkalis Island - Indonesia

2. Methodology

The method used consist of several steps and measurements [10]. There were field survey and laboratory analysis. Field survey were conduct to specify the drilling position and lateral measure section on Selat Baru Beach. Drilling has been carried out at 3 well locations using hand auger on 2 metres depth. The lateral cross section was conducted from the sea tide limit up to lowest position of the sea water [2] and the samples of sediment were taken every 10m. Availability of the data in this study can be seen in Table 1.

Laboratory analysis is carried out by performed sieve analysis (the sieve size in 2.38 cm, 1.19 cm, 0.6 cm, 0.297 cm, 0.149 cm, 0.074 cm, 0.04 cm, and 0.02 cm). This analysis has to be done to obtain the dominant of grain size, sorting, skewness, and kurtosis of the core drilled and lateral measure section samples.

Table 1. Avalability of research data

Data	Name of Data	Location	Coordinate		Depth	Length
			Long	Lat		
Core	#SB 1	Selat Baru Beach	01°33'37.00"	102°14'45.00"	300 cm	-
	#SB 2	Selat Baru Beach	01°33'44.00"	102°14'44.00"	200 cm	-
	#SB 3	Selat Baru Beach	01°33'47.00"	102°14'37.00"	200 cm	-
Measure Section	#SNB 1	Selat Baru Beach	01°33'47.00"	102°14'37.00"	-	185 m
	#SNB 2	Selat Baru Beach	01°34'08.6"	102°11'12"	-	180 m
	#SNB 3	Selat Baru Beach	01°34'07.45"	102°11'12.34"	-	210 m

3. Result and Discussion

3.1 Granulometry Analysis

Granulometry analysis has been performed to trenching data with 30 cm depth at line measure section. The grain size distribution of sediment in Selat Baru Beach is dominated by medium sand, good to excellent sorting and spreading towards finer granules with very platykurtic kurtosis value (Table 2).

Table 2. Summary of Granulometry Analysis

Data	Name of Data	Location	Coordinate		Length	Granulometry Analysis			
			Long	Lat		Mean	Standard Deviation	Skewness	Kurtosis
Measure Section	#SNB 1	Selat Baru Beach	01°33'47.00"	102°14'37.00"	0 - 2.7 m	0.271	0.423	0.334	0.112
					3.7 - 4 m	0.212	0.218	0.007	0.009
					33.3 - 34 m	0.241	0.404	0.48	0.07
					56 - 57 m	0.132	0.222	0.145	0.025
	#SNB 3	Selat Baru Beach	01°33'47.00"	102°14'37.00"	0 - 2.7 m	0.286	0.485	0.534	0.112
					67 - 70 m	0.201	0.279	0.128	0.0203

3.2 Core Data Analysis

Sediment of core data from 3 locations had been analysed, the result shows several types of sediment such as mud, silt and fine sand (Figure 2). Most of these sediment contain Terrestrial Organic Matters (TOM) with varying percentage. In addition, several sediment structures were found in core data. For example: ripple mark, cross lamination, flaser, lenticular, and burrow. Based on core data analysis, characteristic of sediment obtained as followed:

1. Finesand with burrow
2. Finesand with Terrestrial Organic Matter (TOM)
3. Finesand with ripple mark and fine size of TOM
4. Mud with fine size of TOM
5. Laminated mud with lenses of TOM
6. Cross laminated Silt to Mud with TOM
7. Cross laminated Silt to Mud with Burrow

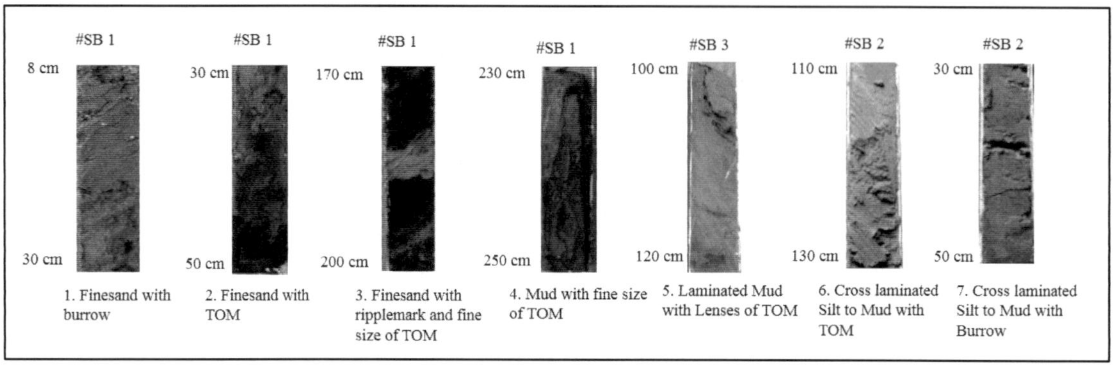

Figure. 2 Coastal Sedimentary Fasies based on Core Data in research area.

3.3 Lateral Measured Section Analysis

Based on the description of measured cross section by observing the physical character in the field, several characteristics of the sediment structure were found. There were massive, ripple mark, burrow, lenticular and flaser. Fine sand sediment with fine sized organic material and ripple mark with a slope angle of 40° and some location have low angle (< 20^0). The types of ripple mark are symmetrical, transverse ripple and barchan-shaped becoming transverse ripple.

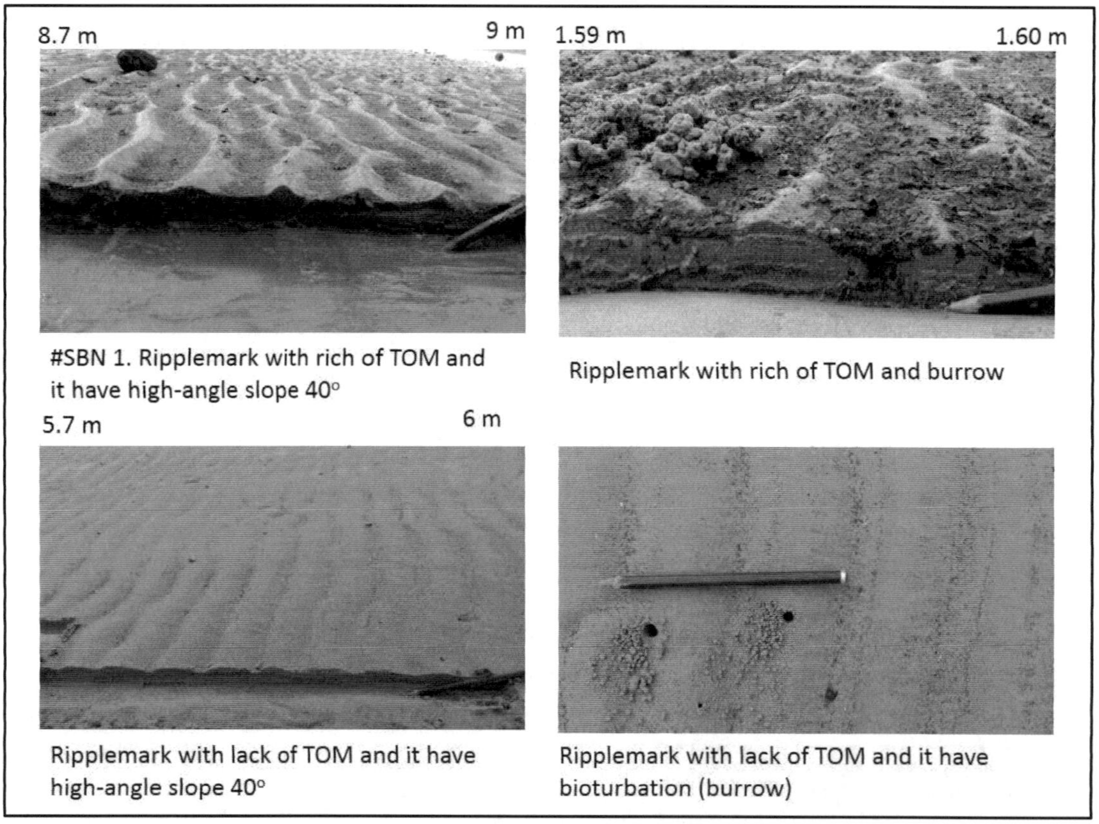

Figure. 3 Sedimentary structures based on Measure Section Data in research area.

Ripples can be generated by waves, currents or both, very small waves and weak currents are unable to move the sand, so that it cannot create or modify the ripples. The orientation of the ripples is aligned with the forcing waves or current, and changes (with a time-lag) as the forcing direction changes, ripples can be flattened by biological action in a few hours. The ripple also will be damaged by very large waves or strong currents wash out ripples, when the strengths of the waves and currents diminish.

Coastal sediment in the research area dominated by well sorting - fine sand. The influenced of land sediment material such as Terrestrial Organic Matter (TOM) given the specific characteristic to the coastal sediment of the research area. The coastline that located closed to estuary would have thicker TOM deposit compared to the beach that far from the tidal mouth. Estuary brought the material from the land to the ocean with various percentage. Selat Baru beach located away from the estuary, therefore the TOM deposit was not dominant in this area. However, high erosion rate in the coastal area, especially in the mangrove conservation area, would bring the mangrove material to the shoreline.

Sedimentary structures that developed in this coastal area were ripple mark, cross lamination, bioturbation, flaser and lenticular. Characteristic of coastal sediment could be used as an analogue in determination of depositional environment, notably coastal and transitional environment that effected by the ocean and occurrence of inland TOM. Ripple mark formation had been influenced by the grain size of sediment, current and waves.

4. Conclusion

Selat Baru beach has several sedimentary facies that dominated by well-very well sorting fine sand (0.125-0.25 cm), TOM content, ripple mark, cross lamination, bioturbation (i.e. burrow, track and trail), flaser and lenticular. In addition, clay and silt deposit could be found. The sedimentary deposit changed vertically and laterally. There are seven sediment facies such as fine sand with burrow; fine sand with terrestrial organic matter (TOM); fine sand with ripple mark and fine size of TOM; mud with fine size of TOM; laminated mud with lenses of TOM; cross laminated silt to mud with TOM and cross laminated silt to mud with burrow.

Acknowledgment

This research was supported by Department of Geological Engineering, Universitas Islam Riau. Also thanks to Rani Amalia, Anang Putranto, Aldila, M. Revanda that assisted the field activity for coring using hand-auger. Furthermore, appreciation given to local authority of Bengkalis district that provide the permit and facilitating data collection activity.

References

[1] Thia-eng C, L Gorre I R, Adrian Ross S, Regina Bernad S, Gervacio B and Corazon Ebarvia M 2000 The Malacca Straits GEF/UNDP/IMO Regional Programme on Partnerships in Environmental Management for the Seas of East *Mar. Pollut. Bull.* **41** 160–78

[2] Yuskar Y, Choanji T, Putra D B E, Suryadi A and Ramsof R A 2018 Coastal Deposit Characteristic Influenced by Terrestrial Organic Matter and Its Sedimentary Structure at Jangkang Beach, Bengkalis District, Riau Province—Indonesia BT - Proceedings of the Second International Conference on the Future of ASEAN (ICoFA) 201 *Proceedings of the Second International Conference on the Future of ASEAN (ICoFA) 2017 – Volume 2* ed R Saian and M A Abbas (Singapore: Springer Singapore) pp 909–16

[3] Putra D B E, Yuskar Y and Hadian M S D 2017 Hydrogeology Assessment Using Physical

Parameter in Bengkalis Riau *Proc. 2nd Join Conf. Utsunomiya Univ. Univ. Padjadjaran* 274–9

[4] Putra D B E and Yuskar Y 2016 Pemetaan Airtanah Dangkal Dan Analisis Intrusi Air Laut *Seminar Nasional ke-3 Fakultas Teknik Geologi Universitas OAdjadjaran*

[5] Hedges J I, Keil R G and Benner R 1997 What happens to terrestrial organic matter in the ocean? *Organic Geochemistry* vol 27 pp 195–212

[6] Burdige D J 2005 Burial of terrestrial organic matter in marine sediments: A re-assessment *Global Biogeochem. Cycles* **19** 1–7

[7] Hopmans E C, Weijers J W H, Schefuß E, Herfort L, Sinninghe Damsté J S and Schouten S 2004 A novel proxy for terrestrial organic matter in sediments based on branched and isoprenoid tetraether lipids *Earth Planet. Sci. Lett.* **224** 107–16

[8] Cameron N R 1983 The Stratigraphy of The Sihapas Formation in the Northwest of the Central Sumatra Basin *Indones. Pet. Assoc. Proc. 12th Annu. Conv. Jakarta, 1983* **I** 43–66

[9] Yuskar Y, Choanji T, EP D B, Suryadi A and Ramsof R A 2018 Coastal Deposit Characteristic Influenced by Terrestrial Organic Matter and Its Sedimentary Structure at Jangkang Beach, Bengkalis District, Riau Province—Indonesia BT - Proceedings of the Second International Conference on the Future of ASEAN (ICoFA) 201 *Proc. Second Int. Conf. Futur. ASEAN 2017 – Vol. 2* 909–16

[10] Yuskar Y and Choanji T 2017 Uniqueness Deposit of Sediment on Floodplain Resulting From Lateral Accretion on Tropical Area: Study Case at Kampar River, Indonesia *J. Geosci. Eng. Environ. Technol.* **2** 14–9

IOP Conf. Series: Materials Science and Engineering **797** (2020) 012004 doi:10.1088/1757-899X/797/1/012004

Characteristics of Quaternary Deep Sea Sediment in the Sumba Strait based on Grain Size and LoI (Lost on ignition Analysis)

Zulhikmah[1], Yuniarti Yuskar[1*], Purna Sulastya Putra[2], Septriono Hari Nugroho[3], Tiggi Choanji[1]

[1]Dapartment of Geological Engineering, Engineering Faculty Universitas Islam Riau
[2]Research Centre for Geotechnology, Indonesian Institute of Sciences (LIPI), Bandung, Indonesia
[3]Research Centre for Deep Sea, Indonesian Institute of Sciences (LIPI), Ambon, Indonesia

E-mail: zulhikmah@student.uir.ac.id

Abstract. This research was conducted in the Sumba Strait, southwest of Sumba East Nusa Tenggara. The Purpose of this study is to determine the depositional facies characteristics of the Quaternary deep sediment. The sample was taken at a depth of 1858 m by gravity core in board of RV Baruna Jaya VII by the geology team of Ekspedisi Widya Nusantara LIPI 2016. The length of the sample was 246 cm. Sedimentology analysis was conducted to measure grain size distribution, organic material and carbonate content in the core. The characteristics of the sediment were dominated by coarse silt, poorly sorted, symmetrical skewness and mesokurtic kurtosis. Generally, the percentage of carbon and carbonate is 20.09% carbon and 9.67% carbonate; it shows that grain size is very influential on organic material in the sediment. The finer the grain of sediment the higher the organic material, which in the sediment results from the grain size analysis indicates the dominant grain size of the fine (Coarse silt).

1. Introduction

Marine sediments are an accumulation of mineral and fragment of rock mixed with the destruction of shells and bones from marine organisms and several other particles formed through chemical processes that occur at sea [1]. This sedimentology study of the Quaternary deep-sea sediment was carried out in the Sumba strait, in the southwest of Sumba, East Nusa Tenggara.

Granulometric analysis is an analysis of sediment grain size. All the size frequency of sediment is sensitive to the processes of the depositional environment [2][3][4]. Analysis of spatial changes in grain size parameters (mean, sorting and skewness) is one method that used to identify the transportation routes. According to [5], there are three factors that influence the grain size of sedimentary rock, namely the variation of the original sediment grain size, the transport process and the precipitation energy.

Loss on ignition (LoI) is a common and widely used method for determining the levels of organic material and carbonates from sediment [6][7]. Determine the percentage level of weight of organic material and carbonates in sediment with LoI based on sample hearting using muffle furnaces [6]. Many research, refined LoI method [6] to provide more accurate determination of organic matter content in terrestrial soils. The focus of the research is to analyze the characteristic of Quaternary deep sea sediment through a grain size and LoI analysis of the sediment samples taken by conducting gravity core from the Sumba strait.

2. Regional Geology

The Sumba strait is separated Flores Island (in the north) form Sumba island (in the south). The subduction activities cause volcanic activities along the island of Bali, Lombok Sumbawa, Flores to Alor to form active volcanic arc. Active deformation zones are located adjacent to the Jave and East trenches forming ridge bounded by the face arc in the north [8][4,9–14]

3. Materials and Methods

3.1 Materials

Sediment samples were taken by the geology team of the Ekspedisi Widya Nusantara 2016 LIPI. Sediment samples were taken by conducting a gravity corer on the northern part of Sumba Island (Figure 1). The total length of the samples was about 257 cm. Then the sample is sampled per 1 cm every 5 cm for an analysis of grain size and LoI. Total samples 49 to analysis grain size and LoI.

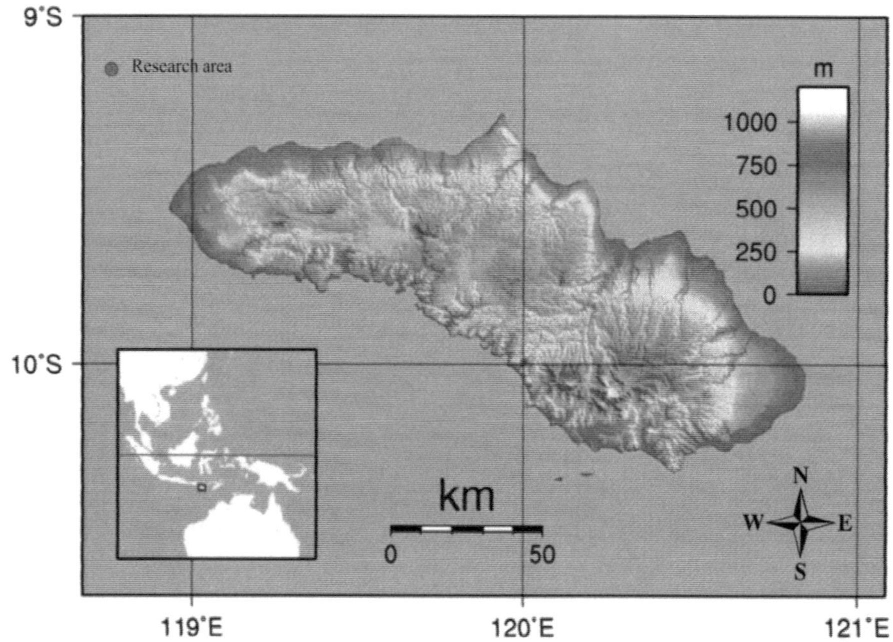

Figure 1. Location of the sample taken in the Sumba Strait.

3.2 Method

Grain size analysis was carried out using the Malvern Mastersizer (grain size analysis) in the Micropaleontology Laboratory, Research Center for Geotechnology, LIPI, Bandung. Every sample needs approximately 2 gram of sediment for this grain size separation process. The grain size parameters were (mean (x), skewness (Skø), kurtosis (Kø), and sorting (σø)) were calculated by using Gradistat v.4.0 software[15]. The grain size classification is defined by Gradistat classification [15]. After drying the sediment sample with an oven with a constant weight (usually 12-24 hours at 105 °-C in the first stage the organic material burns to ash and carbon dioxide at a temperature of 550 °C. LoI can be calculated using the following equation:

$$LOI550 = ((DW105\text{-}DW550)/\ DW105) * 100 \quad (1)$$

LoI_{550} a LoI at 550 °C (%), DW_{105} is a dry sample weight before combustion and DW_{550} is the sample weight after being heated at a temperature of 550 °C (grams). The weight loss must be proportional to the amount of organic carbon content in the sample and [15] shows a strong correlation between LoI at 550 °C and organic carbon content [16]. In the second stage, the carbon dioxide into carbonates, oxides and LoI leave can be calculated with the following equation:

$$LOI_{950} = ((DW_{550}\text{-}DW_{950})/\ DW_{105}))* 100 \quad (2)$$

Where the LoI_{950} is the LoI at 950°C (%), DW_{550} is the sample dry weight after burning organic material at 550 ° C, DW_{950} shows the dry weight of the sample after heating at 950°C, and DW_{105} is the initial weight of the dry sample before burning organic carbon (grams)[17].

4. Result and Discussion

Grain-size characteristics of the Quaternary deep-sea sediment in the Sumba strait were showing a dominant of coarse silt with bad sorting, symmetrical skewness and mesokurtic kurtosis. The LoI indicates that carbon material content was bigger than the carbonate material content. Table 1 shows the result of the grain size data analysis. Figure 5 Bivariate plots are used to show the relationship between grain size and sorting, grain size and skewness, grain size and kurtosis, sorting and skewness.

Table 1. Result on grain size statistical analysis

Statistics	Mean	Sorting	Skewness	Kurtosis
Max	5.997	1.820	0.110	1.116
Min	5.484	1.591	-0.031	1.011
Average	5.798	1.692	0.012	1.056

4.1 Stratigraphy

Based on the result of grain size analysis, it can be concluded that there are two sedimentary layers, the silt layer A and the silt layer B.

4.1.1 Silt Layer A

The silt layer A identified at a depth of 0-65 cm. This layer has a brownish gray color with coarse silt grain size. There was thin intercalation of very fine sand depth of 15 cm, 50 cm, and 65 cm. There were mollusk shells at a depth of 10 cm and increasingly to the lower layer of smooth fossil shell with multiple.

4.1.2 Silt Layer B

The silt layer B silt identified at depth of 65-240 cm. This layer has a gray color, with coarse silt grain size. There was a fine sand intercalation at depth of 200 cm. There are fossils of fine shells and white patches that are thought to be granules of foraminifera fossils.

4.2 Granulometric analysis

Steward diagram analysis, the sorting value is plotted with the value of M (Median). The result of cross plot is slow deposition form calm water wherein the area of sedimentary Sumba strait sediment with a slow sedimentation and show that precipitation occurs in the deep sea environment (Figure 3). CM diagram is used to identify sediment deposition mechanisms in Sumba strait, C (value of first percentile) is plotted with M value (median) in phi [18]. The result of CM diagram shows that the current transforms sediment material is uniforms suspension currents (Figure 4). This shows that the material has been transported far from its origin.

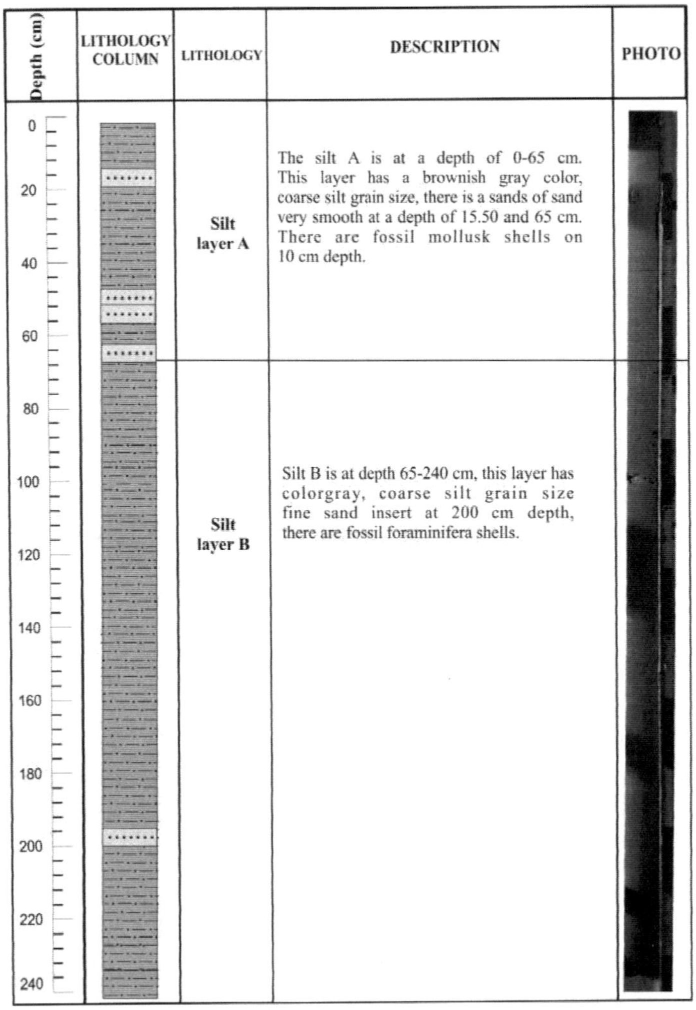

Figure 2. Stratigraphy description of silt layer A and silt layer B on research location.

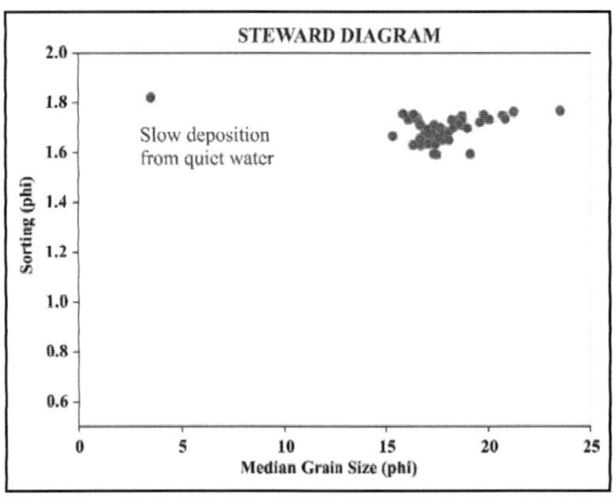

Figure 3. Steward diagram shows that all sediments in Sumba straits are affected by slow deposition from quiet water.

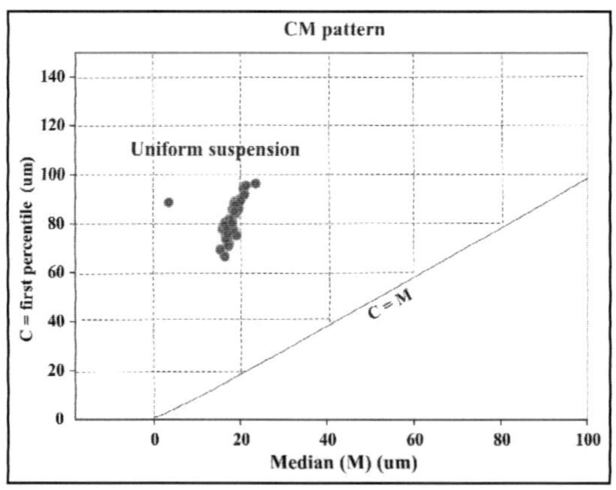

Figure 4. The relationship between C and M show that all sediment in Sumba strait was obtained on uniform suspension.

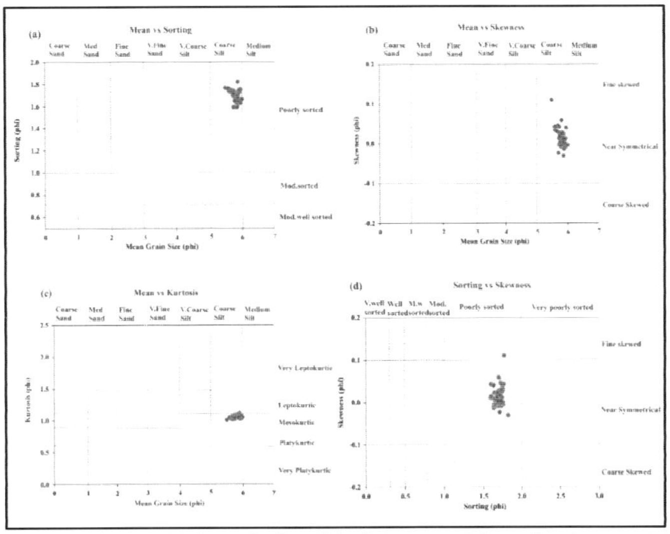

Figure 5. The bivarian plot shows the relationship between (a) grain size vs. sorting (b) grain size vs. skewness (c) grain size vs. kurtosis (d) sorting vs skewness.

Figure 5a shows the relationship between grain size and sorting in Sumba strait. So the result obtained in the Sumba strait have a coarse silt with poorly sorted. Figure 5b shows the correlation between the grain size with kurtosis. The value of kurtosis describes the condition of sediment stored in certain energy (Friedman 1965), Where in the area Sumba strait have a coarse silt grain with mesokurtic to leptokurtic kurtosis. So almost all sediment deposited from the medium-low energy environment. Figure 5c show the relationship between the grain size with skewness wherein the are Sumba strait have a coarse silt with skewness near symmetrical - fine skewed. Figure 5d shows the relationship between sorting with skewness. Sorting analysis show that the deposition area is an oceanic area, wherein the oceanic area predominantly poorly sorted. The result of skewness analysis shows that the deposited material has fine particles, so that the skewness value obtained is positive (near symmetrical - fine skewed).

4.3 Loss on ignition (LoI) analysis

The result of lost on ignition value indicates that the total carbon content is higher than the total carbonate content (Table 2). Sediment comes from biogenous sediment (plants and animals). This was confirmed by the finding of many foraminifera shell fragment and root fibers in the sample.

Table 2. Result on lost of ignition statistical analysis

Colum	Total Carbon (%)	Total Carbonate (%)
Min	7.81609762	2.647977135
Max	20.09639201	9.67044809
Average	16.6506931	7.742016963

5. Conclusion

Analysis of grain size shows the dominance of coarse silt, poorly sorted sediment with symmetrical skewness and mesokurtic kurtosis. Sediment in the Sumba strait is deposited from the medium-low energy environment. CM diagram shows that all sediment is deposited in uniform suspension. Steward diagram shows that slow deposition from quiet water influences deposited sediment. The LoI (lost on ignition) analysis show that the total carbon is greater than carbonate. Many sources of sediment come from land from the sea.

6. Acknowledgment
We would like to thank the LIPI for permission to used data Expedition Widya Nusantara 2016 and also to technicians in the geotechnology research laboratory for assistance LoI analysis.

7. References
[1] Aprilia S and Zakir M Age Determination of Marine Sediment in Spermonde Archipel Via Activity Measurement of 14c by Lsc (Liquid Scintillation Counting) *Mar. Chim. Acta* **15**
[2] Nugroho S H and Putra P S Spatial distribution of grain size and depositional process in tidal area along Waikelo Beach, Sumba *Mar. Georesources Geotechnol.* **36** 299–307
[3] Friedman G M 1967 Dynamic processes and statistical parameters compared for size frequency distribution of beach and river sands *J. Sediment. Res.* **37** 327–354
[4] Suryadi A, Choanji T and Wijayanti D 2018 Infiltration Rate of Quarternary Sediment at Rumbio Jaya, Kampar, Riau *J. Geosci. Eng. Environ. Technol.* **3** 57–62
[5] Folk R L 2002 *Petrology of the sedimentary rocks* vol 4
[6] Agricultura D D E 2009 Soil Organic Carbon in the Soil Scapes of Southeastern Tanzania
[7] Yuskar Y, Choanji T, Suryadi A, Putra D B E and Ramsof R A 2018 Coastal Deposit Characteristic Influenced by Terrestrial Organic Matter and Its Sedimentary Structure at Jangkang Beach, Bengkalis District, Riau Province—Indonesia *Proceedings of the Second International Conference on the Future of ASEAN (ICoFA) 2017--Volume 2* pp 909–16
[8] Watkinson I M and Hall R 2016 Fault systems of the eastern Indonesian triple junction: evaluation of Quaternary activity and implications for seismic hazards *Geol. Soc. London, Spec. Publ.* **441** SP441.8
[9] Kausarian H, Choanji T, Karya D, Kadir E A and Suryadi A 2018 Geological Mapping of Silica Sand Distribution on the Muda Island and Ketam Island, Estuary of Kampar River, Indonesia *Proceedings of the Second International Conference on the Future of ASEAN (ICoFA) 2017--Volume 2* pp 973–82
[10] Choanji T, Rita N, Yuskar Y and Pradana A 2018 Analog Study of Fluid Flow on Deformation Band at Petani Formation, Riau, Indonesia *MATEC Web of Conferences* vol 159 p 1007
[11] Yuskar Y, Choanji T, EP D B, Suryadi A and Ramsof R A 2018 Coastal Deposit Characteristic Influenced by Terrestrial Organic Matter and Its Sedimentary Structure at Jangkang Beach, Bengkalis District, Riau Province—Indonesia BT - Proceedings of the Second International Conference on the Future of ASEAN (ICoFA) 201 *Proceedings of the Second International Conference on the Future of ASEAN (ICoFA) 2017 – Volume 2* ed R Saian and M A Abbas (Singapore: Springer Singapore) pp 909–16
[12] Lubis M Z, Anurogo W, Kausarian H, Surya G and Choanji T 2017 Sea Surface Temperature and Wind Velocity in Batam Waters Its Relation to Indian Ocean Dipole (IOD) *J. Geosci. Eng. Environ. Technol.* **2** 255–63
[13] Yuskar Y and Choanji T 2017 Uniqueness Deposit of Sediment on Floodplain Resulting From Lateral Accretion on Tropical Area: Study Case at Kampar River, Indonesia *J. Geosci. Eng. Environ. Technol.* **2** 14–9
[14] Yuskar Y and Choanji T 2016 *Sedimentologi Dasar* (Pekanbaru, Indonesia: UIR PRESS)
[15] Blott S J and Pye K 2001 Gradistat: A grain size distribution and statistics package for the analysis of unconsolidated sediments *Earth Surf. Process. Landforms* **26** 1237–48
[16] Veres D 2002 A comparative study between loss on ignition and total carbon analysis on minerogenic sediments *Geologia* **47** 171–82
[17] Santisteban J I, Mediavilla R, López-Pamo E, Dabrio C J, Blanca Ruiz Zapata M, José Gil García M, Castaño S and Martínez-Alfaro P E 2004 Loss on ignition: a qualitative or quantitative method for organic matter and carbonate mineral content in sediments? *J. Paleolimnol.* **32** 287–99
[18] Passega R 1964 Grain size representation by CM patterns as a geologic tool *J. Sediment. Res.* **34** 830–47

IOP Conf. Series: Materials Science and Engineering **797** (2020) 012005 doi:10.1088/1757-899X/797/1/012005

Technical Feasibility Analysis of Waode Buri-Lelamo Port as Local Feeder Port in North Buton Regency Southeast Sulawesi Province

Sachrul Ramadan[1], Ishak Kadir[1], Muhammad Zakaria Umar[1*], L M N Arsyad[2]
[1]Department of Architecture, Faculty of Engineering, Universitas Haluoleo, 93232 Kendari, Sulawesi Tenggara, INDONESIA
[2]Department of Civil, Faculty of Engineering, Universitas Haluoleo, 93232 Kendari, Sulawesi Tenggara, INDONESIA

[*]Corresponding author's e-mail: muzakum.uho@gmail.com

Abstract. Wa Ode Buri-Lelamo Port is located in North Kulisusu District, North Buton Regency, Southeast Sulawesi Province, Indonesia. This study can be information about the prospect of developing Lelamo Port, so that it can support economic acceleration in the islands of Southeast Sulawesi Province. This study is aimed at increasing the port of Lelamo to become a local feeder port. Local ports are adjusted to eligibility standards. This feasibility standard is based on the regulation of the Minister of Sea Transportation, No. PM 146, 2016. Data obtained by observation and literature. Data related to port planning. Data is analyzed in a quantitative way. Data is given scoring against eligibility requirements. Technical feasibility requirements include land topog, bathymetry and hydro oceonography. The results of this study concluded that the Waode Buri-Lelamo port is considered feasible to be a local feeder port. This study recommends that the port needs to be made trestell with a size of 6 m x 22 m, a pier made with a size of 8 m x 70 m, and the lowest apron height of 5 m from tides.

1. Introduction

Waode Buri-Lelamo Port is located in North Kulisusu District, North Buton District, Southeast Sulawesi Province, Indonesia. This port is 10 km from the city. This port is the gateway to North Buton. This port has problems, as follows: 1) port support facilities are inadequate; 2) minimal port infrastructure components, such as parking, waiting rooms and causeways; 3) this port only serves small-scale vessels (2500 GRT). Buri-Lelamo Port is designated as a local feeder port [1]. This Ministerial Decree concerns the National Port Master Plan for the Waode Buri-Lelamo port. Thus, this port needs to be upgraded to the study of local feeder port facilities. This study analyzes the value of the technical feasibility of the port area (facilities, infrastructure, and marine areas) against the local feeder port standard. This study is important to do, as follows: 1) this study is a reference for local feeder ports in the archipelago; 2) this study can be a reference for economic development in the island and coastal regions. This study is aimed at increasing the port of Lelamo to become a local feeder port. Local ports are adjusted to eligibility standards. This feasibility standard is based on the regulation of the Minister of Sea Transportation, No. PM 146,2016.

2. Literature Review

The aspects of the technical feasibility of the port are discussed, as follows: 1) the type and characteristics of the ship adapted to the minimum standards of local passenger ports; 2) port facilities adapted to the minimum local port standards, such as ship circulation, port loading and unloading, and passenger circulation.

2.1. Ship Principal Size

If the ship's plan is unknown, the pier length is rounded up. The length of the ship is 5 meters. The depth of the pier pool is rounded up. The pier pool is obtained with 0.5 meters and is calculated by LLWL elevation (Table. 1).

Table 1. Typical dock size for pioneer ships [2].

DWT	GT	Dock length	Depth of Pier	Apron width
(t)	(t)	(m)	(m)	(m)
350	445	60	5	15
500	745	65	5	15
750	980	75	5	15
1000	1200	80	5	15

2.2. Ship Flow and Wave Height in Port Pool

The area of the swivel pool is used to change the direction of the minimum vessel. The area of the swivel pool is obtained by means of a circle radius 1.5 multiplied by the total length of the vessel. The length of the ship used the length of the largest ship. The area of the minimum swivel pool is obtained by means of the circle radius equal to the total length of the vessel (Loa). The minimum swivel pool area is required with the help of anchors and tugs. The depth of the pond is obtained by 1.1 x draft of the ship. The terms of the ship's draft are fully charged and below the water level. The cruise lay out is in a straight line and the current is opposite to the direction of the ship. The turn of the ship is in the shape of a curved curve. Critical wave height is required in the port pool of 0.3 m - 1.5 m.

2.3. Criteria for Local Feeder Ports

Local ports serve sea transportation activities and have a local scale. This is based on PP 61/2009 on Shipping and KM 53/2002 concerning the National Port Order. Local port criteria, as follows: 1) the local port acts as a feeder port; 2) the local port acts as a passenger service area; 3) the local port acts as a place for sea transportation services; 3) local ports are not traversed by regular sea transportation routes; 4) the local port has a minimum depth of -1.5 m LWS; 5) local ports have mooring facilities; 6) local ports have a distance of 5 - 20 miles; 7) the local port has a low loading and unloading volume <8,588,000; 6) the local port has a long distance classification (International / ALKI lines> 179 miles); 7) local ports have a classification of docks to the size of short ships <103 m; 8) the local port has a depth of <5 m.

3. Research Methode

Port feasibility analysis is carried out using Quantitative methods. The criteria of this study refer to KP 414 of 2013 [3]. KP 414 of 2013 is a master port plan for the sea and the Decree of the director general of sea transportation. This regulation is about technical guidelines and procedures for the construction of sea ports and has been modified. building design at the port can reduce wave height [4]. The eligibility criteria are given scoring (Table 2), as follows:

Table 2. Indicator of Technical Feasibility Assessment (Source: KP 414 of 2013 concerning sea port master plan and Director General of Sea Transportation Decree. This regulation is about technical guidelines and procedures for the preparation of sea port development 2015 and has been modified).

Criteria	Weight	Feasibility Indicator	Value
Topography and Slope	25%	• Topographic modeling has a low value and hilly contours of 100 - 500 Meters of MPA.	1
		• Topographic modeling has a moderate value and quite hilly contour 0-100 meters DPL.	5
		• Topographic modeling has high and contour values obtained from 0-100 meters DPL.	10
Bathymetry	50%	• Bathymetry modeling has a low value. This is based on a minimum standard of depth of sea portsThe local local port hierarchy is equated with local feeder ports. The hierarchy of local feeder ports is a value of -4 LWS or 4 m depth (RIPN 2013).	10
		• Bathymetry modeling has a moderate value. This is based on a minimum standard of depth of sea ports. The hierarchy of local feeder ports is a value of -4 LWS or 4 m depth (RIPN 2013).	5
		• Bathymetry modeling has a high value. This is based on a minimum standard of depth of sea ports. The Local Feeder Port hierarchy is obtained at a value of -4 LWS or 4 m depth (RIPN 2013).	1
Hydro - Oceonography	25%	• Hydroeconographic modeling has a low value. This is influenced by tides, monsoons, and high waves.	10
		• Hydroeconographic modeling has a moderate value. This is influenced by tides, monsoons, and medium waves.	5
		• Hydro-oceanography modeling has a high value. This is influenced by tides, monsoons, and low waves.	1

4. Result and Discussion

4.1. Port Area Location
The geographical position of the Waode Buri - Lelamo port, as follows: latitude 4o39'31,740s and longitude 123o11 ', 170e. This port is a distance of ± 10 km from the City Center. This port has facilities, such as waiting rooms, parking lots and causeways. This port has no tresteel yet. This port can only be visited by public boats. Pioneer ships are still anchored outside the Port (Fig. 1).

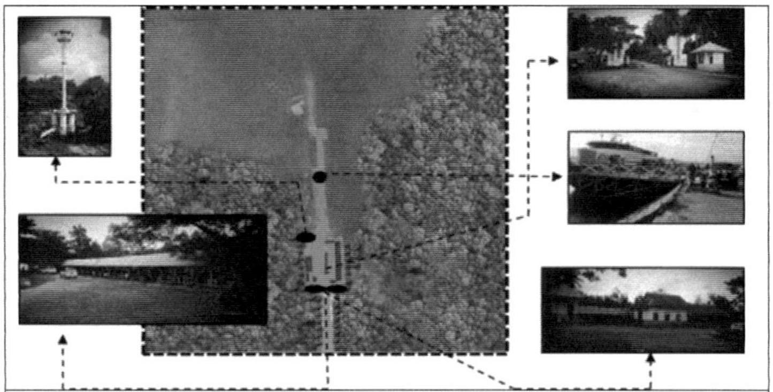

Figure 1. Port Area existing.

4.2. Topography and Slope
Topographic analysis uses data and information. Data and information are obtained in the Regional Spatial Planning document, North-Southeast Buton District, 2014-2034. This document is completed with observations in the field

The location of the Waode Buri port has a slope of 8-15% (hilly) and height at the port plan of ± 4 meters DPL (Table 3). The data processing results above are compared with indicators. Assessment indicators use topographic and slope technical aspects, as follows:

Table 3. Assessment indicators use topographic and slope (Source: Topographic FS Analysis Results, 2017).

Aspect	Weight	Assessment Indicator	Value
Topography and slope	25%	• Topographic Modeling results have a high value and have a sloping value of 0-100 meters DPL	10

4.3. Topography and Slope
Waode Buri Port is deeper than the standard. This is based on a minimum standard of depth of sea ports. The standard hierarchy of local feeder ports is a value of -4 LWS or a depth of 4 meters (RIPN 2013) (Table 4).

Table 4. Bathimetry analysis of the port area based on coastline distance (Source: Results of Bathimetry FS Analysis, 2017).

Bathymetri estimates		Interval	Assessment	Value
Coastline Towards Deep Waters (m)	Aquatic Depth (m)			
0 – 50	20	0 – 100	The port is located in a group interval of 0-100 meters. Bathymetry modeling has a high value.	10

4.4. *Hydro Oceanography*

North Buton Regency has a flat coastline. The coastline ranges from 0-100 meters to the sea. Waode Buri Port is located inside the bay and has a depth of 22 m from the 50-meter coastline. The Port location is located ± 1350 m and leads to the open sea. Thus, the impact of wind and waves is eliminated by the shape of the bay.

4.5. *Tidal Data*

HARI KE	WAKTU PENGAMATAN (JAM)																							
	0.00	1.00	2.00	3.00	4.00	5.00	6.00	7.00	8.00	9.00	10.00	11.00	12.00	13.00	14.00	15.00	16.00	17.00	18.00	19.00	20.00	21.00	22.00	23.00
1	60	30	40	100	120	150	190	200	210	170	120	90	70	60	90	130	150	180	220	210	180	150	130	110
2	100	50	30	60	80	130	180	200	210	200	180	160	130	110	90	60	90	110	120	160	200	220	210	180
3	130	80	40	30	60	80	110	150	170	190	180	170	160	150	120	110	80	120	150	170	190	190	200	170
4	160	90	60	40	50	80	100	120	150	160	170	160	130	110	90	80	90	120	150	170	180	190	190	180
5	150	110	100	80	60	80	100	130	150	160	170	160	150	140	130	110	100	100	110	130	150	170	190	180
6	160	140	120	110	90	70	80	90	120	130	150	160	150	140	130	120	110	100	120	120	130	130	140	150
7	145	150	160	150	130	120	100	90	95	110	120	130	140	150	160	150	140	130	120	110	130	135	140	150
8	140	160	155	140	130	110	90	85	80	90	100	110	140	160	170	160	155	140	120	110	100	110	120	130
9	130	150	160	170	160	150	140	130	120	90	85	110	135	150	160	180	170	160	150	130	120	100	90	100
10	100	140	150	160	170	170	150	140	120	110	90	110	120	135	160	170	185	170	160	120	100	90	70	80
11	90	120	135	155	180	170	160	140	120	100	90	80	85	90	120	160	190	180	140	120	100	90	80	55
12	80	120	130	140	150	190	195	150	140	120	90	80	90	120	140	160	180	210	190	160	140	120	90	60
13	50	90	120	130	155	185	110	170	160	140	100	95	70	100	120	140	190	200	190	180	160	80	70	55
14	50	65	100	110	175	190	200	190	170	140	120	100	80	90	110	140	180	200	190	170	150	120	110	80
15	40	30	35	90	110	130	170	180	200	190	150	130	120	80	110	120	140	170	200	180	160	150	120	90

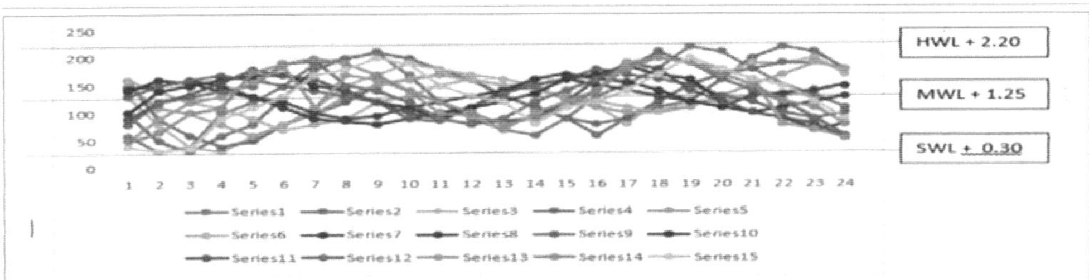

Figure 2. Tidal FS Oceonographic Survey Results at Waode Buri Port (Source: Tidal Location Survey, 2017).

4.6. *Sea Wave Data*

Table 5. Number and Frequency of Wave Events (Source: The results of the FS 2017 survey analysis).

Jumlah Frekuensi Kejadian Gelombang							Persentase Frekuensi Kejadian Gelombang (%)					
Arah	Kecepatan Angin (m/det)					Jumlah	Kecepatan Angin (m/det)					Jumlah
	< 0,1	0,1-0,2	0,2-0,3	0,3-0,4	> 0,4		< 0,1	0,1-0,2	0,2-0,3	0,3-0,4	> 0,4	
Utara	< 0,2	0	0	0	0	0	0,00	0,00	0,00	0,00	0,00	0,00
Timur Laut	< 0,3	0	0	0	0	2	1,39	0,00	0,00	0,00	0,00	1,39
Timur	< 0,4	0	0	0	0	6	4,17	0,00	0,00	0,00	0,00	4,17

Tenggara	< 0,5	0	0	0	0	4	2,78	0,00	0,00	0,00	0,00	2,78
Selatan	< 0,6	0	0	0	0	3	2,08	0,00	0,00	0,00	0,00	2,08
Barat Daya	< 0,7	0	0	5	5	10	0,00	0,00	0,00	3,47	3,47	6,94
Barat	< 0,8	0	19		2	59	0,00	0,00	13,19	26,39	1,39	40,97
Barat Laut	< 0,9	0	1	0	0	1	0,00	0,00	0,69	0,00	0,00	0,669
Jumlah Bergelombang						85	Persentase Bergelombang					59,03
Jumlah Tidak Bergelombang						59	Persentase Tidak Bergelombang					40,97
Total						144	Total					100

4.7. Wind Velocity

Wind speed is measured at the dock point. The results of wind speed measurements are as follows: Most winds are between 1 - 11.5. m / sec (22 Knots) (Table 6).

Table 6. Oceonographic analysis of tides, wave height and wind speed in the sea port area (Source: Results of Tidal FS Analysis, 2017).

Hydro Oceonography Conditions	Assessment	Value
Tidal Conditions: 0 – 0,5 meter Wave Height: 0 – 1,5 Meter Wind velocity: 2 – 20 Kno	The results of hydro-oceanographic modeling have high values	10

4.8. The Width And Flow Of The Ship Turning On The PortArea

The flow of the ship starts from the gate, the port area and the dock. The flow of the ship has a length of 1,310.54 meters. Groove width 01 = 137.15 m, groove width 02 = 220.10 m, and groove width 03 = 488.80 m. The flow of the ship is required with the flow of the local feeder port. The local feeder port has a groove width of at least 114 m (7 x Pl + 30 Length). The maximum ship rotation path is valued at R => 5 Loa (Figure 8 A). The boat rotation path = 220.10 m is the maximum value of R => 3 Loa (Figure 8 B). The distance of the ship's groove starts from the lighthouse gate to the Port = 1.310.54 M. The width of the swivel channel of this port vessel meets the feasibility standard.

5. Conclusion

This Based on the feasibility study above, the Waode Buri-Lelamo port is considered feasible to be a local feeder port. This study recommends that the port needs to be made trestell with a size of 6 m x 22 m, a pier made with a size of 8 m x 70 m, and the lowest apron height of 5 m from tides.

References

[1] The decision of the Minister of Transportation No KP 901 2016
[2] Badan Standar Nasional, SNI xxxx-xxxx tentang kapal Perintis
[3] The decision of President KP 414 of 2013, Tentang Rencana Induk Pelabuhan Laut
[4] Kramadibrata S 1985 *Perencanaan Pelabuhan* Bandung Ganeca Exact
[5] The decision of sea transportation minister No 53/2001, Tentang Tatanan Pelabuhan Nasional
[6] Keputusan Jenderal Perhubungan Laut Tentang Petunjuk Teknis dan Tata Cara Penyusunan Pembangunan Pelabuhan Laut 2015
[7] The Rules of President No 61/2009, It Is About The Nationality of Sailing

IOP Conf. Series: Materials Science and Engineering **797** (2020) 012006 doi:10.1088/1757-899X/797/1/012006

Community Based Tourism for the Concept of Village Tourism Facility Development in Papan Island

Irdinal Arief[1*], Harifuddin Thahir[2]
[1]Department of Architecture and Planning, Faculty of Engineering, State University of Tadulako, Soekarno Hatta KM.9, Tondo, Mantikulore, Palu, Sulawesi Tengah 94148.
[2]Department of Management, Faculty of Economics, State University of Tadulako. Soekarno Hatta KM.9, Tondo, Mantikulore, Palu, Sulawesi Tengah 94148.

*Corresponding author's e-mail: irdinalarief@gmail.com

Abstract. Community based tourism can create employment opportunities for local communities, and reduce poverty, where tourism income is from tourist services to tourists. In the Planning of facilities and infrastructure is needed the concept of development with the location of the local natural environment. It to be developed must be integrated. At the stage of development of tourism facilities, as natural resources and man-made resources are absolutely needed by tourists in their travels in tourist destinations, such as: roads, electricity, water, telecommunications, terminals, bridges and so forth. With the pattern of community-based tourism does not mean that people will run their own tourism business. The level of tourism implementation should be seen as part of integrated development planning that will be carried out in Papan Island of Kadoda Village Talatakoh Subdistrict of Togean Islands, which is designated as the national tourism development and the regional tourism master plan of Tojo Una-Una Regency. One of the strategies is development of marine tourism attraction in Strategic area of National Togean Archipelago, consisting of 50 national tourism destinations. The following spatial concepts that actually include many studies in planning and architecture but this process has included analysis of community-based tourism.

1. Introduction

Togean Islands is geographically located in the middle region of Tomini Bay which consists of large and small islands. It has their own uniqueness from the standpoint of biogeography, as one of the island's ecosystems located in the transition zones of the Wallacea and Weber lines (See Figure 1). Togean Islands is a natural tourist area that is highly favored, because it has the potential of fishery resources, biodiversity of marine and terrestrial biota and the richness of local culture (customs). One of the islands that became the tourist attraction of the Papan Island. The administration of Papan Island is located in Talatako sub-district, Kadoda village, which is part of the Togean archipelago, from existing data of Papan Island inhabited ± 200 KK, which is part of Malenge island and is a Bajo tribe (See Figure 2). Initially the island was only a place of worship and sanitation while houses were made around the island with still be pegged on the seabed. Their house looks very simple because it only consists of boards and mangroves that hold the air as a buffer. Most of the Buildings on the Papan Island is a building with a household typology. Based on the width, the Island Board's width is 1.2 Hectares, with the distribution of 70.2% housing infrastructure at sea and 29.7% land, which is a public facility and housing (See Figure 4 and 5).

IOP Conf. Series: Materials Science and Engineering **797** (2020) 012006 doi:10.1088/1757-899X/797/1/012006

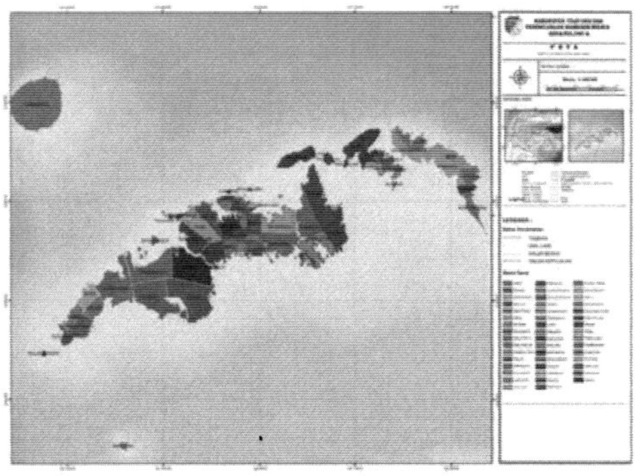

Figure 1. Map. Togean Islands National Tourism Area

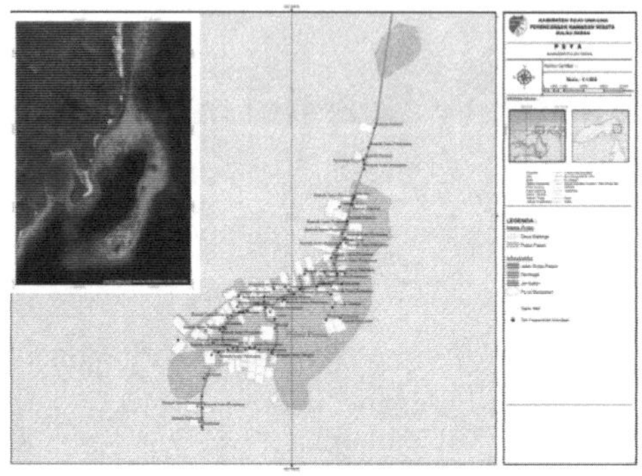

Figure 2. Map Papan Island

Figure 3. Atmosphere of Settlements on Papan Island

On the Papan Island can find a long bridge about one kilometer connecting Papan Island with Malenge Island which is the center of the village of the local area. In the middle of the village on Papan Island there is a place called "Rock Peak". This is usually used as a place for the children of they to play with the tourists. From the top of this peak can be seen the entire length of the bridge and also the view of the sea around the Papan Island (See Figure 3).

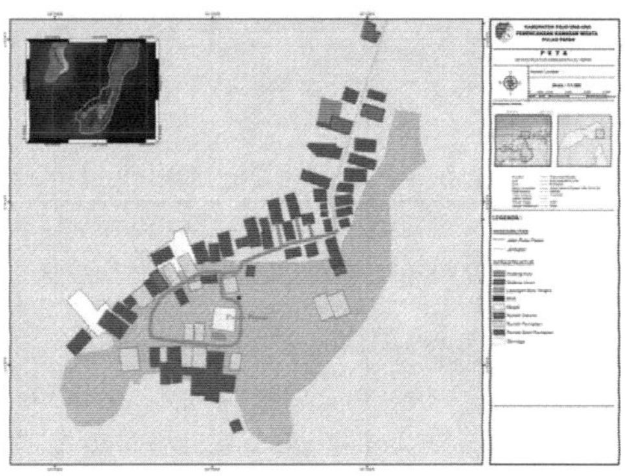

Figure 4. Map of Infrastructure around Papan Island

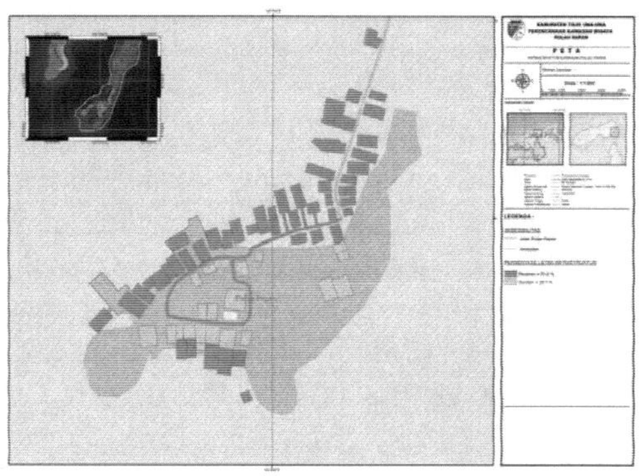

Figure 5. Map Infrastructure Percentage Buildings at sea 70.2% & terrestrial 29.7%

2. Purposes of research

The goal is how to create content from the community that can be used on Pulau Papan. For example, how do people who can work in managing and constructing. We know they are known for their excellent fishing skills. Work as a fisherman. One of them is their ability while enjoying breath in the air with a long duration. This ability is of course an extraordinary ability. Everyone in this village works as fishermen while women work at home or help fish around the house.

The main livelihood of the Bajo tribe is to fish in a way that is still fairly traditional, such as fishing, archery, and crawling fish. The fish will be sold to residents around the coast or nearby islands. The life of theirs is still very simple. Establishing permanent settlements may not have occurred to them if they were not called upon by the local government. The close ties of it and the sea can be seen from the marine environment around the village. They are very guarding the conservation of the sea because the sea is the source of their livelihood so that the Bajo people are very guarding the aquatic ecosystems around the Papan Island. Some of them even already familiar with certain marine product cultivation techniques, such as lobsters, grouper, shrimp, and so forth. They call the cultivation area as a floating pond that is usually located not far from the settlement. A small number of their have even made permanent houses using cement and glass windows.

3. Theory of the Community-Based Tourism Concept

3.1. Conceptual framework of regional development

This concept is the basis for constructing a simple structural framework, which can provide a new foundation for more creative tourism design and regional development. To create a draft concept requires application specifically from design principles. Therefore, humans as users, especially for their travel activities, and the environment requires a functional space for tourism activities.

3.2. The Concept of Regulating the Movement of Regions

Two basic things in the arrangement of the tourist area are:

- First, the attraction group (based on environment, coverage, magnetism, benefits and creation) is the raw material where visitor satisfaction is obtained. For a fun trip, the attractions reveal a whole series of activities.
- Secondly, all attractions, from the most remote to the most urban, are linked to community groups / service centers.
- Third, the fundamental thing that determines the accessibility of tourists is linkage. The linkage between attraction and community and the interconnection between community and access systems that play a role in tourism attractions.

These three elements-attraction groups, community service centers, and linkages deserve special design treatments for specific land uses. In other words, systemic linkage is built is an important thing that will determine the degree of accessibility to a location of attractions.

The concept of it is commonly used by designers of tourism development strategies to mobilize communities to actively participate in development as a tourism industry partner, along with its main focus and the following in the picture below (See Figure 6):

- Strengthening community economic enterprises (related to efforts to increase the value of economic benefits for the community)
- Strengthening of community institutions (related to capacity building and the role of the community to actively participate in tourism development activities)
- Strengthening "tourism awareness" (related to efforts to improve awareness and the role of the community in supporting the development of tourism as well as being a tourism / tourism agent
- Utilize optimally local resources. aspects of local community involvement include 3 (three) areas, i.e. planning stage, implementation stage or implementation stage, and in terms of sharing benefits both socio-cultural.

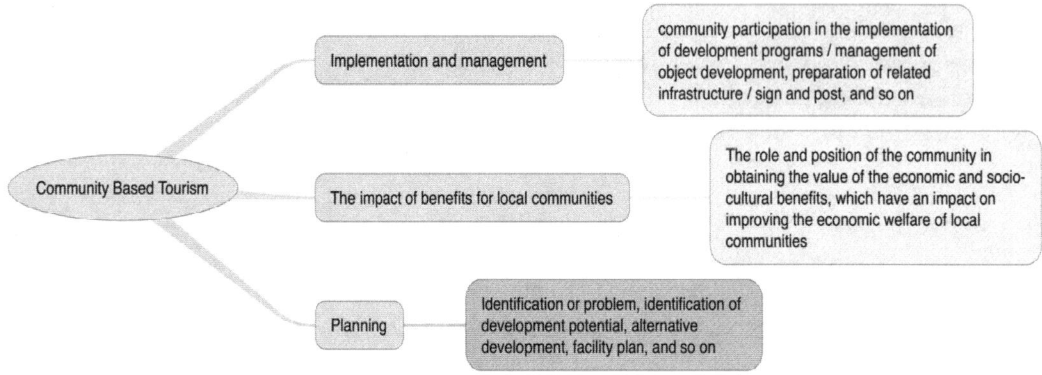

Figure 6. Community Based Tourism Development Scheme

4. Method

The approach taken in this research using rationalistic approach. Data collection is done through primary and secondary surveys, media reviews and literature studies. In the sampling is done by using purposive sampling method to get the respondent competent or influential in achieving the target obtained by using stakeholder analysis.

Analytical techniques that will be used to achieve the objectives and targets, among others, by using descriptive qualitative analysis method, Delphi analysis aims to determine the consensus group for factors supporting the development and analysis of triangulation, FGD (Forum Group Discussion).

5. Result of Village Tourism Facility Development

The development of tourism facilities in Papan Island, which has a land area of 1.2 Hectares and surrounded by the density of Bajo tribe house, becomes a challenge in its development, it is also based on the results of FGD (Forum Group Discussion). Therefore, it is necessary to conceptualize the development of related areas to the community-based eco-tourism, using variables with the results of the analysis below:

- The daily activities of the Bajo community in fishermen, fishing traditionally, usually the local government makes fishing races every year.
- Maintaining the customs of Bajo tribe and commemorate the big day of Islam by holding reading Al-Qurán events in the life of them, as the tourist attraction of the village
- Making education in fisheries, studying the cultivation of fisheries, especially fish cultivation Tuna that often appear in the area around the Papan Island
- Increased accessibility to Papan Island
- Provide supporting facilities and infrastructure such as public toilets and praying rooms at tourist sites
- Provide Floating Villa concept accommodation, Bajo's home adaptation on the water
- Provide souvenir shops from Raising shells and craft skills
- Provide a place of entertainment and recreation without changing the values, culture and norms that exist in Papan Island.
- Improving the quality of human resources (HR) provides training in tourism to increase community participation in activities
- Increase homestay as the home of the village community for tourists
- Rent of tourist facilities, so residents engage in tourist villages

- Additional skills and waste in recycling, also made landfill and the presence of a parent garbage fleet, waste officers for the village. Created by village rules with the fine, in every house or public the existence of trash cans.

After the descriptive analysis of the variables and generated the above factors then performed a Delphi analysis involving respondents. Based on the exploration stage, finally the consensus of all respondents on the factors supporting the development of the tourist village area in Papan Island. And in can factor new findings, among others:

- Promotion media as an identifier of the tourist village area on Papan Island to the outside community
- Management of the local community to take an active role in the development of the tourist village area
- Government policy in controlling development of tourist village area on Papan Island

Concept of Development of Village Area Tourism Papan Island

At the stage of formulation of the concept of development of tourist village area is done by triangulation technique with the data source used in the analysis is the factors supporting the development of tourist village area and policies related to the concept of development of tourist village area on Papan Island. The concept is divided into two, namely the concept of spatial and non-spatial concept. Here's a spatial concept:

- Provide travel routes that surround the area of the tourist village that shows the daily activities of the village community.
- The availability of special sea transportation to the attractions, for the convenience of travel tourists to the attractions.
- Provision of supporting facilities and supporting tourism in every tourist attraction that has no facilities that support the development of undeveloped attractions.
- Provision of lodging facilities concept floating villa / cottage which also makes the typical home Bajo Tribe.
- Provide a souvenir shop that sells fishery products, or souvenirs that are characteristic of the tourist village area so that it can be recognized by the outside community.
- Provision of floating restaurant facilities as well as galleries and entertainment that provide a rustic atmosphere, maintained cleanliness and presents a menu related to the essence of the sea.
- Provide recreation areas that still utilize existing conditions in Papan Island
- Providing alternative access from Papan Island to Malenge Island, via Flying Fox, it can also be a tourist attraction.

While non-spatial development is:

- Making it as a rule to tourists visiting and doing tourism activities in the Region
- Developing a fishery-based tourism village area that also has other attractions such as how to maintain and maintain a sustainable ecosystem of marine habitat, by diving and guarding the coral reefs.
- Improving the quality of human resources by providing training, increasing the insight about tourism, how to treat tourists and also fluent in English so that people can feel directly the impact of the development of tourist village area
- It takes a media campaign by creating a web about the area of tourist village that is directly connected to the web Tojo Una-Una Regency and also work together with existing promotional media
- In the process of developing a tourist village area on Papan Island also involves the local community or community leaders from Kadoda Village in the process of development.
- Required regulations or government policies governing the development process.

6. Implementation

The current implementation that has been done in terms of human resources is the stage of introduction and discuss. Local economics and entrepreneurship are included in village regulations that are obliged to bring up village-owned enterprises or village co-operatives for the development and management of village income independently, while infrastructure development is included in the first phase of construction assisted by the village Ministries and disadvantaged regions (KEMENDES). The following spatial concepts that actually include many studies in planning and architecture but this process has included analysis of community-based tourism, this concept has also been approved by the Village Kadoda in Papan Island and Research Planning Research Agency Tojo Una-Una.

7. Conclusions

Based on the results of the analysis that has been done before, it can be concluded that the daily activities of the community that is the characteristic of the Papan Island very potential to be developed into a tourist village based on community based tourism which has other tourist attraction is how to care and in order to sustain the ecosystem of marine habitat, by diving and preserving coral reefs, then the concept of spatial development in priority on improving the economy and local resources as follows that provides a travel route that surrounds the tourist village that shows the daily activities of the village community, Provision of lodging facilities concept floating villa / cottage which also makes the typical home Bajo tribe, and Provision of floating restaurant facilities as well as galleries and entertainment that provides a rural atmosphere, maintained cleanliness and presents a menu that is related to an seefood. Provision of supporting facilities and supporting tourism in every tourist attraction that has no facilities that support the development of undeveloped attractions.
While the concept of non-spatial development should be done to develop the area of tourist villages in Papan Island in order to develop, among others:

- Provide training, add insight on tourism, how to treat tourists and also fluent in English so that people can feel directly the impact of the development of tourist village area
- Creating a web about the tourist village area that is directly connected to the web Tojo Una-Una Regency and also work with existing promotional media
- Involve the local community or community leaders from Kadoda Village in the process of developing the tourist village area.

Acknowledgements

This study is supported by Development Planning Agency and regional research, Regency Tojo Una- Una, Central Sulawesi.

References

[1] Gunn, Clare A. (2002). *Tourism Planning.* (New York City : Taylor and Francis)
[2] Jones, Samantha. (2005). *Community-Based Ecotourism: The Significance of Social Capital". Annals of Tourism Research. Volume 32*, Number 2. (Great Britain: Pergamon, Elsevier) P 303 – 324
[3] RIPARDA. (2010). *Research Planning Research Regency of Tojo Una-Una.* (Bappeda Tojo Una-Una)
[4] Ristianti, N, S. (2015). *Revitalization of Bajo Trib's Settelement in Kabalutan Village, Togean Islands.* (Indonesia : Semarang)
[5] Sugiyono. (2006). *Educational Research Methods: Quantitative Approach, Qualitative and R&D.* [Bandung: Alfabeta]
[6] Prasiasa, Putu Oka (2012). *Tourist destination, community based,* (Jakarta : Salemba Empat)

Characterization of Liquid Volatile Matter Rice Husk Using Gas Chromatography

Wa Ode Sitti Ilmawati[1*], Muhammad Jahiding Mashuni[2], Kinanti[1]
[1]Department of Physics, Universitas Haluoleo, 93232 Kendari, Sulawesi Tenggara, INDONESIA
[2]Department of Chemistry, Universitas Haluoleo, 93232 Kendari, Sulawesi Tenggara, INDONESIA

*Corresponding author's e-mail: waode.ilmawati@uho.ac.id

Abstract. Rice husk is one type of organic waste that has not been used optimally. Rice husks contain cellulose, hemicellulose and lignin, so they are potentially processed as liquids volatile matter (LVM). The purpose of this study was to determine the effect of pyrolysis temperature on the density and chemical composition of rice husk LVM. This study began with the collection of rice husks. Rice husk is dried and then heated with high temperature 400°C, 500°C, and 600°C. The heating temperature is maintained for 15 minutes, respectively. The process of pyrolysis of rice husk produces charcoal, tar, and smoke. Condensed smoke is called rice husk LVM. The density of rice husk LVM at each pyrolysis temperature is different, namely0.416 g/cm^3, 0.714 g/cm^3, and 1.139 g/cm^3. The rice husk LVM is characterized using gas chromatography. Characterization results showed that the chemical composition of rice husk LVM consisted of ammonia, propanone, acetic acid, hexane, furfural, phenol, etc.

1. Introduction

Rice is the main food crop that can grow in all parts of Indonesia. Indonesian rice production from 1993 to 2015 has always increased and reached more than 75 million tons in 2015 [1]. Rice grains are ground to separate rice granules and husk. The weight of rice husks is 20-30% of the weight of rice seeds. If rice production is high, rice husk production will also be high. Meanwhile, only a small portion of rice husks were used, among others, to burn red bricks, pedestal chicken hens, and others [2].

Rice husk contains 42.2% cellulose, 18.47% hemicellulose, 19.4% lignin, 17.33% ash, and 2.6% water [3]. The lignocellulosic material is potentially processed into more useful materials such as char [2], briquettes [4,5], liquid smoke [6,7], and tar [8]. Biomass processing is carried out by heating at high temperatures above 300°C [4,6,7]. Biomass heating methods include carbonation method [4], pyrolysis [6,7], and microwave [9].

The advantages of pyrolysis method compared to other methods, pyrolysis can produce three products at once, namely charcoal, tar, and smoke. During the pyrolysis process, biomass compounds are gradually degraded. Hemicellulose is degraded at 220°C-315°C, cellulose is degraded at 315°C-400°C, and lignin is degraded in a wider temperature range than 160°C-900°C[10]. Hemicellulose is degraded into acetic acid, furfural, furan, furanone, methanol, formaldehyde, acetone, acetone, lactone, etc. Cellulose is degraded to levoglucosan, hydroxyl acetaldehyde, furfural, hydroxyl methylfurfuralfuran, methanol, etc. Lignin is degraded to 2-methoxyphenol, 2,6-in methoxyphenol, cathecols, phenol, alkyl phenol, methanol, furfural, acetic acid, formaldehyde, acetone, acetol, lactone, etc. [11].

Pyrolysis is a thermo-chemical process in which organic matter is converted into solid carbon and volatile substances by heating without oxygen [12]. Pyrolysis products include charcoal, tar, condensed smoke, and uncondensed smoke [13]. Smoke is a suspension of solid and liquid particles in the gas medium. Smoke is produced by incomplete combustion of the decomposition of constituent polymers into low molecular weight organic compounds due to heat effects including oxidation, decomposition, polymerization, and condensation [14]. Condensed smoke is referred to as liquidvolatile matter (LVM). LVM compounds were identified using gas chromatography (GC) [6,7,11,15,16].

LVM rice husks have been tested as organic insecticides in cocoa plants. LVM phenol content of rice husk gives a distinctive aroma that causes the death of insects in cocoa plants [17]. LVM is also potential for a variety of applications, including preservative fishery products [18, 19], control of cabbage pests [20], as fertilizer [21], and others.

2.Materials and Method

2.1 Materials and Tools

The raw material used to make LVM is rice husk. The main tool used to produce LVM rice husks is a pyrolysis device. Rice husk mass and LVM mass were measured using digital scales, LVM rice husk volume was measured using a measuring cup. Rice husk LVM compounds were identified using the Gilent GC brand, MSD Chemstation type 1431 Ms. 5975.

2.2 Sample Preparation

Rice husks were taken in Lambusa Village, Konda District, South Konawe Regency, Southeast Sulawesi Province. Rice husk is dried in the sun for 3 days to reduce water content.

2.3 Pyrolysis process

The pyrolysis process is carried out by the following procedure: 1). Rice husk is measured as 400gram, 2). It is inserted into the pyrolysis reactor tube, 3). Pyrolysis devices are connected to an electric current source, 4). Thermocouples on pyrolysis devices are set from room temperature to reach the desired pyrolysis temperature of 400°C, 500°C, and 600°C. Each process lasts for 15 minutes. The pyrolysis device used in this study is integrated with the condenser. Smoke that has been condensed into condensate called LVM rice husk. LVM rice husks are stored in bottles, 5). Pyrolysis reactor cooled.

2.4 Measuring Volume and Mass of LVM Rise Husk

The volume and mass of LVM rice husks are measured using measuring cup and digital balance respectively.

2.5 Identification of Chemical Components by GC

Chemical components of LVM rice husk constituents were identified using GC in the Makassar Forensic laboratory. The equipment uses 2 colums, (1). J & W 122-5731: 1721.65322 with a colum temperature of 400°Cwith a diameter of 30 mx 250 μm x 0.1 μm, (2). Frontier Alloy 5Ultra UA - 5 with a temperature of 450°Cwith a diameter of 30m x 0.25μm x 250μm operating conditions at oven temperature160°C /min, injection 250°C, helium carrier gas flow rate of 150 μl /min.

3. Result and Discussion

The pyrolysis process of rice husk in the pyrolysis reactor at a temperature of 400°C, 500°C, and 600°C each produces charcoal, tar, LVM, and the non-condensed smoke. Charcoal is formed as a residue of rice husk pyrolysis in the reactor. Smoke coming out of the reactor towards the condenser was condensed and changed from gas to liquid. Non-condensed smoke is released into the atmosphere.

Parameters of pyrolysis products include mass, m, volume, V, and density, ρ, at each pyrolysis temperature, T, presented in Table 1. LVM volume is an important parameter for determining rice husk LVM density. The density of rice husk LVM at each pyrolysis temperature is shown in the seventh column in Table 1.

Table 1. Parameters pyrolysis of rice husk

T(°C)	m_{char} (g)	m_{tar} (g)	$m_{uncondensed\ smoke}$ (g)	m_{LVM} (g)	V_{LVM} (ml)	ρ_{LVM} (g/cm^3)
400	200	50	130	20	48	0.416
500	100	30	220	50	70	0.714
600	90	20	200	90	79	1.139

Rice husk LVM density increases at higher pyrolysis temperatures. Pyrolysis of rice husk at 600°C produces the largest density of LVM rice husk 1.139 g/cm^3. The mass and volume of rice husk LVM depend on the pyrolysis temperature and the condensation system used, so the density of LVM also depends on both. The approach taken to form LVM is to use water as a coolant so that the heat exchange process takes place relatively quickly. Pyrolysis at high temperature and very long time will cause LVM formation to decrease as the temperature of cooling water increases so that the smoke is not entirely condensed [6,7].

Unscondensed smoke is not measurable, but can be estimated from other parameters. The estimated mass of uncondensed smoke is shown in Table 1 fourth column. Estimated mass of smoke is obtained from the calculation of the rice husk mass raw material of 400 grams minus charcoal mass, tar mass, and LVM mass at each pyrolysis temperature. The pyrolysis temperature also determines the composition of the resulting LVM chemical compound.

Chemical compound LVM rice husk is identified using GC. The characterization results using GC is a chromatogram showing the relationship of retention time and abundance of the compound. Chemical retention times vary depending on the molecular weight. The smallest molecular weight compound will have a small retention time as well. Chromatogram of LVM rice husk produced at 500°C pyrolysis temperatures are shown in Figures 1.

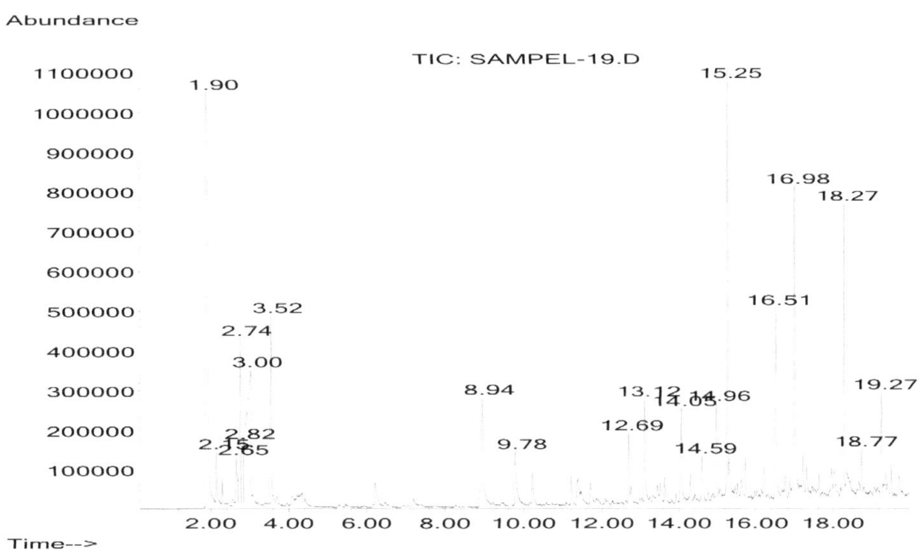

Figure 1. Chromatogram of LVM Rice Husk at Pyrolysis Temperature 500 ° C

IOP Conf. Series: Materials Science and Engineering **797** (2020) 012007 doi:10.1088/1757-899X/797/1/012007

The chromatogram data were supplemented by interpretation of retention time, compounds, and concentration of rice husk LVM compounds. Concentrations of rice husks LVM compounds at pyrolysis temperature 400°C, 500°C, and 600°C were shown in Table 2, Table 3, and Table 4, respectively.

Table 2. LVM compounds of rice husks produced at 400 ° C

Retention (minute)	Compounds	Concentration (%)
1.87	Ammonia	20.18
1.90	Methanol	11.02
2.64	butanedione	0.93
2.68	acetid aced	0.71
2.71	hexane	19.44
2.75	hexane	12.13
2.97	acetid acid	14.17
3.50	propanone	6.66
8.95	furfural	5.65
9.79	furanmethanol	2.41
13.64	furanmethamine	1.06
15.25	phenol	3.08
16.22	phenol	1.32
16.98	benzene	1.42
18.27	guaiacol	1.12
18.36	benzenediol	1.16

Table 3. LVM compounds of rice husks produced at 500 ° C

Retention (minute)	Compounds	Concentration (%)
1.90	Ammonia	2.16
2.15	Propanone	1.07
2.65	pentanone	1.29
2.74	hexane	4.94
2.82	acetid acid	2.2
3.00	acetid acid	12.85
3.52	propanone	4.68
8.94	furfural	5.7
9.78	furanmethanol	1.96
12.69	furancarboxaldehyde	1.57
13.12	phenol	2.94
14.05	cyclopenten	2.61
14.59	phenol	1.47
14.96	benzenemethanol	3.25
15.25	phenol	9.06
16.51	phenol	5.11
16.98	phenol	8.41
18.77	methoxy	0.99
19.27	phenol	2.27

Table 4. LVM compounds of rice husks produced at 600 ° C

Retention (minute)	Compounds	Concentration (%)
1.90	Ammonia	16.67
2.73	hexane	8.34
3.01	acetid acid	10.09
3.52	propanone	4.16
8.94	furfural	5.53
12.69	furancarboxaldehyde	1.83
13.12	phenol	5.15
14.05	cyclopenten	2.52
14.59	phenol	2.51
14.96	benzenemethanol	5.61
15.25	phenol	10.06
16.51	phenol	7.44
16.98	phenol	10.49
18.27	guaiacol	7.12
19.27	phenol	2.48

At pyrolysis temperature 400°C, 500°C, and 600°C there are 13, 12, and 10 compounds respectively. The type of compound varies at each pyrolysis temperature. Compounds where always present at every temperature of pyrolysis are ammonia, propanone, acetid acid, hexane, furfural, and phenol. Propanone highest abundance 6.66% at 400°C, acetid acid 15.05% at 500°C, 31.75% hexane at 400°C, and phenol 38.13% at 600°C. Unfortunately the abundance of its compounds does not show any particular tendency at any pyrolysis temperature. This is thought to be related to the temperature of the pyrolysis and the condensation system used to cause completely non-condensed smoke. The compounds appearing in LVM produced at 400°C partially reappear, partially disappearing, and there is two compounds that only appears at 500°C. Methanol, furanmethamine, and benzene only appear at 400°C. Pentanone and methoxy only appears at 500°C. Compounds at 600°C always appear at 400°C and 500°C. Hexane most abundance at temperature 400°C, meanwhile phenols are most abundant at temperatures of 500° C and 600°C. LVM of rice husk does not contain benzoapiren so that potentially is applied to foodstuffs either as food preservatives [22], fertilizers, and others.

4. **Reference**

[1] Badan Pusat Statistik, Produksi Padi Menurut Provinsi(ton), 1993-2015. Webside:https://www.bps.go.id/linkTableDinamis/view/id/865, accessed 05 Juli 2018.

[2] Badan Pengkajian Teknologi Pertanian Sulawesi Utara. Webside: http://sulut.litbang.pertanian.go.id/eng/index.php/program-litbang/sl-ptt/617-pembuatan-briket-dari-arang-sekam-padi, accessed 05 Juli 2018.

[3] Novia, I. Utami, L. Windiyati. 2014. Pembuatan Bioetanol dari Sekam Padi Menggunakan Kombinasi Soaking In Aqueous Ammonia (SAA) Pretreatment – Acid Pretreatment – Hidrolisis – Fermentasi. Teknik Kimia Vol. 20 No. 1, 46-53.

[4] L. Lestari, V.I.Variani, I.N. Sudiana, D.P. Sari, W.O.S. Ilmawati, E.S.. Hasan. Characterization of Briquette from the Corncob Charcoal and Sago Stem Alloys. Journal of Physics: Conference Series Vol. 846 No. 1 (2017) 012012

[5] E.S. Hasan, M. Jahiding, Mashuni, W.O.S. Ilmawati, W. Wati, I.N. Sudiana. Proximate and The Calorific Value Analysis of Brown Coal for High-Calorie Hybrid Briquette Application. Journal of Physics: Conference Series Vol. 846 No. 1 (2017) 012022

[6] M. Jahiding, Mashuni, W.O.S. Ilmawati, Ermawati, Rahmat, J. Arsyad, S.S. Riskayanti. Characterization of Coconut Shell Liquid Volatile Matter (CS-LVM) by Using Gas Chomatroghaphy. Journal of Physics: Conference Series Vol. 846 No. 1 (2017) 012025
[7] Mashuni, M. Jahiding, W.O.S. Ilmawati, I. Kurniasih, W. Wati, Muzirah, M. Burhan. Characterization of Liquid Volatile Matter (LVM) of Cashew Nut Shell using Pyrolysis and Gas Chomatroghaphy. Journal of Physics: Conference Series Vol. 846 No. 1 (2017) 012026
[8] Ester. 2017. Identifikasisenyawa yang Terkandung Dalam Tar Ampas Sagu menggunakan Metode *Gas Chromatography* (GC). Not published.
[9] F. Bawazir. 2017. Ekstraksi minyak atsiri dari daun dan batang nilam menggunakan pirolisis microwave. *Not published.*
[10] H. Yang, R. Yan, H. Chen, D.H. Lee, and C. Zheng, "Characteristics of *hemicellulose, cellulose, and lignin pyrolysis," Fuel, Vol. 86, (2007). 1781-1788,.*
[11] W.O.S. Ilmawati, M. Jahiding, W.O.S. Musnina. Pyrolysis Temperature Effect on Volume and Chemical Composition of Liquid Volatile Matter of Durian Shell. Proceeding The 1st IBSC: Towards The Extended Use Of Basic Science For Enhancing Health, Environment, Energy And Biotechnology. 2017. 173-275.
[12] Demirbas, A., Arin, G. (2002). An Overview of Biomass Pyrolysis. Energy Sources. 24, 471-482.
[13] S. Sunarsih, Y. Pratiwi, and Y. Sunarto, "Pengaruh Suhu, Waktu dan Kadar Air pada Pembuatan Asap cair dari limbah padat Pati Aren (Studi Kasus pada Sentra Industri Sohun Dukuh Bendo, Daleman, Tulung, Klaten)," ISSN:1979-911X, November 2012.
[14] Girard J P. 1992. Technology of Meat and Meat Products. Ellis Horwood, New York, USA.
[15] John Kenkel, 2013. Analytical Chemistry for Technicians 4th Edition. CRC Press Taylor and Francis Group, London.
[16] W.O.S. Ilmawati, M. Jahiding, W.O.S. Musnina, L.O.M. Arsal. Pengaruh Suhu Pirolisis terhadap Massa Jenis dan Komposisi Kimia *Liquid Volatile Matter* Kulit Durian. Prosiding Seminar Pendidikan Fisika, Fisika, dan Aplikasinya 2016. 2017. ISSN 2085-0379. 139-144.
[17] R. E. Putri, Mislaini, L. S. Ningsih. Pengembangan Alat Penghasil Asap Cair dari Sekam Padi untuk Menghasilkan Insektisida Organik. Jurnal Teknologi Pertanian Andalas Vol. 19, No.2 September 2015. ISSN 1410-1920.
[18] Pemanfaatan Asap Cair Terenkapsulasi Maltodesktrin-Kitosan sebagai Pengawet Ikan Cakalang (Katsuwonus pelamis). R. A. Kurniasih, P. Darmadji, Y. Pranoto. Jurnal Teknologi Hasil Pertanian, Vol. IX, No. 1, Februari 2016.
[19] Zakia. 2010. Pengawetan Ikan Segar dengan Menggunakan Asap Cair (Liquid Smoke) dari Serabut Kelapa. Program Studi Tekhnologi Hasil Hutan Jurusan Pengolahan Hasil Hutan Politeknik Pertanian Negeri Samarinda. Samarinda.
[20] Wiyantono, E. W. Minarni. Kajian Potensi Asap Cair dalam Mengendalikan Ulat Krop Kubis, Crocidolomia pavonana. Jurnal Pembangunan Pedesaan Vol. 9 No. 1, April - Juli 2009: 50-56. ISSN. 1411-9250.
[21] J. D.Jaya, A.Zulmi, D.Wahyudi, Kartika, H.Wati, N.Yuliana, N.Kholis. Optimasi Pembuatan Asap Cair dari Sekam Padi dan Aplikasinya sebagai Pupuk Tanaman Hidroponik. Jurnal Teknologi Agro-Industri Vol. 2 No.2 ; November 2015.
[22] S. Budijanto, R. Hasbullah, S. Prabawati, Setyadjit, Sukarno, I. Zuraida. Identifikasi dan Uji Keamanan Asap Cair Tempurung Kelapa untuk Produk Pangan. J.Pascapanen 5(1) 2008: 32-40.

IOP Conf. Series: Materials Science and Engineering **797** (2020) 012008 doi:10.1088/1757-899X/797/1/012008

Floating Market In Muara Angke

Mieke Choandi[1], **Lourensya Jessica Berlian**[2*]

[1,2]Department of Architecture, Universitas Tarumanagara, 11440 Jakarta Barat, INDONESIA

E-mail: mieke@untar.ac.id[1], jessicajejeyy@gmail.com[2*]

Abstract. Developing the marine economy as the prime mover Indonesian maritime development is the focus of Muara Angke's fishiries as one of many regions with the biggest fishiries market in Indonesia. Infrastructure in fishiries distribution from producer to consumen is still being an obstacle until occur economy and fishiries distribution degradation in Muara Angke. Developing Floating Market is being a solution which focusing in seashore tourism economy, which is one stop recreation, where the tourists can buy both cooked and uncooked fishiries and can enjoy the sea recreation and beautiful Indonesia. The author examines and analyses Muara Angke district, that being the location of the design and research culture study of fishermen and characteristic study of existing Floating Market in the world as a reference in designing. Floating Market must be able to facilitate fishiries distribution infrastructure, either for direct consumption or export outside and creating space that can be felt by society and supporting maritime value of the maritime state that embodies the concept of Indonesia as the global maritime axis.

1. Introduction

1.1. Background

As a capital of a country, DKI Jakarta is a province with the most population with many kinds of tribes in Indonesia. Each of tribe group has different styles and cultures. There are six areas consist of five districts and one sub district, only one area which has physics boundary between land and sea, which is North Jakarta. This condition is very different from the other areas in Jakarta. North Jakarta has the history since the coming Netherlands government to Indonesia, which caused by its boundary to the sea.

Others, when we talk about the existing activity in North Jakarta , related to the sea, is fishiries. This fishiries is about many fishermen who live and have activities as a fisherman catching fishes. The sea products are both marketed in the auction center and the traditional market. It is known that many markets are on the land, because most of Jakarta areas are land. Although North Jakarta is at the sea side, the market trading is still on the land. From the fact, market areas are not merely on the land, they can be also on the sea.

Moreover, when we discuss about Urban Acupuncture, which has the purpose to release the stagnant or degradation area into more vital, both new function and others. One of these seashore areas

[1] Tarumanagara University, mieke@untar.ac.id
[2] Tarumanagara University, jessicajejeyy@gmail.com

IOP Conf. Series: Materials Science and Engineering **797** (2020) 012008 doi:10.1088/1757-899X/797/1/012008

researched is Muara Angke. Since long ago, this area is a fishermen residences, tradisional market is on the land, which enables the acupuncture action is carried out. One of them is building a project to facilitate how the fish are marketed. Muara Angke is able turn into recreation area. This is the aim of the urban acupuncture action.

Muara Angke, a part of North Jakarta, has one harbour which functions as fishiries area. Many of Muara Angke residents work as fishermen and fish mongers. Because of its condition, there are a lot of fishing boats at Muara Angke. Therefore, Muara Angke has the trademark of its fishermen and which has lasted since forty years ago. Most of Muara Angke economy itself comes from trading sea products. Besides the sea products, they also build fish embankments. However, this fishing activity causes basic problems, sucs as the waste of sea shells, which are marketed to Muara Angke Fish Market. The waste of the fish causes bad smell, and the lack of hygiene in Muara Angke. Moreover, because this location is at the sea side, it also gets flood from the sea which flooding most of the areas. The government issued Muara Angke RDTR 2030 to deregulate Muara Angke. In RDTR 2030, Muara Angke will become strategic area because it connect Muara Angke with the reclamation area and Pantai Indah Kapuk. However, it will eliminate some parts of the areas which causes both the loss of fishiries characteristic and the decreasing of the economic.

Through Urban Acupuncture, it is expected to be able to both set back and increase the economic of Muara Angke, as well as create a new job field, and return the trademark of Muara Angke. Therefore, I design a Floating Market project. With this project, I hope it can be one of the solutions and be a part of tourism economic source at the seaside in North Jakarta. Besides, this project is expected to be a magnet to Jakarta residents, as well as Indonesian, and also to the International people.

1.2. Problem Planners

What programs and activites are needed to be suggested to maintain the culture of Muara Angke? What market is needed in Muara Angke? How to carry out the project with local value of Muara Angke without ignoring the surrounding environment? What structure and materials used to construct on the sea?

1.3. The Purposes of The Project

The purpose of each project has a definite architectural achievement. The purpose of the Urban Acupuncture project is: As a form of safeguarding local values against the deregulation of the Muara Angke RDTR 2030. As a Jakarta Government Program development project as a place of recreation and economy. To increase the income of the people of Muara Angke because of the lost fish market as the biggest economic source. Become a more accessible place. Making Muara Angke a magnet for both local and international tourists. Become a means of information activities and economic, social, cultural and recreational development.

2. Methods

In executing the project research, the author do literature studying, the datum collection through electronic media (article, website, journal) which discuss about Urban Acupuncture, culture, market, and space, especially outdoor. To be able to comprehend the project and its problem of local residents, the author does the field survey to collect existing datum about the area and the chosen spot, so that can be both analysed and become a guide in doing the design

3. Discussion and Result

3.1. Comprehension

Urban Acupuncture is a philosophy or approach to address social, economic and urban culture, and improve environmental quality of the city. Urban acupuncture uses small-scale interventions to transform urban contexts into larger ones. In designing the Floating Market, Urban acupuncture chosen is a local socio-economic culture. Culture is communication. The culture of the Muara Angke community is mostly fishermen, catches are marketed to local communities and outside the Muara

IOP Conf. Series: Materials Science and Engineering **797** (2020) 012008 doi:10.1088/1757-899X/797/1/012008

Angke area, both in small and large quantities. The women work as peel shells, shells will be sold to the Muara Angke Fish Market. Therefore, the design project was taken by marine trade activities. The container needed is the market. Floating Market that can answer all of the above needs.

3.2. The Location of The Project

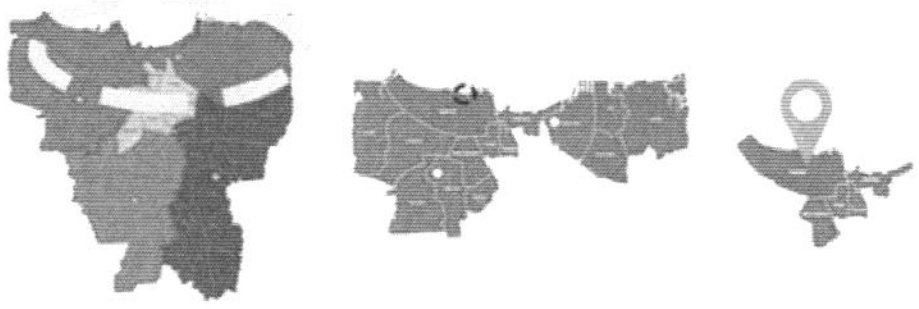

Figure 1. Jakarta – Penjaringan District - Pluit Map

Muara Angke is located on the coast of North Jakarta, a fishing village whose activities are fishing and marketing fish catches. Indonesian BMKG research, waves coming from Northwest, West and Southwest. Wave height of 0.25 m '- 0.75 m', maximum limit of one meter. Therefore, the North coast can be used for beach activities. The wind blows from the Southeast to the Southwest, speeds of around 4-46 km / hour. Generally in the South from East to Southeast around 7-37 km / hour. The temperature in Muara Angke is 30.20 ° C, air pressure is 1009.20 mbar, air humidity is 66.80%. The growth of the fishing population from year to year is higher.

Problems in Muara Angke, survey results and location data collection: The point of pollution is very high. Increased sea levels often cause flooding, disrupt community activities. Irregular infrastructure causes pollution and smell and decreases the fish economy. RDTR 2030, deregulating several parts of the Muara Angke region, caused the following changes: The loss of the fish market as a marine economic resource for the economy of the local cultural community. Job loss. With the new RDTR, making the Muara Angke reclamation area become a more strategic area that connects Muara Angke to PIK. Can increase the economic potential as a tourist area and fishing trade.

3.2.1. Wave Location

In the research of Indonesian BMKG, the fetch that can be arisen the wave coming from North West, West, and South West. The wave is between 0,25 meter and 0,75 meter to the maximum limit of one meter. Therefore, the Northern seashore can be use for seaside activities.

3.2.2. Wind Direction and Speed

The wind blows from Southeast to South West with the speed about 4 to 46 km/hour. Whereas generally in the South from East to Southeast is about 7-37 km/hours. The temperature in Muara Angke is about 30.20°C, and the air pressure is about 1009.20 mbar with the air humidity about 66,80%.

3.3. The Growth Population of Fishermen

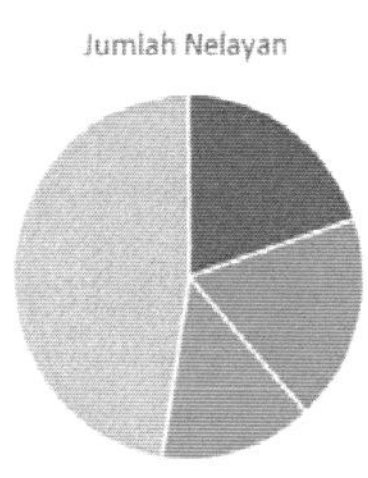

Figure 2. The growth population of fishermen graph

The quota of the growth population of fishermen from year to year is getting higher. It is proved that the growth of the fishermen economy is increasing every year.

3.4. Site Analysis

The transportation to direction from Bandara Soekarno-Hatta to the site by car. The transportation convenience is able to bring in the tourists from the overseas and domestic to the site. Many buses routes to Muara Angke prove that Muara Angke is a public node transportation. The location is in the TOD area, 500 m pedestrian distance from the public transportation node to the location is still convenient if included greening is provided. The road in Muara Angke is a local road and very strategic, because it is on the border between Muara Angke and PIK which is a reclamation area, so tourists can find out the path to the location without traffic signs. LRK 2030, the Muara Angke port near the floating market location, will be designed as a fish distribution to the cold loading and storage building. The operational time for loading docks in the morning and evening, so as not to interfere with the floating market activities that will be designed.

Position of the location of the North Sea coastline, then in the design of buildings pay attention to the conditions and local climate (tropical). Aspects of air circulation is very expected to reach a level of comfort for the visitors to enjoy the beauty of the sea, view and wind. The aspect of the sun, using a saddle-shaped roof, so that in the room get indirect light. Achieving coastal tourism destinations and visitors can enjoy the beauty of the Java sea, some buildings are directly above the sea and along the coastline are designed promenades.

The Basic Massing Form observes macro and micro locations. At first, the quadrilateral form was arranged in a row along the coast, so that it was in the form of a rectangle. utilizing nature with water elements, the public area as a place for water shopping and recreation. The market is a floating market, so the shopping pattern is on the promenade path above the water, at the end of the promanade is made a node to gather the end, the roof cover is made waves like waves in the sea. Promenade along the coastline is more used for eating, drinking and watching if there are performances.

3.5. The Floating Market Design Concept

The Muara Angke Floating Market is closely related to the sea level, so that both the height of the floor piel and the building structure must be resistant to ocean waves. the data obtained, the tide level reaches 1.5-2 meter. Therefore, all buildings use implant structures into the ocean floor. Depth along the coast of Muara Angke only reaches about 2 meters (the result of reclamation).

Floating Market is designed for visitors to shop on the water, can also enjoy seafood, the beauty of the sea scene and recreation. This Floating Market Design can combine Muara Angke market culture with the Floating Market design. The location is near the sea, for visitors every step can see and enjoy the sea. the roof is saddled and shaped like waves crashing against the reef, and using the space frame as a roof frame. This is used to control the effects of sunlight entering the building and the flow of rainwater. Buildings are more outdoor to maximize natural air. Aeration in buildings using cross ventilation. natural lighting of the building by using wide openings and can see the beautiful panoramic sea. Therefore, natural light is given a hollow canopy, so that the beam of sunlight on the canopy provides a sense of aesthetics in space.

The concept of this project focuses on outdoor concepts that rely on natural aeration where the air most comes from the sea. To maximize the concept of floating, the concept of out-door and air circulation is the most important. for that, the front and back of the building are open so the air flows well. The left and right sides of the building are made of hollow glass to give a spotlight effect that comes from the sun be-cause the roof uses an open space frame. The floor to floor le-vel is 5 meters, so the spotlight is not directly on the visitors.

Because it has several functions both as an area of shopping, eating and recreation. then the dining zone and some shopping are inside the building, while the market and recreation zones are in the middle of the sea using canopy and membrane construction. The area for trading marine products directly from boats is carried out in the water recreation zone, so that visitors can enjoy the sea and attract visitors to the water recreation node.

Saddle-shaped roof with a long overstek, sunlight does not directly enter the building but still get natural lighting. The entrance is divided into two parts; pedestrian entrances and vehicles. The pedestrian entrance is at the west side of the sea area, so it can attract pedestrians to enter the building. Drop-off point for the car, located on the south side, the dropoff canopy is made together with the second floor, forming a pattern which is a space. Warehouse zone near the parking area, facilitate loading and unloading. Through the main enterance, visitors can reach the dining area, promenade to the recreation area.

Figure 3. The outdoor atmosphere

The outdoor atmosphere, on the floating deck, visitors can directly buy sea products from boats that come from the sea, and can sit freely while enjoying food and seeing the panoramic views of the sea and the surrounding buildings. Visitors can take photos, a mediator area for jet ski recreation and fishing. This area is surrounded by water so that visitors can feel the natural atmosphere, especially the sea. In this area, visitors can see all the buildings and activities held in the building, especially in the building promenade.

Figure 4. The Promenade at daylight

Promenade area, piel building with promanae floor has a difference of ± 1.20 meters high, serves as a casual seating area while enjoying panoramic views of the sea and sea recreation activities while enjoying food from food/restaurant kiosks. There is a stage in front of the sitting area, functioning as an amphitheater, can be used live music performances or performances presented at night. also provided a playground or children's play area adjacent to the amphi-theater, parents while enjoying food and watching their children

Figure 5. Floating Market at night

The atmosphere of Promenade at night, there is live music with spot-lights as a decoration from the back stage, visitors can dance and sit. Lighting from inside the building gives a beautiful reflection on the surface of the sea water. A spotlight that leads to the sky gives the excitement that accom-panies live music at night. Building interiors. Restaurants on the first and second floors. the roof is made high, thus maximizing air circulation and lighting. With the height of the roof, the lighting and natural air flow can enter the maximum, coupled with the many holes in the roof. Dining activities are carried out on the second floor, eating while enjoying views of the sea. Dining room The restaurant on the second floor, using a canopy with a glass cover, aims to provide natural vibrations in space. Lighting during the day uses natural lighting and at night uses artificial lighting attached to the roof structure.

IOP Conf. Series: Materials Science and Engineering **797** (2020) 012008 doi:10.1088/1757-899X/797/1/012008

Figure 6. Building Structure

Floating Market Structure uses a floating structure foundation. The structure of buildings on land uses skeletal structures and chicken claw foundations. The roof of the building uses a frame of space and roof plana, and the roof of the knot uses a membrane frame with a WF steel column structure.

4. The Conclusion and The Suggestion

a. The Conclusion

Urban Acupuncture is held in areas of stagnation, death, or degradation. In this case, Muara Angke underwent economic degradation and threatened to lose the culture of fishiries because of the relocation of areas on the coast to Tanjung Lesung.

The Floating Market Project is the solution to answering urban acupuncture in Muara Angke by using fishiries culture in Muara Angke as part of the project. Activities in this project, using fishiries and coastal culture, such as boats, fishing, enjoying the sea, eating and drinking, and others. To achieve the objective of Urban Acupuncture, a floating market is made, which is expected to bring the region to life. The building is built on water and the mass of the roof is made waves so as to make the impression attractive and dynamic, the activities are expanded outside which is connected to the sea. Floating Market uses an outside concept created to maximize the atmosphere that gives the impression of being united with nature.

b. The suggestion

Determining the urban acupuncture project in Muara Angke, first review the problems in Muara Angke. Conduct interviews with local residents. Understand the local culture and see the potential in the region, so that the Muara Angke area can be conserved, developed and revived. To be able to revive the activity of the area, the idea arose to build a Floating Market in Muara Angke, considering that Indonesia was a Maritime State. Buy and sell fish can use boats (see the city of Pelembang, Banjarmasin). Traders may be from local fishermen or from outside the region. The buyer is a shopping area while recreation. Thus creating Floating Market building with an interesting atmosphere, beautiful, clean and comfortable, so Muara Angke alive and victorious. The author realizes, still has many shortcomings, to sharpen the design of projects that can be built, enjoyed and improve the economic return of the fishermen .

References

[1] Kaye, L. 2011. Urban Acupuncture Community. United Kingdom : Localised Renewal Project.
https://www.theguardian.com/

[2] Prospek Gelombang Laut di Laut Jakarta Utara. 2018. Badan Meteorologi, Klimatologi, dan Geofisika (BMKG). Jakarta : gelombang dan kecepatan tiupan angin.
http://maritim.bmkg.go.id/

[3] Dinas Tata Ruang Provinsi DKI Jakarta. 2014. Undang Lembar Rencana Kota Jakarta 2030. Pemerintah Provinsi DKI Jakarta. Jakarta.

[4] Dianawati. 2011. Skripsi Tugas Akhir : Maritime Airport at Panimbang, Pandeglang/Banten. Jurusan Arsitektur, Universitas Tarumanagara. Jakarta.

IOP Conf. Series: Materials Science and Engineering **797** (2020) 012009 doi:10.1088/1757-899X/797/1/012009

Hybrid Method Analysis of Variance (ANOVA) and Loss Function to Analyzed Optimizing Composition of Fiber Baggase and Fly Ash

Lilis Laome[1], **Muh. Hasbi**[2]
[1] Departement of Statistics at Mathematics and Natural Science, Universitas Haluoleo, 93232 Kendari, Sulawesi Tenggara, INDONESIA
[2] Department of Mechanical Engineering, Universitas Haluoleo, 93232 Kendari, Sulawesi Tenggara, INDONESIA

E-mail: lhi2slaome@gmail.com[1], muh.hasbi74@gmail.com[2]

Abstract. Research on the utilization of baggase fiber as a composite material has been done, but still has shortcomings in the composite material brilliance. Therefore, the use of baggase fiber combined with coal ash can be an alternative to make a composite material. The purpose of this research is to know the effect of baggase fiber composition and fly ash composition to tensile stress and bending stress. This paper discusses the application of Analysis of Variance (ANOVA) on composite material data. There are two variables measured are variable of tensile stress and variable of bending stress. The result obtained that there is influence composition of baggase fiber and composition of fly ash at tensile stress and bending stress. Furthermore, further tests using Tukey method showed that the different composition pairs significantly were 10% -30%, 10% -40% and 20% -40% in both compositions of the two variables studied. In the determination of loss function with Signal-to-Noise (SN) Ratio with larger-is-the-better criteria, the composition of 40% baggase fiber, 10% fly ash composition and 50% matrix can produce optimal tensile and bending strength.

1. Introduction

Research on the utilization of baggase fiber as a composite material has been done, but still has shortcomings in the composite material brilliance. Therefore, the use of bagasse fiber combined with coal ash can be an alternative to make a composite material. This material can later be used as material on making desks, wardrobe, and others. One reason for choosing such materials is because the baggase and fly ash are relatively cheap and easy to obtain [1].

Seeing the problem, [1] made a new breakthrough in exploiting baggase fiber and fly ash as combined with polyester matrix to make better composite material. [1] has conducted tensile and bending tests on the resulting composite material. To see the optimization of the material will be analyzed by analysis of variance (ANOVA) with the aim to see there is no difference treatment of baggase fiber

[1] lhi2slaome@gmail.com, [2] muh.hasbi74@gmail.com

composition and the composition of fly ash against tensile strength and bending strength in each observation.

ANOVA is a statistical technique used to decide whether samples from a population have the same averages [2]. In this study, expansion of ANOVA is used by calculating the loss function with knowing the value of Signal-to-Noise (SN) ratio at each treatment level. It is expected to provide more information in choosing the optimum composition on the composition material.

2. Review of Literature

2.1. Analysis of Variance (ANOVA)

Suppose there is a treatment or one factor with different levels will be compared. Observation response of each treatment is random variable. Characteristics of analysis of variance with one factor are [2]:

- Level factor of at least 3 groups / levels
- Intergroup assumptions are homogeneous
- The residual distribution is normal

2.2.Linear Model

$$y_{ij} = \mu + \tau_i + \varepsilon_{ij},\ i = 1, 2, ..., k \quad j = 1, 2, ..., n \tag{1}$$

where:

y_{ij} = observation to *i-th,* treatment to *j-th*

μ = general average effects

τ_i = treatment effects to *i-th*

ε_{ij} = residual of observation to *i-th,* treatment to *j-th*

$N = \sum_{i=1}^{k} n_i = nk$ [3].

2.3. Fixed Effect Model

In this model, the treatment effect τ_i, for $i = 1, 2, ..., k$ is defined as a deviation from the general mean so

$$\sum_{i=1}^{k} \tau_i = 0 \tag{2}$$

Statistical Hypothesis **:**

$$H_0 : \tau_i = 0,\ \forall i = 1, 2, ..., k$$

$$H_1 : \tau_i \neq 0, \text{ at least for one } i$$

Testing Criteria **:** Rejected H_0 if $F_{hit} > F_{tabel}$

2.4. Tukey Test

Tukey has introduced a double-comparison procedure that is also based on statistic of student range. The procedure requires the use of $q_{\alpha(k,f)}$ to determine the critical value of all comparison pairs. So the Tukey test specifies that mean is significantly different if the absolute value of the sample mean difference is greater than $T_\alpha = q_{\alpha(k,f)} S_{\bar{y}_i}$

Statistical Hypothesis:

$$H_0 : \mu_i = \mu_j \text{ , untuk } i \neq j$$
$$H_1 : \mu_i \neq \mu_j$$

Test Statistic:

$$T_\alpha = q_{\alpha(k,f)} S_{\bar{y}_i}. \tag{3}$$

where:

$$S_{\bar{y}_i} = \sqrt{\frac{KTE}{n_i}}$$

KTE = Mean Square Error

Testing Criteria: Reject H_0 if $|\bar{y}_i - \bar{y}_j| > T_\alpha$ [5].

2.5. Loss Function

The loss function is a new approach to assessing process capability a product [4]. The quality characteristics are devided into the following three characteristics:

a. *nominal-is-best*
the process under study has a finite target value.

b. smaller-the-better
the process become worse if the value of the measure increases.

c. *larger-the-better*
the process that the quality is judged to be increasing.

In the field of communication engineering a quantity called the signal-to-noise (SN) ratio has been used as the quality characteristic of choice. Two of the applications in which the concept of SN ratio is useful are the improvement of quality via variability reduction and the improvement of measurement. There are several SN ratios available depending on the type of characteristic: continuous or discrete; nominal-is-best, smaller-the-better or larger-the-better. Here is the loss function and the SN ratio formula for n observations in Table 1 [4].

Table 1. Loss Function and SN-Ratio for n observations

Type of Characteristic	SN-Ratio Loss Function Approach
Nominal-is-best	-10 log $[\frac{1}{n}\sum_{i=1}^{n}(y_i - m)^2]$
Smaller-the-better	-10 log $[\frac{1}{n}\sum_{i=1}^{n}{y_i}^2]$
Larger-the-better	-10 log $[\frac{1}{n}\sum_{i=1}^{n}\frac{1}{{y_i}^2}]$

2.6. Composite Materials

The fiber of bagasse in the test as a composite material is usually mixed with hard materials such as epoxy resin, polyester resin, cement, asbestos and concrete which result in press test, bagasse fiber as composite material not yet fulfill requirement as reinforcement material. On the other hand, in the elasticity test using lamination plating technique as an asbestos reinforcement material, bagasse fiber proved to have better elasticity than using plastic fiber [1].

IOP Conf. Series: Materials Science and Engineering **797** (2020) 012009 doi:10.1088/1757-899X/797/1/012009

Fly ash is a solid waste resulting from the combustion process in the furnace on the steam power plant which is then carried out by the burning remains. Fly ash is a mineral residue in a fine grain produced from burning coal that is smoothed on a power plant. Fly ash consists of inorganic materials contained in coal that have undergone fusion during combustion. This material solidifies while inside the exhaust gases and is collected using an electrostatic precipitator. Because these particles solidify during suspension in the waste gas, the fly ash particles are generally spherical [1].

3. Data and Methodology

The data used are tensile stress and bending stress data on a composite material with matrix material, bagasse fiber and fly ash [1]. The composition of each material is 50% matrix and 50% bagasse fiber and fly ash. Because the composition of the matrix is fixed, the material observed is only in the composition of the bagasse fiber and fly ash. Tests with ANOVA were performed separately on each composition of bagasse fiber and fly ash that divided by 10%, 20%, 30% and 40% levels.

4. Estimation Result

The first variable observed is the variable tensile stress of a composite material. In this paper we will see the effect of bagasse fiber and fly ash on four levels are given: 10%, 20%, 30% and 40%. Based on the data obtained [1], there is a relationship between the composition of bagasse fiber and tensile stress and relationship between the fly ash composition and the tensile stress in Figure 1.

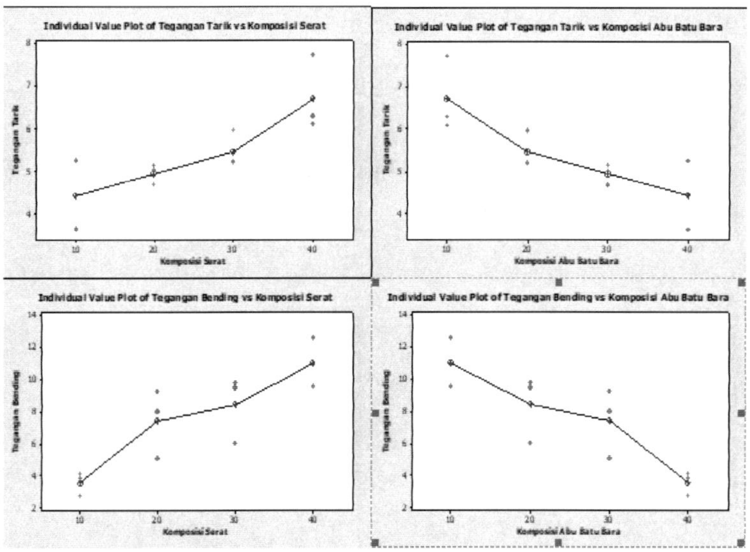

Figure 1. The Plot of bagasse fiber and fly ash composition to tensile stress and bending stress

According to Figure 1, the composition of 40% of bagasse fiber provides the highest tensile stress. This is inversely proportional to the provision of fly ash composition. In the composition of fly ash decreases the tensile stress if the fly ash composition increases at 40% level.

The second variable is the bending stress variable of a composite material. In the same way in the variable tensile stress, the level of bagasse fiber composition and fly ash composition are also given at the same level. Next, Figure 1 about plots the relationship between the composition of the bagasse fiber and the bending stress and the relationship between the fly ash composition and the bending stress. Figure 1 shows the relationship between the bagasse fiber composition and the fly ash composition against the bending stress equal to the tensile variable. The higher of the bagasse fiber composition the higher value of the bending stress. Conversely on the fly ash composition, the higher of the fly ash composition the lower the bending value.

To perform the ANOVA, one of the assumptions that must be met is the assumption of homogeneity. This means that each group should be homogeneous, so it must be tested homogeneity between levels of bagasse fiber composition at tensile stress, homogeneity test between level of fly ash composition at tensile stress, homogeneity test between level of bagasse fiber composition at bending stress and homogeneity test of fly ash composition at the bending stress. Based on levene test for all four test performed, shows that p-value obtained more than 0.05. So it can be continued ANOVA test in Figure 2

Based on Figure 2 shows that the ANOVA test results on the influence of the composition of bagasse fiber and fly ash is significant. This is seen from the p-value generated less than 0.05. This means that there is an effect of giving the composition of the bagasse fiber and giving the composition of fly ash to the magnitude of tensile stress and bending stress.

One-way ANOVA: Tegangan Tarik versus Komposisi Serat

```
Source             DF      SS     MS     F      P
Komposisi Serat     3   8,530  2,843  6,65  0,014
Error               8   3,420  0,427
Total              11  11,950

S = 0,6538   R-Sq = 71,38%   R-Sq(adj) = 60,65%
```

One-way ANOVA: Tegangan Tarik versus Komposisi Abu Batu Bara

```
Source                   DF      SS     MS     F      P
Komposisi Abu Batu Bara   3   8,530  2,843  6,65  0,014
Error                     8   3,420  0,427
Total                    11  11,950

S = 0,6538   R-Sq = 71,38%   R-Sq(adj) = 60,65%
```

One-way ANOVA: Tegangan Bending versus Komposisi Serat

```
Source             DF      SS     MS     F      P
Komposisi Serat     3   86,33  28,78  9,66  0,005
Error               8   23,86   2,98
Total              11  110,19

S = 1,727   R-Sq = 78,35%   R-Sq(adj) = 70,23%
```

One-way ANOVA: Tegangan Bending versus Komposisi Abu Batu Bara

```
Source                   DF      SS     MS     F      P
Komposisi Abu Batu Bara   3   86,33  28,78  9,65  0,005
Error                     8   23,86   2,98
Total                    11  110,19

S = 1,727   R-Sq = 78,35%   R-Sq(adj) = 70,23%
```

Figure 2. Output ANOVA on Variable of Tensil Stress and Bending Stress

Therefore the resulting ANOVA test is significant, it can proceed to further test with Tukey method to see which pairs of compositions differ significantly. Given a summary of Tukey test results in Table 2.

Tabel 2. Summary of Tukey Test Results

Variable	Treatment	Pair of Treatment Significantly
Tensile Stress	Baggase fiber	10% dan 40%
	Fly ash	10% dan 30%; 10% dan 40%
Bending Stress	Baggase fiber	10% dan 30%; 10% dan 40%
	Fly ash	10% dan 40%; 20% dan 40%

To see the loss function used SN-ratio criteria with larger-the-better type. The value of each SN Ratio is presented in Table 3.

Table 3. *SN Ratio* Value on on Variable of Tensil Stress and Bending Stress

Variable	Baggase fiber composition		Fly ash composition	
	Level	*SN Ratio*	Level	*SN Ratio*
Tensile Stress	10%	18,64675	40%	18,64675
	20%	19,8736	30%	19,8736
	30%	20,69669	20%	20,69669
	40%	22,40224	10%	22,40224
Bending Stress	10%	16,51256	40%	16,51256
	20%	22,56626	30%	22,56626
	30%	23,85788	20%	23,85788
	40%	26,67851	10%	26,67851

Based on Table 3, the strength of tensile strength and bending stress is influenced by the large bagasse fiber composition of 40% and the small fly ash composition of about 10%.

5. Conclusion

Based on ANOVA test, it was found that the composition of bagasse fiber and fly ash composition had significant effect on tensile stress and bending stress variables. At variable tensile stress there is one pair of composition of bagasse fiber which is significant is 10% and 40% composition of bagasse fiber while the composition of fly ash contains two significant couples that is 10% and 30%; 10% and 40% fly ash. The bending stress variables are two pairs of sugarcane fiber composition ie 10% and 30%; 10% and 40% while the composition of coal ash contained four significant couples ie 10% and 20%; 10% and 30%; 10% and 40%; 20% and 40%. Furthermore, based on SN-ratio shows 40% bagasse fiber composition, 10% fly ash composition and 50% matrix can produce optimal tensile and bending strength.

REFERENCES

[1]. Akbar, H. 2018. *Studi Eksperimen Kekuatan Tarik dan Tekan (Bending) pada Material Komposit Serat Ampas Tebu Dipadukan Abu Batu Bara (Fly Ash) yang Diperkuat Resin Polyester.*Skripsi. Jurusan S1 Teknik Mesin FT UHO.
[2]. Yamin, S. dan Kurniawan, H. 2009. *SPSS Complete, Teknik Analisis Statistika Terlengkap dengan Software SPSS.* Jakarta: Salemba Infotek.
[3]. Montgomery, D.C. 2001. *Design Analysis of Experiment.* New York: John Wiley & Sons.
[4]. Park, S.H. 1996. *Robust Design and Analysis for Quality Engineering.* New York: Chapman & Hall.
[5]. Suwanda. 2011. *Desain Eksperimen untuk Penelitian Ilmiah.* Bandung: Alfabeta.

IOP Conf. Series: Materials Science and Engineering **797** (2020) 012010 doi:10.1088/1757-899X/797/1/012010

Analysis of Longshore Sediment Transport at Theestuaries of Jeneberang River and Tallo River Caused by Waves on Coast of Makassar

Riswal Karamma[1,3]*, M. Saleh Pallu[1], M. Arsyad Thaha[1], Mukhsan Putra Hatta[1], A. Subhan Mustari[1], Ahmad Syarif Sukri[2]
[1] Department of Civil Engineering, Hasanuddin University Jalan Poros Malino Km. 6 Bontomarannu, 91271, INDONESIA
[2] Department of Civil Engineering, Universitas Haluoleo, 93232 Kendari, Sulawesi Tenggara, INDONESIA

*Corresponding author's e-mail: riswalchiwal@gmail.com

Abstract. The length of the Makassar Coast reaches ± 35.52 km. In the northern coastal area of Makassar City, the Tallo River Estuary and the south side of the Jeneberang River Estuary, both estuaries can affect longshore sediment transport on the coast of Makassar City.The purpose of this research was to determine the direction of longshore sediment transport on the coast at the estuaries of the Jeneberang River and Tallo River caused by waves on the Makassar Coast. The analysis carried out in this research is the analysis of wind direction, wind speed, effective retrieval, wave forecasting, wave return analysis, breaking wave analysis and longshore sediment transport analysis. The result of this research indicates that the longshore sediments transport in the Tallo River Estuary is relatively heading north or away from the coast of Makassar and longshore sediments transport in the Jeneberang River Estuary is relatively heading north or heading to the coast of Makassar.

1. Introduction

Makassar City is one of the cities in Indonesia that has a coastline. The length of the coastal city of Makassar reaches ± 35.52 km.The coastal area of Makassar city provides productive natural resources both as a source of food, mineral and energy mining, recreational media or tourism [1]. To the north of the coastal area of Makassar City is the Tallo River Estuary and to the south there is the Jeneberang River Estuary, these two estuaries can affect sediment transport along the coast of the Makassar City coast. Triadmodjo [2] explained that there are several ways commonly used to predict sediment transport along the coast, one of which is to use empirical equations based on the wave conditions of the area being reviewed. The CERC equation [3] is one of the equation of the empirical sediment transport along the coast. The CERC equation is used to calculate the amount of sediment transport along the coast in conditions without structure [4].

2. Purpose of research

The purpose of this research was to determine the direction of sediment transport along the coast at the estuary of the Jeneberang River and Tallo River caused by waves on the Makassar Coast.Sediment material near the coast will be driven by longshore current (Sorensen 2006). Thus resulting in longshore drift. To calculate sediment transport along the coast, the following equation can be used.

$$K = 1.4\, e^{(-2.5\, D_{50})} \quad (1)$$

$$Q_1 = K\left(\frac{\rho\sqrt{g}}{16\, K^{\frac{1}{2}}(\rho_1 - \rho)(1 - n)}\right) H_b^{\frac{5}{2}} sin(2\, a_b) \quad (2)$$

3. Research method

The method used in this research is descriptive qualitative, the results of the research and analysis are described in a scientific paper in the form of narrative, then from the analysis that has been carried out, a conclusion is drawn.The data used in this research are wind direction and velocity data from 2008 to 2018 as well as data on research locations sourced from ECMWF and satellite imagery (Google Earth).Data analysis techniques used in this research is in the form of wind direction and velocity analysis, effective fetch analysis, wave forecasting, breaking wave analysis and sediment transport analysis.

The thinking flow of this research begins with the literature research or looking for problems that often occur, then data collection in the form of wind direction and velocity and making a map of the research location. After conducting data collection and data analysis, result and discussion are obtained, then draws a conclusion accompanied by recommendations.

4. Result and discussion

In forecasting the wave used wind data sourced from ECMWF with a span of about 10 years from the beginning of 2008 to the end of 2017. The location of wind data retrieval and examples of wind velocity and direction data used in this research can be seen in Figure 1. and Table 1.

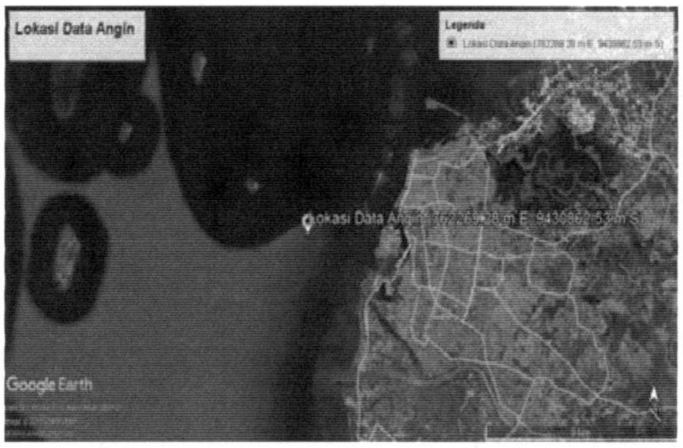

Figure 1. Location of wind data retrieval

Based on the results of wind data analysis, wind events at the research locations as shown in Table 1 were obtained.

Table 1. Percentage of wind events based ondirection

Direction	Data	Percentage (%)
0	329	2.05
45	358	2.23

Direction	Data	Percentage (%)
90	3466	21.56
135	5355	33.31
180	856	5.32
225	612	3.81
270	2381	14.81
315	2719	16.91
Jumlah	16076	100

Based on the table above, it can be seen that the highest amount of wind events in the research location came from the southeast with a total data of 5355 data and a percentage of wind events around 33.31% followed by east, west, northwest, south, southwest, northeast, and north . Meanwhile, wind rose at the researchlocation can be seen in Figure 2.

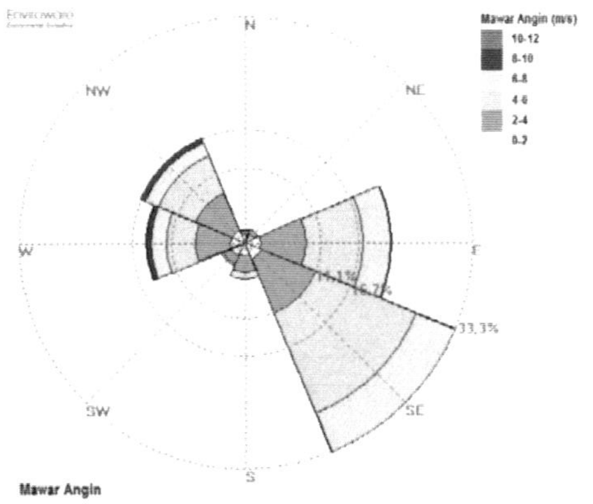

Figure 2. Wind rose in the reserach location

The map of measurement results and fetch calculations can be seen in Table 2.

Table 2. Effective fetch in the research location

No.	Direction	Effective Fetch (km)
1	North	179.64
2	South West	340.36
3	West	447.62
4	North West	419.05

Based on the results of the effective data analysis and fetch above. The height and period of deep sea waves at the research location can be seen in Table 3.

Table 3. The height and period of deep oceanwaves are significant at the reserachlocation

Time	Wave Direction	T (s)	H (m)
2017-06-28 T12:00	270	2.9	5.83
2017-06-28 T18:00	270	2.84	5.79

Time	Wave Direction	T (s)	H (m)
2017-06-29 T00:00	270	2.77	5.74

Based on the results of wave data analysis, wind events were found in the research locations as shown in Table 4.

Table 4. Percentage of wave occurrence based on direction

Direction	Data	Percentage (%)
0	329	5.45
45	0	0
90	0	0
135	0	0
180	0	0
225	612	10.13
270	2381	39.41
315	2719	45.01
Jumlah	6041	100

Based on the table above it can be seen that the highest number of wave occurrence data in the research location comes from the northwest with 45.01 data and the percentage of wave occurance is around 2719% followed by west, southwest and north. Meanwhile, there is no potential for waves from the northeast, east, southeast and south because it is obstructed by the land area of the research location. The wave rose at the research location can be seen in Figure 3.

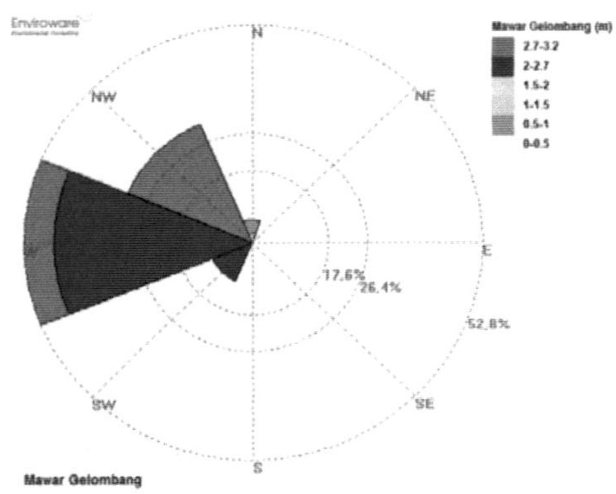

Figure 3. Wave rose in the research location

Table 5. Height and period of ocean waves with certain return times

Return Period	North		North West		West		South West	
	H	T	H	T	H	T	H	T
2	0.40	2.01	1.31	4.58	1.38	4.54	0.99	3.74
5	0.58	3.03	1.79	5.83	2.05	6.30	1.53	5.40
10	0.70	3.71	2.11	6.66	2.50	7.46	1.90	6.50

Return Period	North		North West		West		South West	
	H	T	H	T	H	T	H	T
25	0.86	4.57	2.51	7.71	3.07	8.93	2.35	7.90
50	0.97	5.20	2.81	8.48	3.49	10.02	2.69	8.93
100	1.08	5.83	3.11	9.25	3.91	11.11	3.03	9.96

The determination of the height and depth of breaking waves can also be done by making superposition graphs between wave heights that are affected by shoaling and refraction processes with graphs of wave height affected by water depth. In this case, it is considered the height of the breaking wave is proportional to the depth of the water, i.e. the breaking wave height is 0.78 from the depth of the water.After analyzing the breaking waves the breaking waves at the reserach location can be seen in Table 6.

Table 6. The height and depth of the wave rupture at the research location

No	Direction	Tallo Estuary		Jeneberang Estuary	
		Hb	Db	Hb	Db
1	North	1.05	1.3	0.9	1.2
2	Northwest	3.3	4.3	3.2	4.3
3	West	3.6	4.6	4	5.2
4	Southwest	-	-	2	2.65

After analyzing the breaking waves, the direction of sediment transport along the coast at the research location using equation 1 and equation 2. The direction of sediment transport along the coast in the research area can be seen in Table 7. and Table 8.

Table 7. Longshore sediment transport of Tallo river estuary

Direction	Wave Percentage	K	Cb (m/d)	ab (o)	Pl (Nm/d/m)	Qs (m^3/hari)	Explanation
North	5.45	0.4	3.57	41	2.45	-5.4	Toward South
Northwest	45.01	0.4	6.49	4	6.19	-111.65	Toward South
West	39.41	0.4	6.72	49	54.18	856.3	Toward North
						739.3	Toward North

Table 8. Sediment transport along the coast of Jeneberang river estuary

Direction	Wave Percentage	K	Cb (m/d)	ab (o)	Pl (Nm/d/m)	Qs (m^3/hari)	Explanation
North	5.5	0.4	3.4	62.0	1.5	-3.2	Toward South
Northwest	45.0	0.4	6.5	17.0	23.4	-421.8	Toward South
West	39.4	0.4	7.1	28.0	59.5	941.0	Toward North
Southwest	10.1	0.4	5.1	73.0	7.2	29.1	Toward North
						545.1	Toward North

IOP Conf. Series: Materials Science and Engineering **797** (2020) 012010 doi:10.1088/1757-899X/797/1/012010

For more details, the direction of sediment transport along the coast in the Tallo River estuary and the Jeneberang river estuary towards the Makassar coast can be seen in Figure 4.

Figure 4. Longshore sediment transport directions in the Tallo riverestuaryand the Jeneberang river estuary towards the Makassar coast

5. Conclusion and recommendation

Conclusion from the reserach of Analysis of Sediment Transport Tracing the Coast at the Estuaries of Jeneberang River and Tallo River Caused by Waves on the Coast of Makassar are:

a. Sediment transport along the coast in the Tallo River Estuary is relatively heading north or away from the coast of Makassar.
b. Sediment transport along the coast in the Jeneberang River Estuary is relatively towards the north or leads to the coast of Makassar.

For further research it is recommended to conduct numerical modeling and/or laboratory modeling as a comparison material.

References

[1] Spatial Plans of Makassar City 2005-2035, 2005, Regional Development Agency of Makassar city, Makassar.
[2] Triadmodjo, B, Coastal Engineering, 1999, Beta Offset, Yogyakarta. 184-188
[3] USACE, Coastal Engineering Manual, 2002, Washington, D.C.
[4] Umar, H, et al., Sediment Transport Reduction Along the Coast Case Study of Glagah Beach, Kulon Progo, Central Java, 2016, JPE-UNHAS, Vol. 20, No. 2, Makassar.

Analysis of Torobulu - Tampo Port Service Performance to Improve Mobility of People and Good Movements

Adris A. Putra[1*], Edward Ngii[1], Rudi Balaka[1], Mappa Nashrun[1], Harsawan Harsawan[1]

[1]Transportation Engineering, Department of Civil Engineering, Faculty of Engineering, Universitas Haluoleo, 93232 Kendari, Sulawesi Tenggara, INDONESIA

*Corresponding author's e-mail: putra_adris@yahoo.com

Abstract. South Konawe and Muna Island are strategic locations to make the ASDP Port one of the sea centers in Southeast Sulawesi, but the current performance of the Torobulu port is not maximized in providing services to users. The purpose of this study is to study the performance of the Torobulo Konawe Selatan Port for now. and Provide solutions to the problems of the Torobulu-Tampo port in terms of service to the government. Performance analysis using the Importance Performance Analysis (IPA) method of the analysis method is done by considering the factors of interest and level of satisfaction. The results showed that of the 15 indicators that can be used in quadrant 1 (low priority) is the attitude of officers in serving jobs 3.52 and a value of 4.35, the area of scope with performance of 3.50 and the value of 4.26, supporting facilities in the room wait with a performance value of 3.70 and a value of 4.24, cleanliness of the port area with a performance value of 3.71 and a value of 4.35. Support facilities in the waiting room with a performance value of 3.66 and a value of 4.07.

1. Introduction

Sea transportation has a role as a means of distributing cargo (goods and passengers), as a supporting tool and at the same time driving the dynamics of development, supporting regional development, and further strengthening the development of community life, nation and state in order to realize the Archipelago's Insights and enhance international relations.

The development of transportation is directed to bridge the gap between regions and enhanced equitable distribution of development outcomes. Inter-regional transportation will raise opportunities for trading; reducing price differences; increasing the mobility of goods, services and labors in order to reduce the concentration of skills and abilities in some areas.Thus, encouraging the creation of opportunities for development, opening isolation, increasing mobility and social contact between the population could maintain activities and facilitate access to the use of technology as well as improving social aspects. The process is capable to support the economy aspect of the local citizen and increase local revenue [2].

ASDP Torobulo-Tampo Port is located in South Konawe District, Southeast Sulawesi Province, which is useful for connecting the islands of South Konawe and Muna Island, with a strategic location to make the ASDP Port one of the marine centers in Southeast Sulawesi, but there are still components and services that are less than optimal.

2. Literature review

2.1. Port

According to [5], the port is a place consisting of land and water and surrounding areas with certain limits as place of Government activities and service activities. Besides, port is where the ship is leaned, anchored, lifted up and down passengers and/or loading and unloading of goods equipped with shipping safety facilities and port supporting activities as well as canal and inter-modal transportation. Meanwhile, said that port is a place consisting of land and surrounding waters with certain limits as a place of government activity and economic activity which is used as a place of ship leaning, anchored, up and down passengers, and/or loading and unloading of goods which is equipped with shipping safety facilities and supporting port activities and as well as a place inter-modal transportation. The public port is a port that is organized for the benefits of public services [7].

Meanwhile, said that the ports are a watershed protected against storms, waves and currents [6]. So, it can be said that the ships can hold movement, lean back, throw anchor in such a way that loading and unloading of goods and passenger movements can be done well. The main function of port is displacement function of the industry in terms of port entrepreneurs, completing the facilities for the activities of ships at the port such as the shipping channel for in and out of the ships from and to the port, mooring equipment, loading and unloading activities at the dock, checking goods, warehousing, and provision of local transport links in the port area.

Said that ports can play a role in stimulating the growth of economic, commercialand industrial activities from their spheres of influence but they do not create such activities [8]. They only serve to grow and develop these activities such as activities in the increasing role of ports from just as a ship place to the center of economic activity. Said that, in terms of regional importance, port area has an economic meaning that the port has import and export functionof other economic activities that are interconnected by cause and effect [3]. The transportation system is an interaction that occurs between three components of a system that are interrelated and influence, namely the system of activities,The port as a marine transportation infrastructure has a very important and strategic role for industrial and trade growth and is a business segment that can provide a constituent for the national economy and development because it is part of the chain of transportation and logistics systems [1].

2.2. The Role of Ports in Port Infrastructure Development

Infrastructure development is an integral part of national development. Infrastructure is the driving force of economic growth. Transportation sector activity is the backbone of the distribution pattern of both goods and passengers. Other infrastructure such as electricity and telecommunications related to the nation's modernization and its supply is one of the most important aspects to increase productivity of the production sector. The availability of housing and settlements, including drinking water and sanitation so as widely and equitablysustainable management of water resources, will determine the level of community welfare. In this context, future approaches to the development of region-based infrastructure are increasingly important to note and consider as major development.

Port infrastructure has a very strong linkage to economic growth, social welfare and even social-politics as indicated that regions with better port infrastructure will have welfare levels with good economic growth. Thus, the development of port infrastructure is one of the main factors of economic growth or the development of a region. Emphasizedthat infrastructure development through capacity expansion will have a positive impact on national and regional economic development [4]. The development of the infrastructure must be acceptable to the community with various considerations such as environmental effects, urban planning and social-economic aspects although the cost of development becomes more expensive [4].

2.3. Port Connectivity System

Jinca (2011) In the archipelago region, strengthening the connectivity system is needed to make the inside island and inter-island connections run smoothly in supporting the development of a region's economy. Connectivity in Sislognas is Local Integrated, Globally Connected and has 3 levels of integrated connectivity: Local Connectivity (Intra Island), National Connectivity (Inter Island) and Global Connectivity (International) as shown. In the archipelago region, connectivity scenarios are expected to open up isolated areas linked to economic centers in order to aim at reducing price and service disparities, increasing competitiveness and accelerating poverty alleviation. The aims can ultimately increase social and economic accessibility. The role of ports in the connectivity system is the point of transfer of modes between sea transport and land/road transportationas shown in Figure 1.

Port as the main element in sea transportation system is a supporting factor of economic activities, providing services for mobility of people and goods and being a catalyst for economic growth. Ports can play a driving role in remote areas development as it is known to be pioneer transportation to open isolated areas, increase trade, mobilize population, reduce regional disparities and regional stability. The connectivity or node chain has 4 functions as (composition, conection, interchange, decomposition) in intermodal transport (Rodrigue and Comtois) as shown in Figure 1.

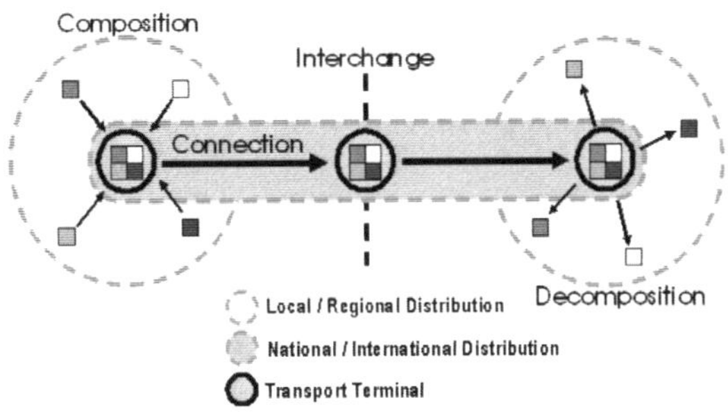

Figure 1. Intermodal Transportation Chain

3. RESEARCH METHODS

This study includes the type of description research with a quantitative approach. Description research is research that attempts to solve current problems based on data. The study was conducted at the Torobulu port of South Konawe Regency, Southeast Sulawesi Province. The sampling method used in the study was Accidental Sampling. Accidental Sampling is a sampling technique that bases itself by chance or where it is found, which meets the requirementsas a passenger (service user) ASDP Torobulu-Tampo port.

4. RESULTS AND DISCUSSION

4.1. Service Performance and Level of Interest/Expectation of the Port of Torobulu.

Based on the sum of the results in Table 5, the suitability level is 80.92%.. Based on the calculation of the suitability level of the 15 indicators, it can be seen the suitability level of each for the quality of the Torobulu Port service, as follows Table 5 the average overall suitability level of the 15 Port Torobulu service indicators.The performance analysis of the service of the Torobulu port using the Importance Performance Analysis (IPA) method of the analysis method was carried out by combining the measurement of importance factors and the level of satisfaction of the visitors and users of port terminals in the graph, making it easier to explain the data. Interpretation of the science chart on four quadrants based on the measurement results obtained from the results of the questionnaire in the field. The equation used to calculate the average score of the level of performance and level of importance is as follows:

Table 1. Average Level of Conformity in Service Quality of Torobulu Port

No	Statement	quality Performance (X)	quality Interests (Y)	Level Suitability (Tki)
1	Attitude of officers in serving passengers	352	435	80.92
2	Ship arrival schedule information	379	450	84.22
3	Security of ports	350	426	82.16
4	Lighting area system at the port (night)	330	402	82.09
5	Availability of supporting facilities in the waiting room	370	424	87.26
6	Cleanliness of the port area	371	435	85.29
7	Ticket purchase system	362	393	92.11
8	Availability of vehicle parking area	380	399	95.24
9	Order hawkers when selling at the port	396	436	90.83
10	Port entry fee	402	423	95.04
11	Cleanliness of the waiting room	385	433	88.91
12	Departure time	399	404	98.76
13	Seating in the waiting room	360	399	90.23
14	Seating in the waiting room	366	407	89.93
15	Ship comfort	395	397	99.50
Average Amount				**89.50**

Source: Questionnaire results

Cartesian diagram is a building that is divided into four sections which are bounded by two lines intersecting perpendicular to the points (X,Y), which X is the average of the average score of the level of implementation or passenger satisfaction on the port terminal of all factors or attributes , and Y is the average of the average score of the importance of all factors that influence passenger satisfaction. There are all 15 factors or attributes. The next formula:

The Cartesian diagram drawings for the service performance of the Torobulu port can be specified in Figure 2. Based on Figure 3 of this Cartesian diagram, it can be seen that the location of the implementation elements of the factors or attributes that influence the satisfaction of passenger services in the South Torobulu Konawe Port is divided into four parts. The interpretation of the Cartesian diagram is explained as follows:

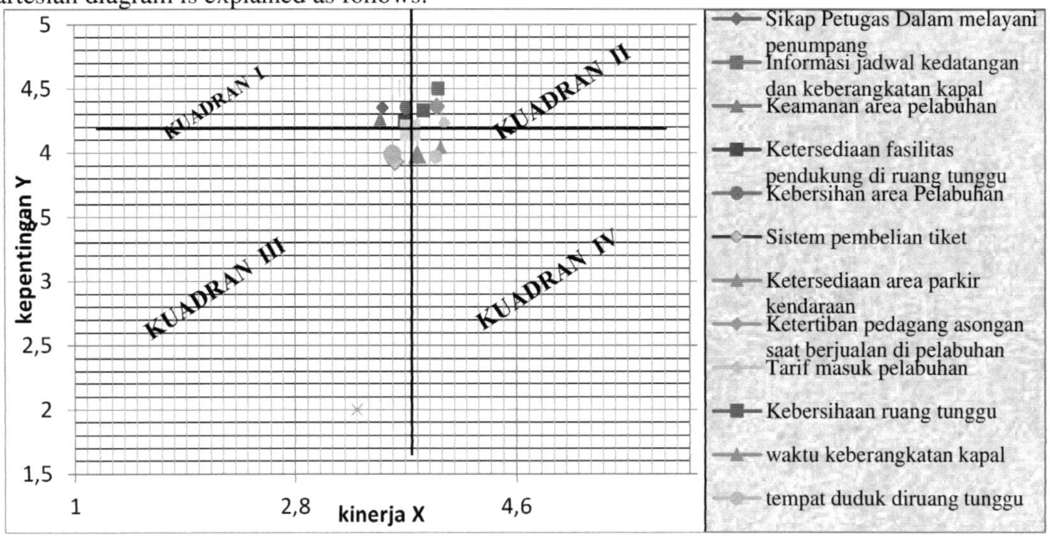

Figure 2. Cartesian Diagram

1) Quadrant I (main priority)
   The factors in this quadrant I are: a) The attitude of officers in serving passengers with an average score of performance levels and interests is 3.52 and 4.35. b) Port area security with performance and interest level scores is 3.50 and 4.26. c) The availability of supporting facilities in the waiting room with performance and interest level scores are 3.70 and 4.24. d) The cleanliness port area with a score of performance and interest levels is 3.71 and 4.35. e). Supporting facilities are shipped with an average score of performance (3.66 and 4.07).
2) Quadrant II (maintain achievement): a) Ship Arrival and Departure Schedule Information with an average level of performance and importance are (3.97 and 4.50). b) The involvement of hawkers when selling in the port area with an average score of the level of performance and level of importance are (3.96 and 4, .36). c) The cleanliness of the waiting room with the average score of the level of performance and the level of importance are (3.85 and 4.33). d) Port entry rates with an average score of performance levels and interests are (4.02 and 4.23)
3) Quadrant III (low priority): a) Seating in the waiting room with an average score of performance level and level of importance are (3.60 and 3.99). b) Ticket purchase system with an average score of performance level and level of importance are (3.62 and 3.93). c) Port Lighting Area (Night) with an average score of performance level and level of importance are (3.30 and 4.02),
4) Quadrant IV (excessive): a) Availability of Vehicle Parking Area with an average score of performance level and level of importance are (3.80 and 3.99). b) The departure time of the ship with an average score of the level of performance and level of importance is (3.99 and 4.04). c) Ship comfort average score of performance level and level of importance are (3.95 and 3.97).

4.2. Problem Solutions

Based on the results of the discussion on the service performance of the Torobulu-Tampo port, the researchers recommend solutions:

1) Attitude of officers in serving passengers
   For indicators of the attitude of officers in serving is very important and can affect the level of passenger satisfaction. The problems related to the indicator are frequent clashes between officers and passengers because the port lacks officers in serving passengers, especially on

holidays.So to optimize the indicators that can affect the level of passenger satisfaction, the need for the officers to be more friendly in serving passengers in any situation and the port must add port officers to anticipate the overflow of passengers that can cause noise within the port.

2) Security of the port area
The indicators related to the port security area are one of the factors that enter the quadrant I where the security of the port area is still unfavorable because the port does not provide security or the like, which results infrequent theft of motorcycles and other electronic devices,
As for suggestions from researchers for the port or instasi related to the need to provide security at the port to prevent loss and to ensure passenger safety.

3) Availability of supporting facilities in the waiting room
The availability of supporting facilities in the waiting room is something that can not be ignored by the port, because the existence of these facilities can provide comfort for passengers. among them there are 2 bathrooms, tv and seating. Based on the problems found in the supporting facilities in the waiting room, it is necessary to repair the bathroom provided by the port.

5. Conclusion and Recommendations

5.1. Conclusion

Based on the description and analysis that have been stated from the previous chapters, the results of this conclusion can be summarized as follows:

1) Performance of the service of the South Torobulu Konawe port, its implementation has been in accordance with the interests and expectations of passengers, and has provided good services including the following: Information on the ship's arrival and departure schedule, Order hawkers when selling at the port, Cleanliness of the waiting room,Port entry fee.
2) While the indicators that need to be prioritized by the port and are still considered unsatisfactory for passengers are as follows: attitudes of officers in serving passengers, port area security, availability of supporting facilities in the waiting room, cleanliness of the port area.

5.2. Recommendations

The recommendations that can be given related to this research are:

1) Pay more attention to the level of service including: Attitude of officers in serving passengers,,port area security, availability of facilities in the waiting room,cleanliness of the port area, supporting facilities in the ship (WC, Canteen)
2) In order to maximize the service performance of the Torobulu port, the need for the port, ASDP or related agencies, pay more attention to facilities at the port and ship which greatly affect the satisfaction of passengers.

References

[1] Putra,A.A and Djalante,S.,2017,*Port Infrastructure Development Strategy In Supporting Development Islands Areas*, international journal of engineering research & technology (IJERT) vol. 6 issue 06.

[2] Putra, A.A.,Ngii, E and Djalante,S.,2018, *Port Development In Supporting Connectivity System Of Archipelago Region*, International Journal of Mechanical and Production Engineering Research and Development (IJMPERD) Vol. 8, Issue 3.

[3] Bintarto, 1989, *Rural-Urban Interactions and Problems*, Jakarta: Ghalia Indonesia

[4] Dekker, Sander, Verhaege, R.J. and Pols A.A.J., 2003, *Economic Impact and Public Financing of Port Capacity Investment*:

[5] Gurning,.O.S and Budiyanto, 2007. *Port Business Management*. Publisher, APE Publishing,

[6] Jinca, M.Y., 2011, *Sea Transportation*, Publisher, Brilliant International, Surabaya

[7] Suranto, 2004. *Management of Sea and Port Transport Operations and Procedures for Import of Goods*. Publisher PT. Gramedia Pustaka Utama, Jakarta

[8] Soemantri, Midian. 2003. *Overview About Automated Marketing Services At PT (Persero) Indonesian Harbour II Cirebon Branch*. Job, University Computer Indonesia, Bandung

[9] Vuchic, Vukan. R., 1981, *Urban Public Transport: Systems and Technology*, New Jersey: Prentince-Hall, Inc.

Potential Use of Pomalaa Nickel Slag as a Substitute for Sand in Brick Making

E Ngii[1], **B Mursidi, and Y Y Umar**
Civil Engineering Department, Engineering Faculty, Halu Oleo University, Indonesia

E-mail: edward.ngii@uho.ac.id

Abstract. The latest variance of nickel slag from PT.Antam Pomalaa has that material gradation size equivalent to sand has, so it has potentially as a sand substitute material. The study aims to determine the use of fine nickel slag as a substitute for sand in brick making. Nickel slag used is variation of 0%, 25%, 50%, 75% and 100% against of total weight of sand. The brick is designed with a cement – sand ratio of 1 : 7. The exprimental is a hollow brick with dimension of 40 cm×18 cm×10 cm. Compressive test is done after brick curing along 28 days. The maximum result of compressive test is obtained in the use of 50 % nickel slag of 19.58 kg/cm^2 compared to the brick without nickel slag of 12.92 kg/cm^2. The conclusion that by using a fine nickel slag type of FeNi 4 up to 50 % on production, it can reduce sand use and improve the strength up to 51%.

1. Introduction

South East of Sulawesi has become one of regions that included in specific Economy Zone and in cement Center of National Mining Industries. The one of the biggest mining in South East of Sulawesi is a nickel mining managed by PT. Antam Pomalaa. Byproduct that is produced is nickel slag to reach a number 1 million tone of weight in 2018. A new variant of nickel slag is FeNi 4, with distribution size passed the sieve number 4 is more than 90%. Based on material size distribution, nickel slag FeNi 4 as equal as sand aggregate [1]. This become apportunity to use nickel slag FeNi 4 of PT. Antam Pomalaa as a fine aggregate to substitute a river sand for many constructions.

Since in 2014, Indonesian government has released regulation that stated that nickel slag is a category of hazardous waste [2]. Yet,level of hazardous of nickel slag has still on level 2, It means that it has a delayed effect and indirect effect to the environment and people [3]. Beside that, some researchers concluded that the level of metal leachate is still below of requirement standart if it is used for concrete aggregate as backfill material and road pavement [4, 5, 6]. This is indicated that nickel slag can be considered as a green environment material on construction made of cement.

The using nickel slag on construction together with cement is called solidification mechanism. This solidification can cause chemical stability and bind the pollutant within solid matrice [7]. Characteristic of nickel slag is latent that can not react if only with water (H_2O).Yet, It has a highly reaction if together with calcium and other compound contained in cement [8,9].

Brick is a building component based on cement that consist in mix of sand, cement, and water with a certain composition. Hollow brick has characteristic of a good heat conductor, its weight lighter than 1/3 of compact brick and more sound absorber [10, 11]. A strength of brick is influenced by a number

[1] To whom any correspondence should be addressed.

IOP Conf. Series: Materials Science and Engineering **797** (2020) 012012 doi:10.1088/1757-899X/797/1/012012

of water cement ratio and a number of sand to cement ratio in the mix. Sand is used commonly a fine gradation [12] with a minimum compressive strength of qualities I,II,III, and IV respectively 65 kg/cm^2, 45 kg/cm^2, 30 kg/cm^2, dan 17 kg/cm^2.

Even though a new variant of nickel slag FeNi 4 sizing likely to sand, but a former researcher result showed that a unit weight of nickel slag is heavier than a normal aggregate [13,14,15]. Based on above result, so the aim of this research is to analyze the level of nickel slag substitution against normal sand used on hollow brick making.

2. Material and Method

The experiment is done into two steps,1).A trial mix step is to find a composition and water cement ratio on a normal brick as control, 2). Step of brick making by using nickel slag

2.1. Material and equipment

Composition of material used are Tonasa cement, sand of Pohara ,FeNi 4 nickel slag from PT. Antam Pomalaa, and water clean from PDAM company. Material characteristic of sand and slag are examined according to SNI/AASHTO/ASTM, whereas material cement and water are observed by visual. Qualification and characteristic of material will be used into mix design.

A main equipment used is a manual compression of brick, a hollow type of 40×10×18 cm in length×wide×hight that is shown on Figure.1. A net volume of hollow brick is 0.0056 m^3.

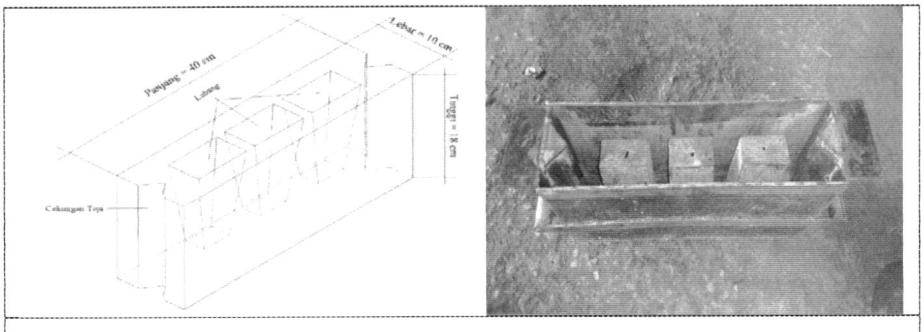

Figure 1. Dimension of brick hollow mould

2.2. Mix design

Mix design of brick is calculated based on ratio of cement and sand weight using volume absolute method. Composition of materials on each variation are shown on Table 1 and Table 2 as following:

Table 1. Mix design of normal brick (BN) for preliminary testing.

Trial	Mix design (cement : sand)	Water/Cement Ratio	Weight of material (kg//m^3) sand	cement	water
1	1: 8	0.4	2028,77	253,60	101,44
2	1: 8	0.5	1978,60	247,32	123,66
3	1: 8	0.6	1930,84	241,36	144,81
4	1: 7	0.4	1977,38	282,48	112,99

Table 2. Mix design of slag nickel brick for ratio cement : sand (1:7) using water/cement ratio 0.4

Variation by types	Code of Samples	Composition (slag : sand)	Weight of material (kg//m^3)			
			Slag	Sand	Cement	Water
Normal brick	Nb	0% : 100%	-	1977,4	282,5	113,0
Slag brick-25	Sb-25	25% : 75%	520,4	1561,1	260,2	104,1
Slag brick-50	Sb-50	50% : 50%	1068,4	1068,4	267,1	106,8
Slag brick-75	Sb-75	75% : 25%	1646,5	548,8	274,4	109,8
Slag brick-100	Sb-100	100% : 0%	2193,7	-	313,4	125,4

2.3. Testing
Each variation of mix is made three test samples so that 15 bricks to be tested. Compressive strength is taken after brick 28 days of age. A peak load accepted when sample test has already broken. However observation of brick quality is done by looking a visual deformation and shape.

2.4. Analysis
Analysis on volume weight and compressive strength of brick is defined by eq.(1) and eq.(2).

$$W = \frac{w}{v} \tag{1}$$

$$f_c = \frac{P}{A} \tag{2}$$

Where W is volume weight of brick (kg/m^3); w is weight of brick (kg); v is volume of brick (m^3); f_c is compressive strength of brick; P is maximum load (kg); A is a surface of stress (m^2).

3. Result and Discussion

3.1. Result of material testing
Expriment result of normal sand and slag sand as following;

Table 3. Properties of normal sand and slag sand.

Properties of material	Pohara Sand	Slag Sand	Specification SNI
SSD specific gravity	2.48	2.83	2.4 - 2.9
Absorption	0.81 %	0.66 %	< 3%
Bulk density	1.65 gr/cm^3	1.40 gr/cm^3	1.4 – 1.9
Clay, Silt Content	1.32 %	0.10 %	< 5 %
Moisture content	0.53 %	1.32 %	< 3 %

Tabel 3 shows that material of sand slag accepted the requirement as material for concrete brick.. Test result of unit weight of nickel sand slag FeNi 4 is higher than normal sand has. A number of unit weight of sand slag is 2.8 indicated that it has in the level of heavy aggregate. Aggregate with unit weight between 2.5 to 2.7 gr/cm3, it will produce concrete that has 2.3 gr/cm3 of unit weight. When aggregate has a high unit weight and high compressive strength as well.

3.2. Exposure of normal brick making

The results of bricks in the preliminary test with the composition as shown in Table 1 are presented as follows:

a) Mix of 1:8, w/c 0.4

b) Mix of 1:8, fas 0.5

Figure 2. A common brick using normal sand (pohara river), mix of 1:8 with 0.4 and 0.5 water cement ratio.

a) Mix of 1:8, fas 0.6

b) Mix of 1:7, fas 0.4

Figure 3. A common brick using normal sand (pohara river), mix of 1:8 and 1:7 with 0.6 and 0.4 water cement ratio.

From the equipment test result of a normal brick using Pohara Sand, we have a composition of mix 1 : 7 and water cement ratio 0.4.

3.3. Exposure of sand slag brick making

In preparing to sand slag brick making, we make a normal brick as control and comparison according to the result of expriment shown in Figure 4.

Figure 4. A normal brick as control, mix 1:7 and water cement ratio 0.4

The exposure of sand slag brick based on composition on Table 2 as following:

a) 25% slag : 75% sand

b) 50% slag : 50% sand

Figure 5. A slag brick with nickel slag of 25% dan 50%.

a) 75% slag : 25% sand

b) 100% slag : 0% sand

Figure 6. A slag brick with nickel slag of 75% dan 100%.

From testing result shown that slag brick 100% is not accepted a visual formation (SNI 03-0349-1989) because of occuring deformation on test object with composition 1 cement : 7 aggregate and 0.4 water cement ratio. So we do not continue a test object of slag 100%.

3.4. Weight examination of slag brick result

The examination of volume weight of each nickel slag variation is shown below.

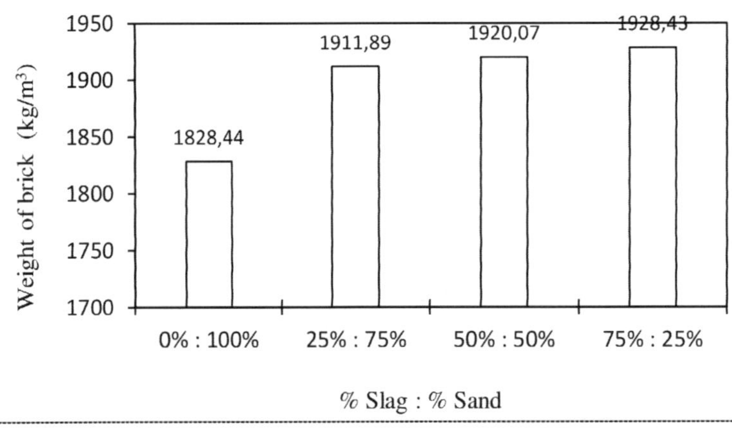

Figure 7. Weight of brick base on slag used variation.

Based on figure as shown, it can be seen that weight of brick is increasing because of using slag in the mix. The more slag using the more weight of brick produced. The number of weight of brick increasing about 5% of normal brick weight (without slag). A increasing of weight is caused of a heavier unit weight of slag than normal sand.

Based on the figure, it can be seen that weight of brick is increasing because of using slag in the mix. The more slag using the more weight of brick produced. The number of weight of brick increasing about

5% of normal brick weight (without slag). A increasing of weight is caused of a heavier unit weight of slag than normal sand.

3.5. Compressive strength examination of slag brick

Compressive strength test of slag brick in variations can be seen below.

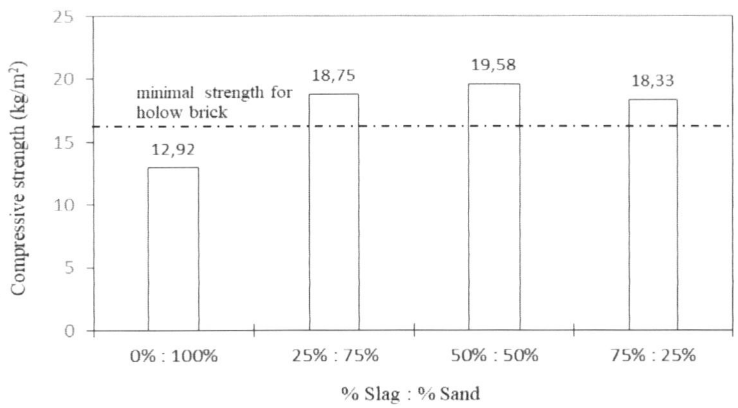

Figure 8. Compressive test of brick base on variation slag in the mixture.

The figure shown that compressive test of normal brick (test object as control) is lower than a minimum requirement number 17 kg/cm2. This condition indicated that composition design of normal brick need more improvement of water cement ratio to meet the requirement of strength.

The using of nickel slag FeNi 4 a number of 25% and 50% will improve the compressive strength of slag brick against a normal brick a number of 45% and 52%. However, by using nickel slag a number of 75% the compressive strength decreased even though it is still higher than a normal brick strength.

An increasing of brick strength using nickel slag FeNi 4 is occurred because of the aggregate texture of nickel slag are rough and sharp. This is in line with the research [14,15] that concluded an increase in strength of concrete due to use of sand slag. A decreasing of brick strength by using of 75% of nickel slag can be caused of higher weight of slag material. Fig. 7 shows that weight of brick increase because of slag nickel addition. For composition of 1 : 7 with water cement ratio of 0.4, using 75% nickel slag FeNi 4 aggregate is not enough optimum to increase compressive strength of sla brick. Mix design modification is needed to find a strength of mix that is higher than 50% of slag using.

4. Conclusion

Using nickel slag FeNi 4 as substitute of sand in concrete brick making can increase its compressive strength, in the same time a volume weight become higher. Best composition and maximum strength we have on 50% slag FeNi 4 : 50% normal sand from Pohara river.

5. References

[1] ASTM C33 2018 *Standard Specification for Concrete Aggregates, ASTM International* (West Conshohocken: PA)

[2] PP No. 101 2014 *Pengelolaan Limbah Bahan Berbahaya dan Beracun* (Indonesia: Jakarta)

[3] Albeto 2017 *Pemanfaatan Slag Nikel dalam Mendukung Pembangunan Nasional* (Edisi XXVIII –November 2017 hal 21-23)

[4] Saha A K and Sarker P K 2017 *Procedia Engineering Conf. on Sustainable Civil Engineering Structures and Construction Materials, SCESCM* (Elsevier) p 689-694

[5] Romy S E, De Schepper M, Gruyaert E, De Belie N 2016 *J. Construction and Building Materials.* **119** 31-44

[6] Tangahu B V, Warmadewanthi, Saptarini D, Pudjiastuti L, Tardan M A M, and Luqman A 2015 *J. Advances in Che m-ical Engineering and Science.* **5** 408-412

[7] Bone B D, Barnard L H, Hills C D 2005 *Stabilisation/Soldification for the treatment of contaminated soil, Taylor & Fancis* (London).

[8] Sun D D, Zhang D, and Lai 2013 *J. of the Air & Waste Management Association* **63** 1469-1478.

[9] Shing T S, Pant 2006 *J. Hazardous Materials* **131** 29-36.

[10] SNI 03-0349 1989 *Bata Beton untuk Pasangan Dinding* (BSN: Jakarta)

[11] Thorat V M, Papal M, Kacha V, Sarnobat T, and Gaikwad S 2015 *International. J. of Modern Engineering Research (IJMER).* **5** 2249-6645

[12] Ukiman, Setio U, Supardjo, Iman N, and Pentardi R 2017 *J. Bangun Rekaprima.* **3** 31-37

[13] Ngii E and Mursidi B 2009 *Proceeding The 1st International Seminar on Sustainable Infrastructure and Built Environment in Developing Countries,* (Bandung, Indonesia) p 20-25

[14] Salain I M A K, Sudarsan I K, and Wayan M 2015 *Proceeding The 5th Environmental Technology and Management Conference "Green Technology towards Sustainable Environment",* (Bandung, Indonesia).

[15] Rini S, dan Ngii E 2017 *Proceeding International Symposium on Concrete Technologi, ISCT* (Makassar, Indonesia).

IOP Conf. Series: Materials Science and Engineering **797** (2020) 012013 doi:10.1088/1757-899X/797/1/012013

The Development of Service Desk Application (SDA) as a Media to Improve Care Services and Asset Improvement in the University

Riyadi Purwanto[1*], Muhammad Yamin[2], Laode Muh. Golok Jaya[3], Yuspian Gunawan[4]

[1] Department of Informatics Engineering, Politeknik Negeri Cilacap, 537992 Cilacap, Jawa Tengah, INDONESIA

[2, 3] Department of Informatics Engineering, Faculty of Engineering, Universitas Haluoleo, 93232 Kendari, Sulawesi Tenggara, INDONESIA

[4] Department of Mechanical Engineering, Faculty of Engineering, Universitas Haluoleo, 93232 Kendari, Sulawesi Tenggara, INDONESIA

email: [1*]adidok_bayu@yahoo.com, [2]putra0683@gmail.com, [3] laodemgj@uho.ac.id, [4]yuspian_gunawan@yahoo.com

Abstract. An asset is a wealth. Assets at universities in the form of facilities and infrastructure are used as learning media. To ensure the assets are in good condition, assets must be managed properly. Universities is an entity that has various types of assets and distributed in various work units. Maintenance technicians have the task of carrying out maintenance and repair of assets. Care and repair history must be recorded. Requests for repairing damaged assets or having problems must be dealt with quickly. However, some universities are still implementing conventional systems. The purpose of this study is to develop Service Desk Application (SDA) as medium to improve maintenance and repair services for campus assets quickly and precisely. SDA is a centralized service application that is used to handle repair requests from customers. The results showed that SDA can monitor the status and progress of asset maintenance and control the performance of technicians so that they can complete work quickly and according to procedures, SDA serves as a useful knowledge-based management to enrich knowledge so that if there are problems with similar assets, then the solution to the problem can be immediately known.

1. INTRODUCTION

Assets are resources or wealth owned by an entity. In general, an asset is an item (thing) or something (anything) that has: economic value, commercial value or exchange value; owned by agencies, organizations, business entities or individuals (individuals)[1]. Physically, the assets purchased need to be managed better, so that adequate administrative equipment is needed so that the assets purchased are maintained[2].

In Educational Institutions assets can be in the form of facilities and infrastructure used as tools and supporting learning processes. To ensure that assets owned can be maintained in good condition, long life, and have a wealth value, then these assets must be managed and maintained properly through maintenance and repair both preventively and corrective.

University is an entity that organizes academic activities which of course has various types of assets including educational laboratory equipment, office equipment, furniture, building buildings, and other supporting infrastructure distributed in various work units. Each University has a Maintenance Technical Implementation Unit which has the task of carrying out maintenance, maintenance and repair of assets as a whole. A list of history of treatments and repairs must be properly inventoryed.

IOP Conf. Series: Materials Science and Engineering **797** (2020) 012013 doi:10.1088/1757-899X/797/1/012013

Service requests for repairs from work units to damaged assets or having problems must be managed and handled quickly. But on the other hand, some universities still implement conventional asset maintenance and repair services. If there is an asset that has a problem, then the head of the work unit will contact the Maintenance Technical Implementation Unit by telephone or delivered orally. Then the technical staff will record the request in the repair inventory list (*paper based).*

The mechanism certainly raises a number of problems, among others, inventory lists for repair requests are poorly stocked and not systematized, maintenance and repair history is not fully recorded, takes a long time to look for similar or similar maintenance and repair history. These problems result in slow response time for maintenance and repair services. Because of that, a Service Desk Application (SDA) is needed that can handle and improve maintenance requests and improve campus assets in a fast, precise and accurate manner. Service Desk Application is a centralized information technology service application that is used to handle requests for repair of an event from a customer. Service Desk is a function in the ITIL process that is responsible for managing incidents and problems and as a bridge between users and service managers[3].

The purpose of this study is to create a service desk information system as a medium to improve care services and repair campus assets. The system is used as a medium of communication and request maintenance services and asset improvements online between asset users and the technical asset maintenance unit at each university.

2 LITERATURE REVIEW

A. Asset

Assets are resources or wealth owned by an entity. In general, an asset is an item (thing) or something (anything) that has: economic value, commercial value or exchange value; owned by agencies, organizations, business entities or individuals (individuals)[4]. Physically, the assets purchased need to be managed better, so that adequate administrative tools are needed so that the assets purchased are maintained[5]. In Educational Institutions assets can be in the form of facilities and infrastructure used as tools and supporting learning processes.

B. Service Desk

Service Desk is a function in the ITIL process that is responsible for managing incidents and problems and as a bridge between users and service managers[3]. Service desk has an important and integrated function in IT service management to support effective management for organizations. Service desk is also a center of contact between service providers and users[4].

Service Desk is a functional unit consisting of a number of employees / employees who are dedicated to handling various services through telephone calls, web interfaces, or infrastructure events that are automatically reported by a system[6]. The purpose of Service Desk is to conduct communication centers between users and IT service management, to handle various events and requests[5]. Task in the Service Desk in general, among others: accepting incidents, recording incidents, classification of incidents based on priorities, classification and escalation, finding solutions, providing information to the end user about the process that takes place[7].

C. Incident & Problem Management

Incident & Problem Management is a handling and prevention of an event / problem. Incident & Problem Management has a role to convince the use of good resources to support the business, to develop, repair and store data relating to trouble as well as possible. Incident & Problem Management has the following responsibilities[5], that is Detect events and record them, Classify each event and initialize the support form, Conduct investigations and diagnostics, Make provisions and fix them, Resolve every problem / event Maintenance

D. Maintenance

Maintenance is an activity to maintain facilities, so that the facility is in a ready-to-use condition as needed. Treatment is generally seen as a physical activity such as cleaning the equipment in question, giving lubricants, repairing damage, replacing components and the like if needed[8]. In general, treatment consists of two types, namely planned care in the direction of prevention and unplanned care (unscheduled) direction to damage.

3. RESEARCH METHODS

The research was carried out in several stages, that is :

A. Research Materials

In this study, the research materials used were as follows:

1) Data obtained from a literature study on Service Desk Application.
2) Information on mechanisms and procedures for maintenance and repair of assets
3) Information regarding inventory list of problems for maintenance and repair of assets

B. Research Instrument

The research tool used in this study is a computer device that has enough specifications to run an operating system and software, namely Windows 7, Macromedia Dreamweaver (software developer), xampp 1.7.3 (web server), and Adobe Photoshop (graphic design software).

C. Road Research

In the process of developing Service Desk Applications, researchers refer to software development methods in general, namely the method of developing the System Development Life Cycle (SDLC) Waterfall model[9]. Method of developing the Waterfall Model consists of several hases / stages. That is Analysis & Requirement Phase, Design Phase, Implementation Phase, Testing Phase Is a stage of testing the system and improving the system developed, and Maintenance Phase.

D. System Design

The process of requesting maintenance assets in the developed system can be shown in Figure 1.

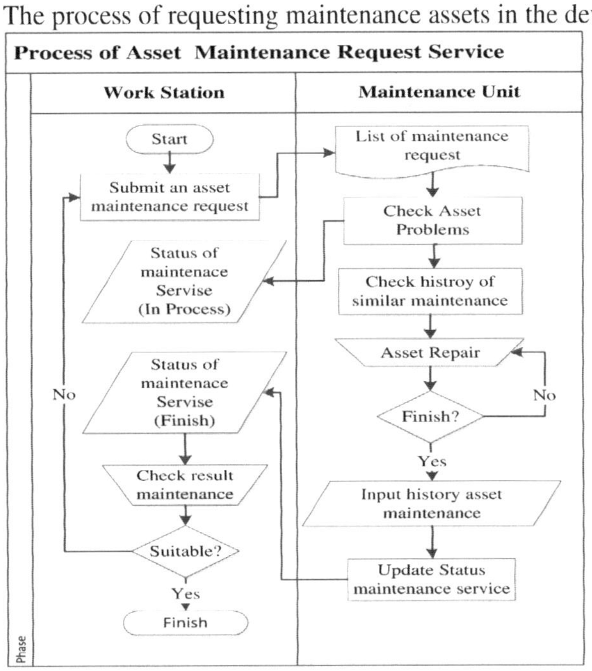

Figure 1. Flowchart of the process of maintenance requests and asset improvements

Figure 1. Explain the flow of the request process and asset repair through the SDA which can be explained as follows:

1. There are 2 (two) actors who play a role in the process of maintenance.
2. The process of maintenance and repair of assets begins with the Work Unit that submits a request to the Maintenance Unit by filling in the request form on the SDA.
3. The Maintenance Unit receives a list of requests for maintenance will then check the contents of the problem that has been written on the request form.
4. Technicians from the Maintenance Unit will check the damaged assets and trace the history of repairs in the system if similar problems occur. Furthermore, the operator will update the status that the asset repair request is in process.
5. If a similar problem occurs, the repair will be easier. But if it has never been done, the technician will repair the asset from the beginning.

System design is made in the form of Data Flow Diagrams (DFD).

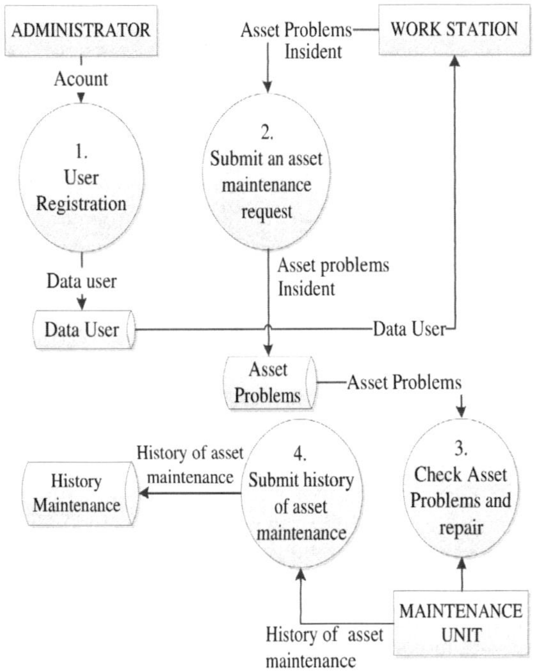

Figure 2. Data Flow Diagram (DFD)

Figure 2, describes the main processes in the Service Desk Application which can be explained as follows:

1. There are 4 main processes including the user registration process, the input process requesting asset repair and maintenance, the process of checking asset problems, and maintenance history input and asset improvement.
2. Administrators only have access rights in managing user data as shown in process 1.
3. The work unit has access rights to input incidents or problems and submit requests for asset repair and maintenance. Requests for maintenance and repair of assets will be stored in the database as shown in process 2.
4. The maintenance unit checks the demand for repair and maintenance of troubled / damaged assets as shown in process 3.
5. After the asset is repaired, the maintenance unit will input the list of asset problems and solve the problem into the system as a history of maintenance assets.

4. RESULTS AND DISCUSSION

The results of the research conducted were the creation of an information system as a medium to improve care services and asset improvements in universities called Service Desk Applications. The system can function to convey incidents, requests for maintenance and repair of damaged or problematic assets delivered by asset users. An overview of Service Desk Application can be shown in Figure 3.

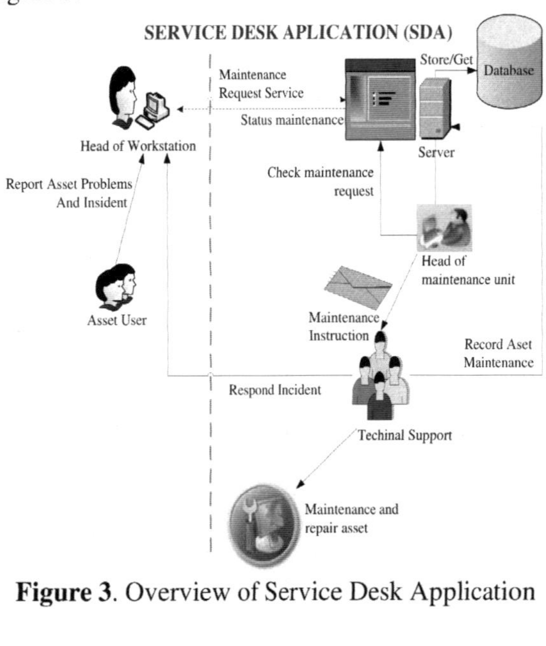

Figure 3. Overview of Service Desk Application

Figure 3, shows the general description of the service desk application work process. From the picture, it is explained that, when an incident or problem occurs in an asset used by the user, it must be reported to the head of the work unit. The next stage, the head of the work unit reports the incident or problem to the maintenance unit through an online application, namely the service desk application to be repaired. The head of the maintenance unit will check the list of asset repair requests from the work unit and forward it to technical support to make improvements. Requests for asset repair and solution to the problem will be stored in the system database as maintenance histories.

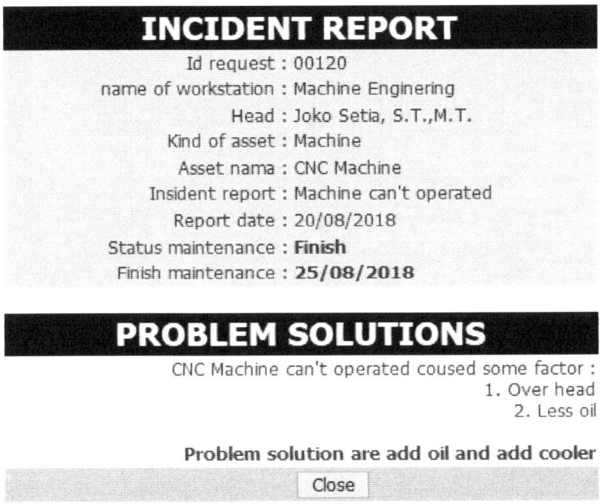

Figure 4. History of asset maintenance

Figure 4, shows information on asset maintenance history as knowladge based. Thus, if similar problems occur on the same machine, then the solution to the problem can be immediately known and the repair of assets will be resolved quickly. Based on the study of the research that has been done, the Service Desk Application can provide benefits as a medium in improving maintenance demand services and improving assets at Higher Education as shown in table 1.

Table 1. Benefits of Service Desk Aplication (SDA)

No	Field	Benefits
1.	Insident report	1. Incident reports on assets that are problematic, damaged, or experiencing well-documented interference in the system. 2. Types of assets in universities that are reported through natural resources can be in the form of educational equipment, office equipment, furniture, buildings and buildings, and other supporting infrastructure. 3. Incident reports can be made by users online
2.	Monitoring maintenance	1. The repair status of assets can be monitored by the reporter and maintenance unit head with the status of waiting, process, and finish so that maintenance progress is more controlled. 2. As a tool to control the performance of technicians (technical support) in maintaining and repairing assets at the University to work quickly and according to procedures.
3.	Time management	1. Incident reports can be done at any time and are not limited to working hours, thus speeding up the process of data collection of complaints against problem assets. 2. Response to asset repair based on report date (queuing system), except for matters that are urgent and very urgent on the basis of orders from the Head of the Maintenance Unit.
4.	Knowlade based	1. SDA has a basic knowladge that is obtained based on maintenance history, so that if similar problems occur, the repair solution can be immediately known and the repair of assets will be resolved quickly. 2. As a media for handling problems for Technicians and asset users. Thus, if there are problems with assets that are of a minor nature, asset users can try to improve themselves

5. CONCLUSION

Based on the results of the research that has been done by developing it can be made as follows:

1. Desk Application Service can receive incident report information on troubled or damaged assets from various work units at the University for maintenance and repair by the Maintenance Unit. Incident reports are done online so that they can be done at any time not limited to working hours.

2. Service Desk Application can monitor asset maintenance status to control the progress of maintenance and repair as well as controlling the performance of technicians so that they can complete the work quickly and according to procedures.
3. Service Desk Application can function as knowladge based management with maintenance history data sources. This information is useful for maintenance unit technicians to enrich knowledge. If there are similar asset problems, then the solution to the problem can be immediately known and the repair of assets will be resolved quickly (responding to maintenance service times faster).

The suggestions that can be made for system development and further research are as follows:
1. Service Desk Application can be developed again by adding online chat features.
2. Service Desk Application can be developed again by adding the function of asset maintenance schedules at Universities on a regular basis in order to minimize the occurrence of asset damage.

Reference

[1] I. Yunita and J. Devitra, "Informasi Manajemen Aset Pada Smk Negeri 4 Kota Jambi," *Anal. Dan Peranc. Sist. Inf. Manaj. Aset Pada Smk Negeri 4 Kota Jambi*, vol. 2, no. 1, 2017.
[2] F. Nugraha, "Rancang Bangun Sistem Informasi Manajemen Aset Perguruan Tinggi Dengan Metode Simple Additive Weighting(Saw)," *ISSN 2252-49483 J. Inf. Syst. Eng. Bussiness Intell.*, vol. 3, no. 1, pp. 1–5, 2014.
[3] L. Aisha, K. Wardani, and L. Ramadani, "Perancangan Tata Kelola Manajemen Layanan Teknologi Informasi Menggunakan ITIL V3 Domain Service Transition dan Service Operation di Pemerintahan Kota Bandung," *Ind. F. R., Oper. S., Transition, S., Publik, S*, vol. 3, no. 2, pp. 3272–3278, 2016.
[4] A. Pramurjadi, "Analisis dan Rancangan Manajemen Service Desk di Instansi Pemerintah (Studi Kasus : Badan Litbang Pertanian) Analysis and Design of Service Desk Management in Governmental Institution (A Case of Indonesian Agency for Agricultural Research and Developme," *J. IPTEK-KOM*, vol. 16, no. 1, pp. 69–88, 2014.
[5] H. Surbakti, "Pendekatan Model Information Technology Infrastructure Library (Itil) Dalam Rekayasa Sistem Informasi Untuk Pencapaian It- …," *Teknol. Inf. Respati*, vol. 3, pp. 97–110, 2013.
[6] A. A. Agustinus Suradi, M. Suyanto, "Analisis kematangan tata kelola informasi service desk dan insiden di yayasan pangudi luhur yogyakarta," *J. Teknol. Inf.*, vol. VIII, no. November, pp. 75–86, 2013.
[7] Hendra, "Penerapan Best Practice It Service Management Dalam Perbaikan It Service Desk : Studi Kasus Pt . Matahari Putra Prima," *Comtech*, vol. 2, no. 1, pp. 514–526, 2011.
[8] A. Purwanto, "Analisis Failure Rate Mesin Reverse Osmosis dengan Perhitungan Evaluasi Sistem Perawatan di PT . XYZ," *j. Indept*, vol. 3, no. 3, 2013.
[9] Y. Bassil, "A Simulation Model for the Waterfall Software Development Life Cycle," *Int. J. Eng. Technol.*, vol. 2, no. 5, pp. 2049–3444, 2012.

IOP Conf. Series: Materials Science and Engineering **797** (2020) 012014 doi:10.1088/1757-899X/797/1/012014

Carbon Stocks Mapping of Mangrove Forest in North Buton Indonesia using Combination of Landsat and ALOS PALSAR Images in the Perspective of Climate Change Mitigation

La Ode Muh. Golok Jaya [1], Muhammad Yamin[1], Ika Purwanti Ningrum[1], Jamhir Safani[2] and Ketut Wikantika[3]

[1]Faculty of Engineering, Universitas Halu Oleo (UHO), Jl. HEA. Mokodompit No. 8 Kampus Hijau UHO Bumi Tridharma Anduonohu Kendari, South East Sulawesi, Indonesia

[2]Faculty of Earth Science and Technology Universitas Halu Oleo (UHO), Jl. HEA. Mokodompit No. 8 Kampus Hijau UHO Bumi Tridharma Anduonohu Kendari, South East Sulawesi, Indonesia

[3]Center for Remote Sensing, Institut Teknologi Bandung (ITB), Faculty of Earth Science and Technology, ITB, Jl. Ganesha 10, Bandung, West Java, Indonesia.

Email: laodemgj@uho.ac.id, putra0683@gmail.com, ika.purwanti.n@gmail.com, jamhirsafani72@gmail.com, wikantika.ketut@gmail.com

Abstract. Mapping of carbon stocks, distribution and its changes become one of the important strategies in climate change mitigation. This study aims to map the carbon stocks of mangrove forests and analyze the changes using a combination of Landsat and ALOS PALSAR images in North Buton, South East Sulawesi Province, Indonesia. Carbon stocks in mangrove forests are known stores several times more than forests in the mainland. Mangrove forest at coast of North Buton is one of the largest area of mangrove forest in south east Sulawesi. The research conducted using backscatter coefficient of HH and HV polarization from ALOS PALSAR and Normalized Difference Vegetation Index (NDVI) from Landsat. The results showed that Landsat image provided information about the distribution, density and change of the carbon stocks during year 2001-2017 while ALOS PALSAR provided the volume of carbon stocks in this region.

1. Introduction

The existence of mangrove in coastal area plays important role for a healthy coastline and able to become a resource for the environment and humans who live around the area. Mangrove forests have various advantages over other mainland natural forests in terms of carbon stocks, environmental functions and disaster mitigation as well as climate change. Mangrove forests are able to store carbon five times more than forests on land [1]. In terms of environmental functions, mangrove forests become spawning areas for several fish species and several other marine species. Regarding to disaster mitigation, mangrove forests can become a kind of strong barrier to ocean currents including tsunamis.

[1] To whom any correspondence should be addressed.

Thus, mangrove forests need to be conserved for the future. However, population today increase and settlement expansion have threatened the existence of mangrove forests.

Mangrove forest and its carbon stocks need to be mapped and monitored periodically to ensure the sustainability of mangrove existence. Mapping of carbon stocks is carried out in various ways with various methods and approaches and with various instruments. One effective way to map and monitor carbon stocks is using remote sensing.

Landsat and also other optical sensor such as SPOT, Ikonos, World-View etc. have been used in biomass or carbon stock estimation since a few decades ago. Landsat has a high spectral resolution for the analysis of biomass and land. However, the optical remote sensing encountered obstacles due to hydrometeorology conditions such as cloud cover, rain, fog and smoke. Thus, today, remote sensing radar namely Synthetic Aperture Radar (SAR) becomes an alternative for mapping land use and land cover. SAR sensor can penetrate rain, clouds, fog and can be operated day and night.

Today, there are several radar satellites can be used to monitor and map carbon stocks, such as ALOS PALSAR and Sentinel-1. Dual polarization ALOS PALSAR (Advanced Land Observing Satellite-Phased Array type L-band Synthetic Aperture Radar) and Sentinel-1 images are available in abundance and distributed free of charge. The easy of obtaining cloud-free image data for carbon stocks mapping purposes in the context of disaster mitigation and climate change needs to be welcomed. The abundance of Landsat image data in the online system especially from https://earthexplorer.usgs.gov/ is very beneficial for biomass mapping today. We also can obtain ALOS PALSAR data (L-band) freely from https://www.asf.alaska.edu/sar-data/palsar/, as well as Sentinel-1 SAR data (C-band) from https://scihub.copernicus.eu/.

It has been hypothesized that the integration of SAR at different frequency or with optical data can improve biomass prediction [2]. Thus, the combination of optical and SAR images to obtain high accuracy biomass or carbon stocks become a good effort in disaster and climate change mitigation. This research aims to analyze combination of optical and radar images to map mangrove carbon stocks in North Buton, South East Sulawesi Province, Indonesia.

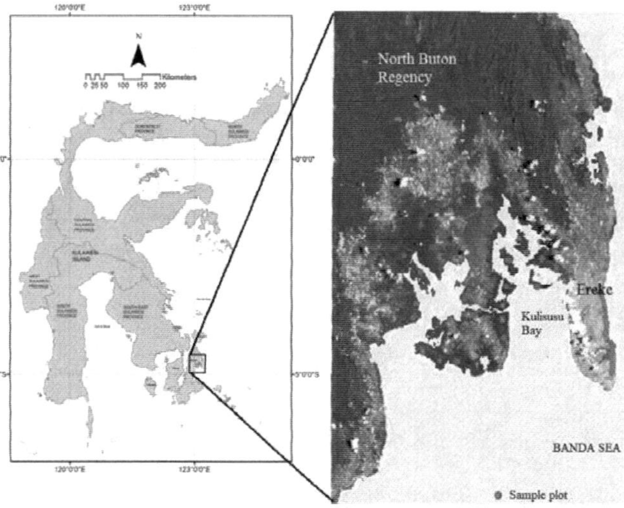

Figure 1. The study location in North Buton, Indonesia.

2. Methods

2.1. Location

This research conducted in mangrove forest in North Buton, Indonesia about 150 kilometers at south east of Kendari, the capital city of South East Sulawesi Province. The mangrove cover is about 13,000 hectare stretches from north to south of North Buton as displayed in Figure 1. The landscape surrounded by mountains in the north and relatively flat in south. The land cover dominated by forest, settlement and agriculture. Mangrove forest covers in some small islands near Ereke (the capital city of North Buton).

2.2. Data

We used optical and SAR image (Table 1) to show the beneficial of these images in mapping carbon stocks.

Table 1. Data used in the research

Platform	Acquisition Date	File Name
Landsat ETM	2001	EPP112R063_7F20010506
Landsat 8	2018	LC08_L1TP_112063_20180614_20180614_01_RT.tar
ALOS PALSAR	18 July 2009	ALPSRP185477090-L1.1
ALOS PALSAR	21 July 2010	ALPSRP239157090-L1.1

2.3. Procedure

Carbon stocks analysis can be conducted through several methods. In this research, we conducted backscatter coefficient analysis for ALOS PALSAR data and NDVI (Normalized Difference Vegetation Index) analysis for Landsat. Shimada et al [3] developed backscatter concept using equation (1).

$$\sigma^{o}[dB] = 10 * log_{10}(DN^{2}) + CF \quad (1)$$

Where DN=Digital Number of radar image and CF=Calibration Factor for ALOS PALSAR data,

While for NDVI, we used general equation (2) from [4]:

$$\mathrm{NDVI} = \frac{\mathrm{IR-R}}{\mathrm{IR+R}} \quad (2)$$

Where NDVI = Normalized Difference Vegetation Index, IR= infrared band, and R = red band.

2.4. Validation

For validation, 20 sampling plots with 25 x 25 square meters grid have been measured in the field during July 2018 as displayed previous in Figure 1. We identified four species of mangrove in the field namely Sonneratia alba, Rizophora apiculata, Bruguiera gymnorhiza, and Xylocarpus granatum. We used allometric equation for mangrove vegetation from [1] which integrate Diameter at Breast Height (DBH) and vegetation height using equation (3).

$$B = 0{,}0825 * (D^{2}H)^{0{,}89966} * \rho \quad (3)$$

where B = biomass (kg), D = DBH (cm), H = height (m) and ρ= vegetation density (gr cm-3).

3. Result and Discussion

Mangrove in North Buton is one of the largest mangrove forest in South East Sulawesi. It consists about 13,000 hectares extend from north to south in Kulisusu Utara, Kulisusu Barat, Kulisusu and Bonegunu Sub Districts. Mangrove ecosystems play an important role in supporting the local communities. Mangroves provide food security and livelihoods through its natural resources, such as mangrove crabs, and fish-spawning habitats. Related to disaster mitigation, mangrove forests also act as storm buffers for coastal communities who may otherwise lose their homes during storm surges. Unfortunately, mangroves in Indonesia are being lost at an estimated rate of 1-2% each year to agriculture and aquaculture [5].

Figure 2. Mangrove forest in North Buton coastal area surrounded by traditional settlement of Bajo, nipa, coconut tree and fishpond (Photo by Jamhir Safani and team)

Figure 2 shows mangrove forest and the existence of Bajo settlement surround the forest. Bajo (or Bajau) is traditional tribe who lives in coastal area. Their life is depending on coastal resources. However, for many years they are expanding settlement and fishpond in mangrove area. These activities release carbon in the atmosphere.

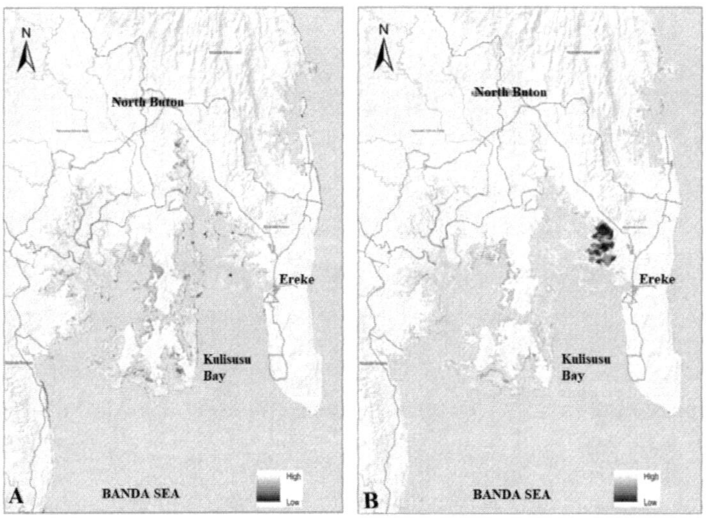

Figure 3. NDVI of mangrove density and distribution in North Buton based on Landsat image: (A) Year of 2001 and (B) Year of 2018

NDVI has been used to analyze carbon stocks in forest and mangrove [6][7]. Middle and high resolution satellite can be used for NDVI analysis. Besides forest density and biomass, we also obtain distribution of vegetation cover in a certain area. In line with this, SAR data become an alternative for biomass mapping. SAR data are acquired in X, C, L and P bands. Each of these bands has their own characteristics in relating to forest stand parameters [8]. The ability of radar images for mapping mangroves cover is very dependent on radar channel and also its polarization. Previous studies have shown that the use of bands with longer wavelengths, for example L Band, will be better in identifying mangroves [9]. While the polarization used is known to polarize HV or VH has the ability that is also better at mapping mangroves due to lower saturation values.

NDVI value normally is between -1 (for non-vegetated area) to +1 (fully-vegetated) [4]. Figure 3 shows NDVI map of mangrove density using Landsat in year 2001 and 2018. The NDVI value between -0.58 to +0.67 in 2001 and become -0.19 to +0.84 in 2018. This informs us that the forest density is increase. However, the area of mangrove forest reduces in the last seventeen years from 12,918 hectare in 2001 become 11,521 hectare in 2018 or 10.8% lost. It may change into fishpond, settlement or agriculture.

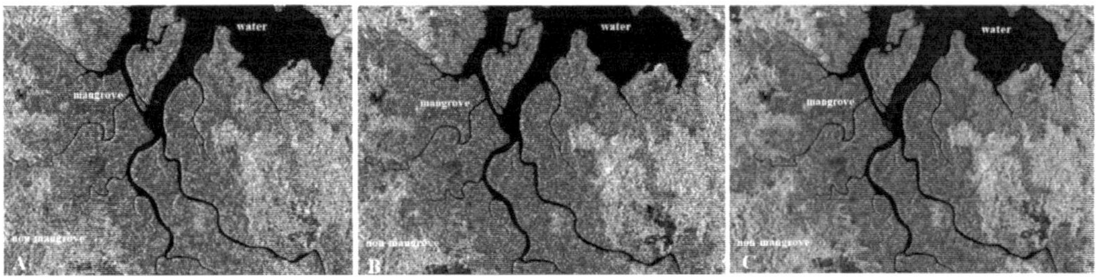

Figure 4. ALOS PALSAR image of mangrove forest in north Buton in multi looking mode: (A) HH Polarization, (B) HV Polarization, and (C) RGB composite (R=HH, G=HV, B=HH/HV) (note: north direction is opposite)

Figure 4 shows ALOS PALSAR image of mangrove forest based on HH and HV polarization and the color RGB composite. The darkness part of image indicates mangrove forest. It can be separated from other land cover such as water and non-mangrove cover.

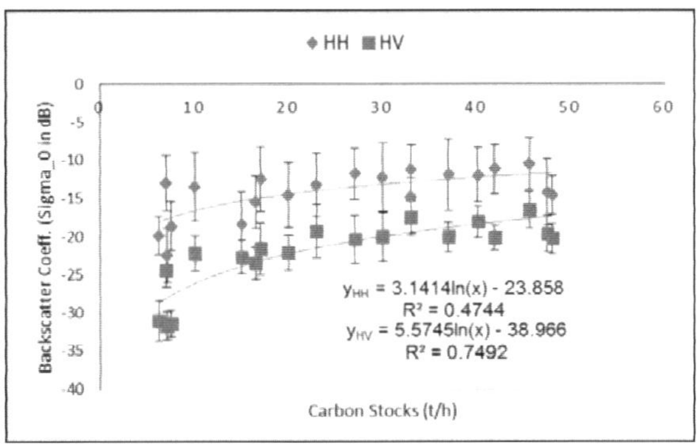

Figure 5. Relation between carbon stocks from sample plot and backscatter coefficient

The backscattering of the L band is more related to the trunk and main branches and less sensitive to the environmental conditions so it is valuable in the biomass estimation [9][10], especially mangrove forest [11]. Using a combination of C and L band has shown better results than using of one band [12]. Figure 5 shows relation between carbon stocks from 20 sample plot and backscatter coefficient of ALOS PALSAR. The maximum carbon stocks among 20 sample plots is 48 ton/ha. We conducted survey only in the coastal line due to difficult accessibility for getting data the older mangrove in the middle part of mangrove cover. We obtained R^2=0.4744 for HH polarization and R^2=0.7492 using HV polarization. This result confirmed some previous researches that HV polarization of SAR image has a good result in mapping biomass.

4. Conclusion

Optical image such as Landsat has advantage to map density and distribution of carbon stocks through NDVI method. We found that the density of mangrove in the study site is increase, but the area of mangrove cover is reduced. In the other hand, SAR image can provide volume of biomass or carbon stocks using backscatter coefficient analysis. We found that using ALOS PALSAR image, HV polarization has a good relation with carbon stocks than HH. The combination both methods can provide a good accuracy of carbon stocks map, distribution and density. The spatial information of carbon stocks can be used to support local and national program related to mitigation of climate change and disaster mitigation as well for example REDD+.

Acknowledgment
This research is part of the research on the application of radar and optical remote sensing images for carbon mapping in the context of disaster mitigation and climate change in Southeast Sulawesi, Indonesia. Thank you to the Directorate of Research and Community Service (DRPM) of the Directorate General of Research and Development (Dirjen Risbang) of the Ministry of Research and Higher Education of the Republic of Indonesia who funded this research through the Halu Oleo University Applied Research Grant in 2018/2019.

References
[1] Murdiyarso D, Purbopuspito J, Kauffman JB, Warren MW, Sasmito SD, Donato DC, Manuri S, Krisnawati H, Taberima S and Kurnianto S 2015 The potential of Indonesian mangrove forests for global climate change mitigation. *Nature Climate Change* **5(12)**: 1089
[2] Laurin GV, Balling J, Corrona P, Mattioli W, Papale D, Puletti N, Rizzo M, Truckenbrodt J and Urban M 2018 Above-ground biomass prediction by Sentinel-1 multitemporal data in central Italy with integration of ALOS2 and Sentinel-2 data *J. Appl. Remote Sens.* **12(1)**: 016008
[3] Shimada M, Isoguchi O, Tadono T and Isono K 2009 PALSAR radiometric and geometric calibration *IEEE Transactions on Geoscience and Remote Sensing* **47(12)**: 3915-32.
[4] Lillesand T M and Kiefer RW 1994 *Remote Sensing and Image Interpretation* 3rd edition (John Wiley & Sons)
[5] Conservation International 2016 *Indonesia's mangroves one of the world's best solutions to climate change*, viewed 11 July 2018 <https://www.conservation.org/NewsRoom/pressreleases/Pages/Indonesias-mangroves-one-of-the-worlds-best-solutions-to-climate-change.aspx>
[6] Hamdan O, Khairunnisa MR, Ammar AA, Hasmadi IM dan Aziz HK 2013 Mangrove carbon stock assessment by optical satellite imagery *Journal of Tropical Forest Science* **1** : 554-65.
[7] Clerici N, Rubiano K, Abd-Elrahman A, Posada Hoestettler J and Escobedo F 2016 Estimating aboveground biomass and carbon stocks in periurban Andean secondary forests using very high resolution imagery *Forests* **7(7)**:138.
[8] Richards JA 2009 *Remote Sensing with Imaging Radar, Signals and Communication Technology* (Springer, ISBN: 978-3-642-02019-3)
[9] Luckman AJ, Frery AC, Yanasse CC and Groom GB 1997 Texture in airborne SAR imagery of tropical forest and its relationship to forest regeneration stage *International Journal of Remote Sensing* **18(6)** : 1333-49
[10] Sun G, Ranson KJ and Kharuk VI 2002 Radiometric slope correction for forest biomass estimation from SAR data in the Western Sayani Mountains, Siberia *Remote Sensing of Environment* **79(2-3)**: 279-87
[11] Jaya LMG, Wikantika K, Sambodo AK and Susandi A 2017 Comparison of PolSAR and PolInSAR Method to Estimate Mangrove Carbon Stocks in Southeast Sulawesi Indonesia using ALOS PALSAR Dual-Polarizations in the Perspective of Climate Change Mitigation *International Journal of Tomography and Simulation* **30(3)** CESER Publication
[12] Lumsdon P, Mercer B and Zhang Q 2008 Estimation and Monitoring of Tropical Forest Biomass Using Polarimetric Interferometric SAR Data *Intermap Technologies Corp.*

IOP Conf. Series: Materials Science and Engineering **797** (2020) 012015 doi:10.1088/1757-899X/797/1/012015

Geometric Accuracy Assesment Of Orthorectification Method Based On Sensor Model Refinement In Open Source System Environment
(Case Study : Sangir Subdistrict, South Solok District, West Sumatra Province)

Andri Suprayogi[1], La Ode Muh. Golok Jaya[2]
[1]Cadastral Survey And Mapping Laboratorium, Geodetic Engineering Dept., Engineering Faculty, Diponegoro University Semarang, Jl. Prof Sudarto, Tembalang, Semarang-Central Java Province, 50275
[2]Geo-information Science and Technology Research Group, Department of Informatics Engineering, Faculty of Engineering, Universitas Haluoleo, 93232 Kendari, Sulawesi Tenggara, INDONESIA

Email: [1]andri.suprayogi@ft.undip.ac.id, [2]laodemgj@uho.ac.id

Abstract. Orthorectification is one of the methods of correcting geometric condition of satellite imagery. Commonly used in high resolution satellite imagery such as IKONOS, Quickbird, and Pleiades, to reduces the effect of relief displacement. The Goal of this study is to implement the sensor model refinement procedures in orthorectification method. In this case we use orfeo toolbox (OTB) version 6.6.0 Installed in linux Mint 19. A true color Pleiades imagery of the area study in 2013, ACE2 DEM, and ground control points obtained by were used to obtain refined sensor model. This Study aimed to obtain the geometric accuracy of the output imager. The sensor model refinement process was done by access the imagery sensor model and changes its parameters using ground control points. The process of systematic orthorectification then carried out to transform the satellite imagery using modified sensor models. The geometric accuracy presented as horizontal accuracy were obtained by plotting several check points in the orthorectification output measured and calculated to obtain the horizontal accuracy. In this study we obtained that the sensor model refinement has an overal process accuracy of 1.664 meters. The model refined the imagery horizontal accuracy to 0.800 meters.

1. Introduction
Satellite imagery with spatial resolution smaller than approximately 1 meters such as Ikonos Quickbird, GeoEye, and Pleiades considered as Very High Resolution (VHR). These satellite imagery used in many applications such as base map production, Land registration, change analysis, and by its "finer" resolution, it may be used in modelling spatial phenomena with detailed characteristics such as hydrologic modelling[4]. Orthorectification is one of the methods for correcting geometric condition of a satellite imagery prior to further use[4]. There are two types of orthorectifications, systematic orthorectification where no GCP were used and using available sensor models, either Rigorous (RG) model or rational polynomial function (RPF), model and a DEM as External data[9]. Another, the rectification that taking GCPs coordinates into account where expected accuracy results were displayed, guiding the operator to correctly placed the GCP[3].

IOP Conf. Series: Materials Science and Engineering **797** (2020) 012015 doi:10.1088/1757-899X/797/1/012015

In an orthorectification process, DEM represents the terrain condition. In some flat coastal areas[1], DEM were not required. Curently there are free DEMs such as ACE2, a combined SRTM with satellite radar altimetry[8]. DEMs referenced to mean sea level (MSL) no longer relies to geometric relationship. As height difference should be vertical distance required by sensor model implementation. Thus, DEMs reference conversion fom MSL to Ellipsoid are required prior to, or within the sensor model refinement stages.

The Goal of this study is to observe the sensor model refinement of an orthorectification process[5]. Orfeo Toolbox (OTB) packages separate orthorectification process into sensor model refinement and (systematic) orthorectification process[6]. The expected accuracy of placed GCPs were unknown since we use external application to collect pixel coordinates. Thus proper interpretation of GCPs placement are required to ensure geometric quality and become one of the major aspect of this research. As open source environment, we executed orthorectification in Linux Mint 19 Cinnamon as the operating system. We also observe some the processesing sessions to collect aspects that may be useful for obtaining a good geometric output such as data condition, application features, and hardware usage, similar to the study of various orthorectification methods[3].

2. Study Area

The study area of this research located in Sangir Subdistrict, South Solok District, West Sumatra Province. This location have approximately 632,99 square kilometers with average Altitude of 453 meters since it lies in the south of Mt. Kerinci. And have 41.237 number of populations[2]. **Figure 1.** shows the depiction of Sangir subdistrict where it lies in the border with Jambi province[2].

Figure 1. Sangir Subdistrict Boundary, North to Mt. Kerinci and lies in the border of West Sumatra and Jambi Provinces.
Source : Kecamatan Sangir In Figures 2017

3. Methodology

In this study, sorthorectification are consists of preparations, GCPs collections and implementation in RPF Sensor Models refinement process. The final orthorectification process are systematic in terms no GCPs were used in the final steps. In order to obtain geometric accuracy, we use the Independent Check Points (ICPs) available in the area study . Both of GCPs and ICPs available from a total of 30 GPS surveying points. No spectral calibration process conducted since the imagery already in pan-sharpened state, results of post processsing stages to any spectral processing.

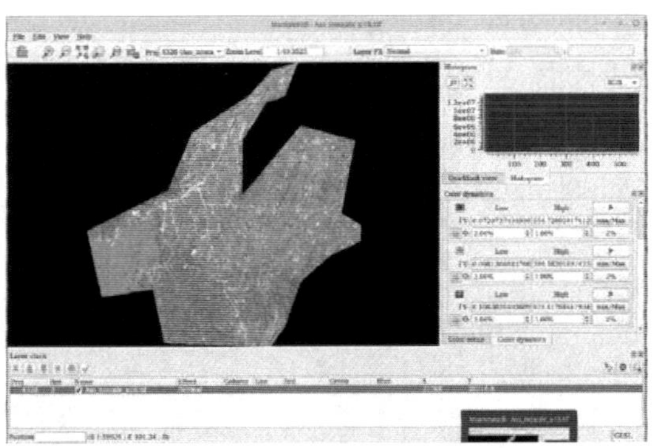

Figure 2. Monteverdi User Interface

For processing the data, a destop PC with Intel i7-3770 with speed of 3.4 Ghz with four cores and 8 functional threads, and 16 GB DDR3. Primary software to orthorectification process are OTB version 6.6.0. via graphical user interface called *mapla*, and *monteverdi* (**Figure 2**), useful for displaying imagery and it's histogram.

3.1. Preparations

The preparations of data started with image assembly where Pleiades imagery converted into OSSIM structures[6] as shown in **Figure 2.** We avoid global contrast stretching that may cause difficulties in imagery interpretation. The Sensor model of the imagery were stored in geom files, A single tile of ACE2 DEM With spatial resolution of 30 meters, that contain study area as shown in **Figure 3**. were downloaded. This local ACE2 have average height of with maximum elevation of 3774 meters and minimum elevation of 150 meters. We also prepare the EGM96 grid data to be used later in sensor refinement process to obtain ellipsoid referenced elevation.

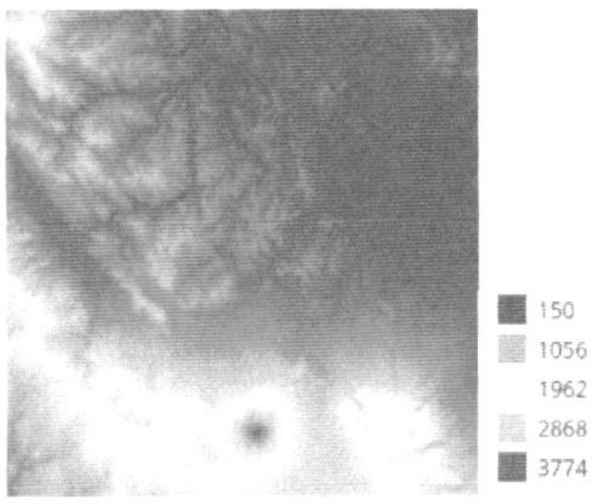

Figure 3 ACE2 DEM within the Pleiades Extent (Height unit in meters)

3.2. GCPs Placement

We use QGIS 3.2.1. To plot the GCP on the satellite Imagery. GPS fields survey coordinates (x,y) in decimal degree were manually entered with fraction no less than 5 digits to match spatial resolution of the imagery pixels. In order to obtain a well interpreted locations, we preserve the imagery radiometric resolution in 16 bit that may cause some locations looks too bright or too dark. Conditional histogram stretching were used in terms of local contrast in QGIS where brightness are adjusted based on the current extent histogram rather than the entire images. ***Figure 4*** shows the comparison of various image contrast in one location.

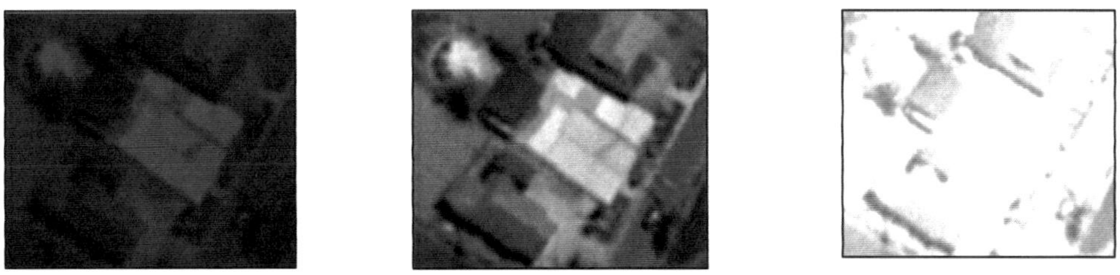

a. Original Histogram (unstretched) b. Local stretch c. Global Stretch

Figure 3. Contrast Variation Of Images, notice in this location local contrast stretch offer a better objects interpretation than global histograms stretch

OSSIM georeferences were unreadable in QGIS. We perform an initial systematic orthorectification to guide GCPs placement. By using the list of GCP Observed coordinates, we create x,y plot and overlaid them to the image to obtain a rough locations of GCPs. Field Photos that have scenery orientation of The GPS were used to refine GCP locations. From this step, we obtained a distribution of 12 GCPs shown in Figure 5. Some of them listed in **Table 1**. Notice that the pixel coordinate colums are shown in negative number, as it origins started from upper left while the image pixel coordinates arrangement increased to maximum at the lower right.

Table 1. Examples of Collected GCP

Points	**row**	**column**	**Longitude**	**Latitude**
GCP 5	20772	-6857	101.3070746	-1.5381736
GCP 7	14553	-8051	101.2784824	-1.544231381
G_ICP 12	4126	-15351	101.2306653	-1.579112036
G_ICP 9	14996	-14386	101.2805192	-1.574076947
GCP 13	1371	-16946	101.2180514	-1.586787769
GCP 14	6733	-20313	101.2425404	-1.602589778
GCP 15	14249	-20975	101.2770069	-1.605367303

Note that some of the GCP were ICP, since the accuracy of GPS survey were equal, the role changes of points are acceptable

Figure 4. Distribution of GCPs Identified in the Pleiades Imagery

3.3. Orthorectification Procedures

After GCPs were collected and stored in text document, we use OTB's refine sensor model tool to modify the coefficients of RPF models. The key to this process is the affine transformation to the models, as shown in equation (1), where $rpf_r(\lambda,\varphi,h)$ and $rpf_c(\lambda,\varphi,h)$ are the original RPF, while set of coefficients a and b are the six estimated affine transformation parameter [3] In refining the sensor models, we use geom file that need to be refined, a GCP list to refine the sensor model, ACE2 DEM, and EGM96. The output of the proces is another geom file that contain refined sensor model, while the achieved accuracies shown in **Table 2**.

$$r_{refined} = a_0 + a_1 \cdot rpf_r(\lambda,\varphi,h) + a_2 \cdot rpf_c(\lambda,\varphi,h)$$
$$c_{refined} = b_0 + b_1 \cdot rpf_c(\lambda,\varphi,h) + b_2 \cdot rpf_r(\lambda,\varphi,h) \quad (1)$$

source : *Hoja, D., Schneider, M., Müller, R., Lehner, M., Reinartz, P., 2008*

Table 2. Estimation of refined Sensor Model final accuracy

Accuracies	**Values (meters)**
Overall Root Mean Square Error	1.63432
X Mean Error	-0.00427
X standard deviation	1.10246
X Root Mean Square Error	1.10247
Y Mean Error	0.013312
Y standard deviation	1.20639
Y Root Mean Square Error	1.20647

Next, we replace the original geom file with refined one in the OTB's systematic orthorectification process, the Pleiades then corrected with the refined sensor model. We change the default RAM size which is 128 MB to 8096 MB (8GB) to improve performance. The orthorectifications process were utilized most threads up to their maximum load as shown in **Figure 6**. Only thread 2 and 6 are were not fully. The output of rectification process are converted to UTM Zone 47 S in to be able to obtain geometric accuracies in metric units.

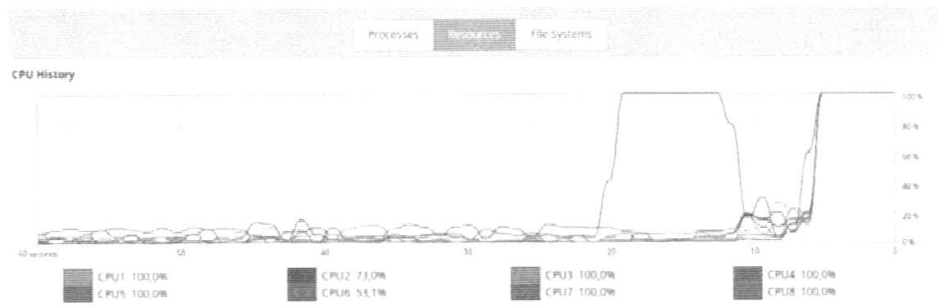

Figure 5. CPU Load When Rectification Process Started

3.4. Accuracy Assesments

To obtain the geometric accuracy presented as horizontal accuracy in the form of horizontal Accuracy. As the Imagery geoometric coordinate are now available, we can now use QGIS to locate ICP locations directly. From the ICP Placement, we obtain 11 Check points shown in Figure 7 where some of them are also points that planned to be GCP.

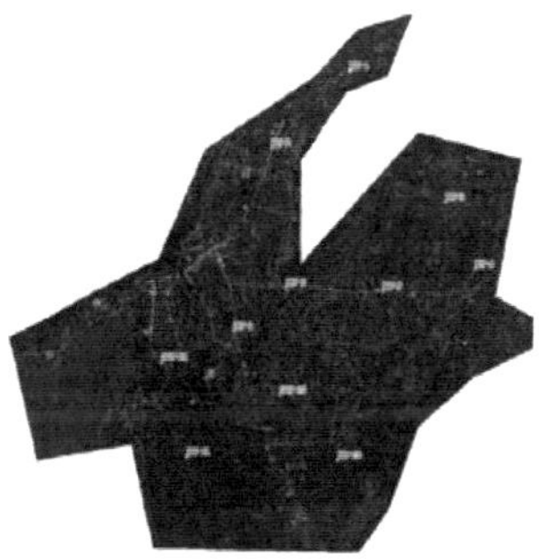

Figure 6 Distribution of ICPs Identified in the Pleiades Imagery

4. Results And Discussion

The topography of the area study are considered high since it is near the Mt. Kerinci with height range of 3672 meters. Thus relief displacement may occur and need to be corrected with DEMs. DEMS used in this study is referenced to MSL. There is no explicit undulation for the EGM 1996 so we let the OTB calculate the correct geometric height within the refinement process.

Not all GPS survey results were used since no identifiable placement after modifying the imagery contrast. Some points also have a vague photograph orientation where its properly oriented. The accuracy of refined sensor model are presented in standard deviation of 1.634 meters by using collected GCPs from its original 1.664 meters. From the ICPs collection stages, we obtain the RMSE of 0.800 meters as the geometric accuracy of Pleiades Imagery

5. Conclusion

The orthorectification of Pleiades in an open source environment opens the possibility of obtaining high accuracy imagery in reduced cost since no licensed softwares. On the other hand, there still a little documentation available that explain the proper OTB's functions such as creating GCP list in a correct structures to be correctly read by OTB's sensor refinement process. The use of tab delimiter show correct reading, but the process of correctly delimit the GCP list still need to be observed.

Although the usage of ACE2 DEM resulting good accuracy, the availability of local DEM such As recently published DEMNAS opens the possibility of obtaining more accurate result for the Pleiades imagery. The availability of undulation of the EGM 2008 in DEMNAS made the change of DEMs elevation reference to ellipsoid prior the sensor model refinement possible.

The usage of CPU's most thread indicates that OTB are optimized in multi-threaded system. Thus, increasing available thread may improve the process performance. Additional memory size, may improve the orthorectification process, however, change of default RAM size can only be done in the application compilation. Thus, knowledge of compiling the OTB is also required

References

[1] Alawy M. and Moh. Sukojo, M. Bangun 2016 Analisis ketelitian geometrik citra satelit Pleiades 1B dan Geoeye untuk pembuatan peta dasar RDTR kawasan pesisir (studi kasus: Kecamatan Bulak, Surabaya, Jawa Timur), J. *Teknik ITS* **5** 2 pp 165-9

[2] BPS Kab. Solok Selatan 2017 Sangir Subdistrict In Figures 2017th edition. *https://solokselatankab.bps.go.id/publikasi.html*

[3] Hoja, D., Schneider, M., Müller, R., Lehner, M., Reinartz, P., 2008 Comparison of orthorectification methods suitable for rapid mapping using direct georeferencing and RPC for optical satellite data, Beijing *ISPRSarchives* **XXXVII** B1 pp 1617-28

[4] Mateo C. M. R., Yamazaki D., Kim H., Champathong A., Vaze J., and Oki T., 2017 Impacts of spatial resolution and representation of flow connectivity on large-scale simulation of floods Hydrol. Earth Syst. Sci. **21** pp 5143-63

[5] Müller R., Krauß T., Schneider M., and Reinartz P. 2011 Automated Georeferencing of Optical Satellite Data with Integrated Sensor Model Improvement *Photogrammetric Engineering & Remote Sensing* **78** 1 pp 61-74

[6] OTB Development Team 2018 The ORFEO Tool Box Software Guide Updated for OTB-6.6.0, CNES,

[7] Santillan J. R. Makinano-Santillan M. 2016 Vertical accuracy assessment of 30-m resolution Alos, Aster, and SRTM global DEMs over Northeastern Mindanao, Philippines Prague *ISPRSarchives* **XLI** B4 149-56

[8] Smith R.G., Berry P.A.M., Benveniste J. 2010 ACE2: The New Global Digital Elevation Model. *Proc. ESA Living Planet Symposium* ESA SP-686,

[9] Widyaningrum E., Fajari M., and Octariady J. Accuracy comparison of VHR systematic-ortho satellite imageries against VHR orthorectified imageries using GCP Prague *ISPRSarchives* **XLI** B4 305-09

IOP Conf. Series: Materials Science and Engineering **797** (2020) 012016 doi:10.1088/1757-899X/797/1/012016

Fuzzy Analysis to Determine Potential Catching Areas of Skipjack Tuna on Southeast Sulawesi Waters

Ika Purwanti Ningrum, Jayanti Yusmah Sari*, Rizal Adi Saputra, Sutardi, Ismail Yushar

Department of Informatics Engineering, Faculty of Engineering, Universitas Haluoleo, 93232 Kendari, Sulawesi Tenggara, INDONESIA

E-mail: ika.purwanti.n@uho.ac.id, *jayanti@uho.ac.id, rizaladisaputra@uho.ac.id, sutardi@uho.ac.id, ismailacc97@gmail.com

Abstract. Southeast Sulawesi is one of the catching fisheries bases in the Eastern Indonesia Region which deals directly with the Banda Sea. Skipjack Tuna (*Katsuwonus pelamis*) is classified as an important economic pelagic fish resource and is one of the export commodities. The existence of skipjack tuna is certainly influenced by the condition of the ocean. The parameters used in the analysis of the determination of skipjack fish catches are Sea Surface Temperature, the distribution of chlorophyll-a and catches. The purpose of this study was to analyze the catching area of skipjack fish around the waters of the Southeast Sulawesi. To determine of skipjack tuna fishing area is supported by Fuzzy Mamdani method. The results showed that the catching area of skipjack fish on the peak catching season in the waters around Southeast Sulawesi May was in position 03008'30.2 S / 122025'17.4 E to 03009'48.8 S / 122042'32.7 E. In June the potential area is at point 03009'47.4 S / 1220 41'02.3 E and on July potential area at point 03005'50.8 S / 122041'16.3 E.

1. Introduction

Southeast Sulawesi is one of the catching fisheries bases in Eastern Indonesia which faces directly to the Banda Sea. The Banda Sea is a vast ocean that separates several large islands and small islands around it. Data from the Directorate General of Capture Fisheries of the Ministry of Maritime Affairs and Fisheries in 2011 stated that the potential for large pelagic fish resources in the Banda Sea is estimated to reach 104,100 tons per year [2].

Skipjack tuna (*Katsuwonus pelamis*) is classified as an important economic pelagic fish resource and is one of the export commodities. The existence of tuna skipjack fish is certainly influenced by the condition of the ocean. The parameter used in the analysis of the determination of skipjack fish catches is Sea Surface Temperature or SST. SST observation to predict the presence of skipjack fish is used because skipjack tuna is a fish species whose swimming layer is found in the surface layer [4]. Another parameter that also affects the distribution area of skipjack tuna is the distribution of chlorophyll-a. Chlorophyll is an ingredient in the process of plant photosynthesis. The chlorophyll-a concentration of a water is largely determined by the intensity of light and the presence of nutrients. Tropical marine waters generally have low chlorophyll-a content due to nutrient limitations and strong stratification of the water column. Furthermore, based on the chlorophyll-a distribution pattern both seasonally and spatially, in some parts of the water a high concentration is found. Chlorophyll-a on the surface of the waters are grouped into three categories: low, medium and high which are contains chlorophyll-a of <0.07 mg/m^3; 0.07 mg/m^3 -0.14 mg/m^3 and > 0.14 mg/m^3 concecutively [6]. The potential for catching skipjack fish is closely related to environmental parameters, especially optimum SST in the range of 29.9 - 31.0^{0}C and optimum chlorophyll-a in the range of 0.12-0.22 mg/m^3 [4].

The Artificial Intelligence method that can be used in this case is Fuzzy Logic. Fuzzy Logic allows membership values between 0 and 1, gray level and also black and white, and in linguistic form which have an uncertain concepts like "little", "pretty", and "very". This logic is related to fuzzy sets and probability theories.

2. Research Methods

Sea Surface Temperature data and chlorophyll-a distribution used in this system are Terra / MODIS satellite image correction data from the NASA website (*oceancolor.gsfc.nasa.gov*), while the Catch Result data is obtained from the data collection from Indonesian Fisheries Port Information Center (*pipp.djpt. kkp.go.id*) using data from May to July 2016.

This research uses Fuzzy Mamdani method which is interpreted in Fuzzy Interference System (FIS). Every data that has been obtained will go through the correction process before being used in system analysis. The Fuzzy Mamdani method requires fuzzyfication process, inference, and defuzzyfication.

3. Results and Discussion

3.1. Fuzzyfication

In this section, each dataset is entered into the system by changing the crisp value to a fuzzy set. The SST (Sea Surface Temperature) dataset has a unique linguistic set of "Cold", "Warm", and "Hot" with a range of values of 20^{0}C - 24^{0}C, 23^{0}C - 27^{0}C, and 26^{0}C-30^{0}C respectively. Descriptions in the graph can be seen in Figure 1 and the degree of membership in mathematics formula can be seen in equation (1).

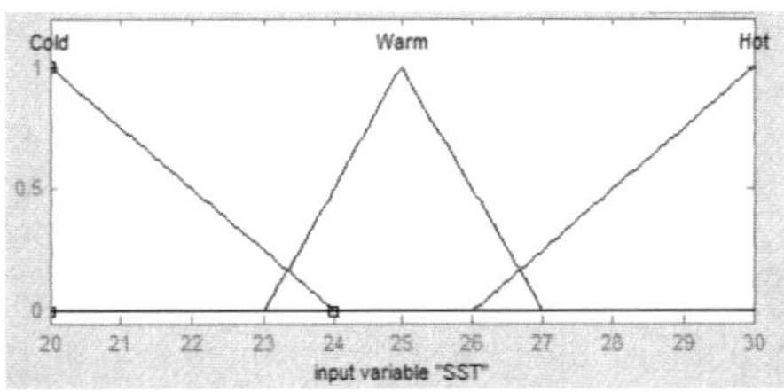

Figure 1. Graph representation for membership degree of SST

$$\mu[x]\begin{cases} \frac{24-x}{24-20}; x = 20; \\ \frac{x-23}{27-23}; \frac{27-x}{27-23}; x = 25; \\ \frac{30-x}{30-26}; x = 30; \end{cases} \qquad (1)$$

In equation (1), μ [x] states the function of the x variable or input which is initialized as *x*. "Cold", "Warm", and "Hot" is a linguistic classification of SST (Sea Surface Temperature) variables. The triangle curve in figure 1 states the value of the cluster set range.

Chlorophyll-a dataset has a linguistic set that is "Low" which is between 0 - 0.075 mg/m^3, "Medium" is between 0.065-0.145 mg/m^3, and "High" has a set between 0.135-0.2 mg/m^3. The drawing with the curve can be seen in figure 2.

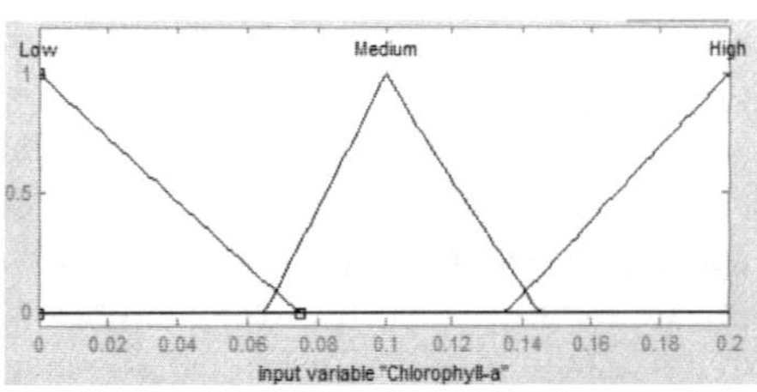

Figure 2. Graph representation for membership degree of chlorophyll-a variable

$$\mu[x]\begin{cases} \frac{0.075-x}{0.075-0}; x = 0; \\ \frac{x-0.065}{0.145-0.065}; \frac{0.145-x}{0.145-0.065}; x = 0.1; \\ \frac{x-0.135}{0.2-0.135}; x = 0.2; \end{cases} \qquad (2)$$

Based on figure 2, the degree of membership can be formulated with the μ[x] function in equation (2), with x represents the input value.

The Catch Dataset are grouped in a linguistic set of "Less" with values ranging from 0-125 amount of fishes, "Sufficient" with values ranging from 100-200 fishes, and "Plentiful" with values ranging from 175-300 fishes, the description of which can be seen in figure 3 .

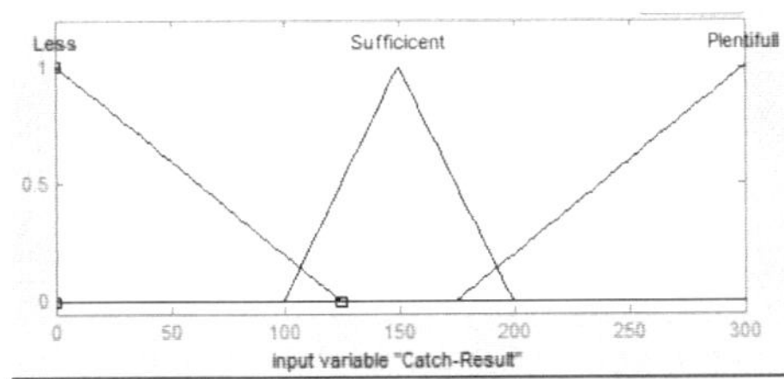

Figure 3. Graph reprsentation for membership degree of catch result

$$\mu[x]\begin{cases} \frac{125-x}{125-0}; x = 0; \\ \frac{x-100}{200-100}; \frac{200-x}{200-100}; x = 150; \\ \frac{300-x}{300-175}; x = 300; \end{cases} \quad (3)$$

Based on figure 3, the membership degree can be formulated with the function μ[x], and x is the initialization of input (processed dataset) by using equation (3).

Based on the purpose of the study, the output variable is determined as "Potential-Area". This variable contains three attributes with successive linguistics, namely "Not-Potential", "Potential-Medium", and "Potential-High". The range of the set of output variables is determined between 0 to 1. The picture can be seen in Figure 4.

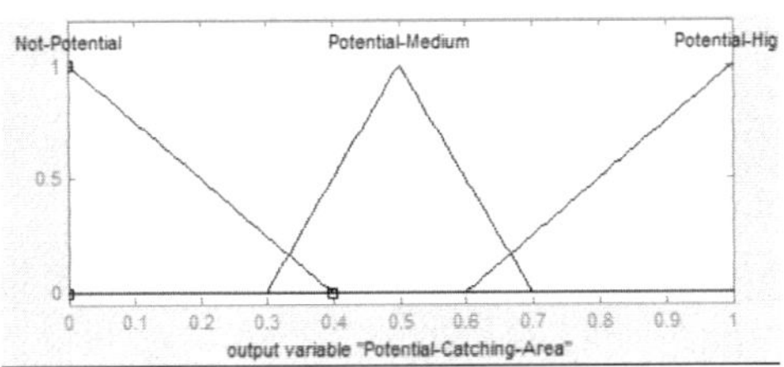

Figure 4. Graph representation of membership degree in the potential catching area cluster of skipjack tuna

$$\mu[x]\begin{cases} \frac{0.4-x}{0.4-0}; x = 0; \\ \frac{x-0.3}{0.7-0.3}; \frac{0.7-x}{0.7-0.3}; x = 0.5; \\ \frac{1.0-x}{1.0-0.6}; x = 1.0; \end{cases} \quad (4)$$

Based on figure 4, the membership degree can be formulated with the equation of the function μ[x], where x is the value of the dataset processed.

Equation (1) to equation (4) is written based on the rules of formulation in the triangle representation which is used to describe the graph of the membership degree of each variable.

3.2. Inference

Inference is part of the process of making a fuzzy system that connects or analyzes the relationship between each variable and is made in sentences called *rule base*. Based on the relationship that has been observed from the three variables and input data set that is managed, the rule base used in this system are listed in Table 1.

By applying substitution of the membership value into the inference, then the solution using the minimum rules formula can be written in equation (5).

$$\alpha = \min\left(\mu[x]_i \cap \mu[x]_j \cap \mu[x]_k\right) \quad (5)$$

Equation (5) states that the three determinants are i, j, k, which are sequentially SST, chlorophyll-a, the catch result, registered as a set of α values and has an AND operator. The AND operation result is the minimum value between the three variables which will then be used in the defuzzification process. After getting the value of the implication function, the rules are compiled using the Max method. The Max method is written by registering each value function implication.

$$\mu[z]\begin{cases} \alpha 3;\ z \leq A1 \\ \frac{A2-z}{A2};A1 \leq z \leq A2 \\ \alpha 2;\ A2 \leq z \leq A3 \\ \frac{A3-z}{A3};\ \alpha 1;\ A3 \leq z \end{cases} \tag{6}$$

In equation (6), μ[z] denotes the set of *z* functions, where *z* is the result of the calculation of the implication function, α denotes the predicate value of each implication function or is the output value of equation (5), while A states the threshold value of each set.

Table 1. Rule Base for Implication Function

No	Rule Base
19	*If (SPL is Warm) and (Chlorophyll-a is Low) and (Catch is Less) then (Potential-Area is Not-Potential)*
20	*If (SPL is Warm) and (Chlorophyll-a is Low) and (Catch is Sufficient) then (Potential-Area is Potential-Medium)*
22	*If (SPL is Warm) and (Chlorophyll-a is Medium) and (Catch is Less) then (Potential-Area is Not-Potential)*
23	*If (SPL is Warm) and (Chlorophyll-a is Medium) and (Catch is Sufficient) then (Potential-Area is Potential-Medium)*
24	*If (SPL is Warm) and (Chlorophyll-a is Medium) and (Catch is Banyak) then (Potential-Area is Potential-High)*
10	*If (SPL is Hot) and (Chlorophyll-a is Low) and (Catch is Less) then (Potential-Area is Not-Potential)*
11	*If (SPL is Hot) and (Chlorophyll-a is Low) and (Catch is Sufficient) then (Potential-Area is Not-Potential)*
12	*If (SPL is Hot) and (Chlorophyll-a is Low) and (Catch is Plentiful) then (Potential-Area is Potential-Medium)*
13	*If (SPL is Hot) and (Chlorophyll-a is Medium) and (Catch is Less) then (Potential-Area is Not-Potential)*
14	*If (SPL is Hot) and (Chlorophyll-a is Medium) and (Catch is Sufficient) then (Potential-Area is Potential-Medium)*
15	*If (SPL is Hot) and (Chlorophyll-a is Medium) and (Catch is Plentiful) then (Potential-Area is Potential-Medium)*
16	*If (SPL is Hot) and (Chlorophyll-a is High) and (Catch is Less) then (Potential-Area is Not-Potential)*
17	*If (SPL is Hot) and (Chlorophyll-a is High) and (Catch is Sufficient) then (Potential-Area is Not-Potential)*
18	*If (SPL is Hot) and (Chlorophyll-a is High) and (Catch is Plentiful) then (Potential-Area is Potential-Medium)*

3.3 Defuzzification

This study uses the centroid method for *defuzzication*. The formulation of *defuzzification* using the centroid method is written in equation (7).

$$Z = \frac{M1+ M2+ M3+M4}{L1+L2+L3+L4} \tag{7}$$

The *M* variables is the implementation of the median value of each composition of the implication function registered, while the *L* value is the implementation of the area value of each composition. The median value is written in equations (8.1), (8.2), (8.3), and (8.4).

$$M_1=\int_0^{A1}(\alpha 3)\ zdz; \tag{8.1}$$

$$M_2=\int_{A1}^{A2}\left(\frac{A2-z}{A2}\right)\ zdz; \tag{8.2}$$

$$M_3=\int_{A2}^{A3}(\alpha 2)\ zdz; \tag{8.3}$$

$$M_4=\int_{A3}^{max}\left(\frac{A3-z}{A3}\right) zdz; \quad (8.4)$$

Value *L* or Area of *A* can be formulated in equation (9.1), (9.2), (9.3) and (9.4).

$$L_1=A1*\alpha3; \quad (9.1)$$

$$L_2=(\alpha3+\alpha2)*(A2-A1)/2; \quad (9.2)$$

$$L_3=(A3-A2)*\alpha2; \quad (9.3)$$

$$L_4=(max-A3)*(\alpha1); \quad (9.4)$$

There are four M and L variables which represent the total number of variables both input and output in the system.

The proposed system combines the Fuzzy Interference System library with Matlab Graphic User Interface. The results obtained from the analysis can be seen that the average potential level of the skipjack fishing area is in the medium potential class, as expressed in Table 2 which clearly reveals the results of the fuzzy system and the manual clustering results. Clustering is done manually by looking at the predetermined threshold value and listed in the pink area in figure 5.

Table 2. Cluster results with fuzzy systems

Month	Position	Potential value	Potential class
May	03^0, 08' 30, 2 S / 122^0 25'17, 4 E	0.5	Medium potential
	03^0, 09'48, 30 S / 122^0, 42'24, 7 E	0.5	Medium potential
	03^0, 09'48, 8 S / 122^0, 42'32, 7 E	0.5	Medium potential
	03^0, 09'16, 15 S / 122^0, 42'49, 2 E	0.658008	Medium potential
June	03^0, 09'16, 6 S / 122^0, 42'49, 1 E	0.173422	Not potential
	03^0, 10'24, 5 S / 122^0, 43'05, 5 E	0.17964	Not potential
	03^0, 10'19, 4 S / 122^0, 41'13, 8 E	0.173422	Not potential
	03^0, 09'47, 4 S / 122^0, 41'02, 3 E	0.5	Medium potential
July	03^0, 07'03, 5 S / 122^0, 43'04, 5 E	0.185667	Not potential
	03^0, 05'50, 8 S / 122^0, 41'16, 3 E	0.5	Medium potential
	03^0, 04'22, 0 S / 122^0, 39'15, 6 E	0.185667	Not potential

In Table 2, the position states the coordinate point of the skipjack tuna fishing area from east to south (E, S). The value obtained in the fuzzy system that has been made is expressed by the potential value in table 2, and the clustering results are expressed in the potential class column. In May, it was entirely above the threshold value of 0.5 potential value, even at the last point it continued to rise to about 0.6 units of potential value, this indicates that these points are in a medium potential class that allows fishermen to get enough skipjack fish in the area. . In June a drastic decline occurred, this is known from the potential value generated by the fuzzy system is very different from the previous month at the same point. Potential values obtained at the points 03009'16.6 S / 1220 42'49.1 E to the point 03010'19.4 S / 1220 41'13.8 E only ranged from 0.17 potential value units and were in the class area which was not potential for tuna fishing, but in point 03009'47.4 S / 122041'02.3 E in June is in the medium potential class because its potential value is above the threshold of 0.5 potential value units. In July, the points 03007'03.5 S / 122043'04.5 E and 03004'22.0 S / 122039'15.6 E had a potential value ranging from 0.18 units of potential value so that the points did not become a potential catching area for skipjack fish, but point 03005'50.8 S / 122041'16.3 E has a potential value of 0.5 units of potential value and is on the threshold of potential "medium" class value of skipjack tuna fishing grounds.

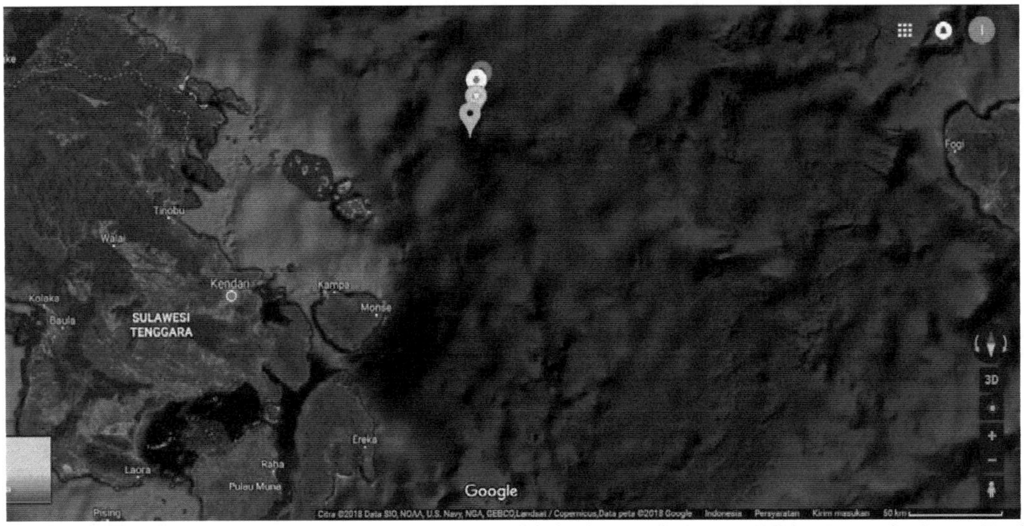

Figure 5. Figure 5. Potential areas for catching skipjack tuna in May (White Round), June (Red Round), and July (Yellow Round)

The system finds that the average potential area is in the "medium" potential class. The appearance of potential tuna fishing areas around Southeast Sulawesi from May to June 2016 is shown in figure 5.

4. Conclusion

The catching area of skipjack fish in the waters of Southeast Sulawesi for May is in the position of 03008'30.2 S / 122025'17.4 E up to 03009'48.8 S / 122042'32.E. In June the potential area is at the point 03009'47.4 S / 1220 41'02.3 E, and in July the potential area is at the point of 03005'50.8 S / 122041'16.3 E. Classification of potential fishing areas for skipjack tuna from May, June, and July 2016 is entirely at moderate potential.

This study uses an old dataset so that the results cannot be implemented at this time. The system also does not process satellite imagery directly so that further research is needed with the ability of the system that can process satellite imagery in order to facilitate the analysis process.

References

[1] Agusliana M, Tadjudda M, dan Mustafa A, 2016. Sebaran suhu permukaan laut dan tracking daerah penangkapan Ikan Cakalang di Perairan Barat Laut Banda. *Jurnal Manajemen Sumber Daya Perairan*, **2(1)** 41-49.

[2] Azzahra N Z, Permata C A D, dan Akhirta N A, 2017. Pemetaan Potensi Penangkapan Ikan Cakalang di Perairan Sulawesi. *Seminar Nasional Penginderaan Jauh*. Sekolah High Meteorologi Klimatologi dan Geofisika.

[3] Hidayah G, Wulandari, S Y, dan Zainuri M, 2016. Studi Sebaran Chlorophyll-a Secara Horizontal di Perairan Muara Sungai Silugonggo Kecamatan Batangan, Pati. *Buletin Oseanografi Marina*, **5(1)** 52 – 59.

[4] Jufri A J, Amran M A A., dan Zainuddin M Z, 2014. Karakteristik Daerah Penangkapan Ikan Cakalang Pada Musim Barat Di Perairan Teluk Bone. *Jurnal IPTEKS PSP*, **1 (1)** 1-10.

[5] Kusumadewi S, 2002. Analisis & Desain Sistem Fuzzy Menggunakan Tool Box Matlab. **1(1)** xii-276. Yogyakarta : Graha Ilmu.

[6] Tangke U T., 2014. Pendugaan Daerah Penangkapan Ikan Pelagis Berdasarkan Pendekatan Suhu Permukaan Laut Dan Chlorophyll-a Di Laut Maluku. *Jurnal Ilmiah agribisnis dan Perikanan*, **7(1)** 74-81.

IOP Conf. Series: Materials Science and Engineering **797** (2020) 012017 doi:10.1088/1757-899X/797/1/012017

Experimental Study on the Performance Characteristics of a Microwave - Solar Heating Dryer

Lukas Kano Mangalla[1*], L O Ahmad[3], A Lolok[2], Jenny Delly [1], Kadir[1], Samhuddin[1], Abdul Kadir[1], B Sudia[1], dan L Pagiling[2]

[1]Mechanical Engineering, Faculty of Engineering, Universitas Haluoleo, 93232 Kendari, Sulawesi Tenggara, INDONESIA

[2]Electrical Engineering, Faculty of Engineering, Universitas Haluoleo, 93232 Kendari, Sulawesi Tenggara, INDONESIA

[3] Department of Chemistry, Universitas Haluoleo, 93232 Kendari, Sulawesi Tenggara, INDONESIA

*Corresponding author's e-mail: lukas.kano@uho.ac.id

Abstract: The use of microwave energy for drying of agricultural products is recently developed due to the advantageous of heating the product much faster and more effective than conventional methods. Moisture content of agricultural product should be evaporated immediately after harvesting in order to inhibit microbial growth and decrease enzymes activity. This work aim to investigate characteristics of microwave-solar hot air drying of agricultural products such as rice grain and potato using a commercial microwave oven and hot air from a can solar collector. A domestic microwave oven was modified in order to operate as a microwave dryer device. The state of the art of microwave-solar heating was developed for investigation on drying performance of agricultural products such as grains and potato slices. Measurements on the temperature of drying chamber and moisture content of the product were conducted under different power inputs of microwave (low, middle and high) and intensity of solar drying. It was observed that microwave-solar air dryer proposed good heating performance and quality of products. Furthermore, temperature, moisture changes and drying rate could be controlled during the drying process.

1. Introduction

Drying is an important preservation processes of moisture removal from an agricultural product. This process be intended to evaporate the moisture content and to reduce water activity, in order to inhibit microbial growth and decrease enzymes activity inside the product [1]. The process can be quality enhancement, ease of handling and transporting, and probably the oldest method of food preservation [2]. Delayed drying of the crops may result in deterioration of products quality [3]. Drying processes is influenced by heat and mass transfer between hot airflow and product. Common traditional way of dehydration process is using hot air that is characterized by ineffective drying. Heat is transferred to the surface of the material by conduction, convection or radiation and into the interior part of material by thermal conduction. In this process moisture is initially evaporated from the surface and remaining water diffuses to the surface. The process time is limited by the rate of the heat flow into the body of the material which is depend on specific heat, thermal conductivity, density and viscosity of the material [4]. To overcome the limited of this scheme drying and to improve quality of the products, combination with the modern drying system is needed.

Microwave heating is an attractive heating scheme recently. Conventional heating and drying methods approach material from the surface, applying heat only to the outside edges. In this heating system the energy transmits from the surface to the interior while the moisture is transferred in opposite way from the inner to the surface. Microwave, however, is not forms of heat but forms of energy as heat that generates from interaction with materials. Electromagnetic wave produced from the magnetron of the microwave penetrates directly inside the material, Heat generated depends on the dielectric properties, shape and thickness of material [5-6]. In microwave drying system, the inner part of material is warmed as heat travels from the outer layers inward. If the material is highly wet and the pressure inside rises rapidly the moisture will be removed from the material due to the difference in pressure. This phenomenon creates a force action on the liquid to push it to the surface. Microwave energy does not heat the room but only the desired materials [7-8]. Heat producing in microwaves processes can be used for several dielectric solid and liquid materials containing some water [9-11].

Modern heating schemes propose high efficient drier as can be done in the microwave heating. Use of microwave energy in drying system offers a shorten time drying and completion the internal residual moisture by conventional drying in later steps [1,12]. Several studies have been using microwave as energy sources and compared with another heating schemes. Beaudry et al. [13] studied the effect of several heating methods including microwave convective (0.7W/g and 62°C), hot air (62°C), freeze and vacuum drying (94.6 kPa) on the quality of cranberries. They reported that all dried samples were acceptable and closest to commercial dried cranberries. Same effect was also studied by Piotrowski et al. [14] using strawberries. They reported that the fastest drying method was microwave convective, and the longest was for hot air drying at 62°C. This work aims to investigate the performance heating of microwave and hot air heating from can solar collector for agricultural products. In this study, preheating of microwave is supposed to bring all the moisture to the surface of the product and later flashed the moisture out by conventional heating-hot air.

2. Material and Methods

Agricultural products used in this study (rice grains and potato) were obtained from local farmer in Kendari during the rainy season 2018. The selected samples of potato were cleaned and sliced into thin layer-about 3mm of thickness. The grains and the sliced potatoes were stored at room temperature for 12hours until drying process. This process is intended to preserve its original quality and condition when the products just harvested.

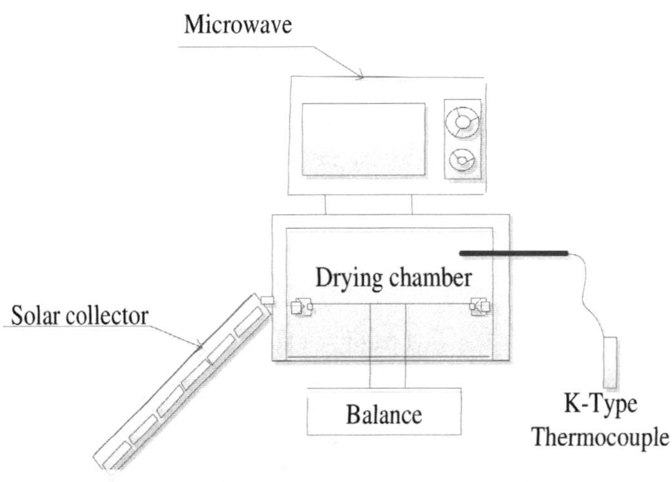

Figure 1. Schematic of the experimental study

The schematics of the experimental investigation of microwave-hot air drying setup is shown in Fig.1. The solar absorber made from soda cans was used as the heat sources. Hot air from solar collector forced through the products placed inside the heating chamber. Aluminum soda cans were holed and arranged in the absorber panel of 60cm width and 120 cm length.

Can collector is used to absorb heat from the sun and heat the air flow throughout the arranged cans. Air is entering to the bottom part of collector and experiencing heating inside the cans collector. Heated air is forced into the drying chamber by a small fan mounted on the top of collector. In order to evaluate the heating performance of the solar collector, temperature of air inlet and outlet were measured using K-Type Thermocouple.

A domestic microwave oven was used as the main energy source in this study. Electromagnetic wave from magnetron of the microwave was guided to the drying chamber by modifying the bottom part of microwave cavity. The driven electromagnetics can interact directly with the product inside the drying chamber to produce heat.

Experimental investigation on performance heating of the microwave-solar collector dryer was evaluated at different sample of materials and power input of microwave. Microwave used in this study is the microwave oven for domestic appliance with frequency of 2450MHz. Three levels of power heating scenarios were used - low, middle and high level. In order to investigate the temperature of the heating room and atmospheric temperature, K-Type thermocouple with 2 sensors cable were used. Moisture losses from samples dried was recorded by a means of digital balance at each 10 second intervals.

Experimental investigations were conducted at three different processes- natural drying at open sun (SUN), hot air coupled with microwave heating (MW+SUN), and microwave (MW). Drying rate is calculated based on the moisture reducing during a specific heating time for drying. Comprehensive analysis and descriptions on the experimental data are presented below.

3. Result and Discussion

Solar collector is one of the main heating sources for drying of the agricultural products in this study. Figure 2 shows the trend of hot air temperature produced from solar can collector (Tin) compared with the environment temperature (To) during experiment. Conditions of the sky was very clear along the time, however the blowing of wind was sometime come and affected the temperature of the hot air. It shows from the figure that the average temperature of hot air can reach around 60°C and it is hot enough for drying the agricultural products. Fluctuated trend at several points can reach 55°C and even lower is probably affected by the speed of the wind entering to the solar collector.

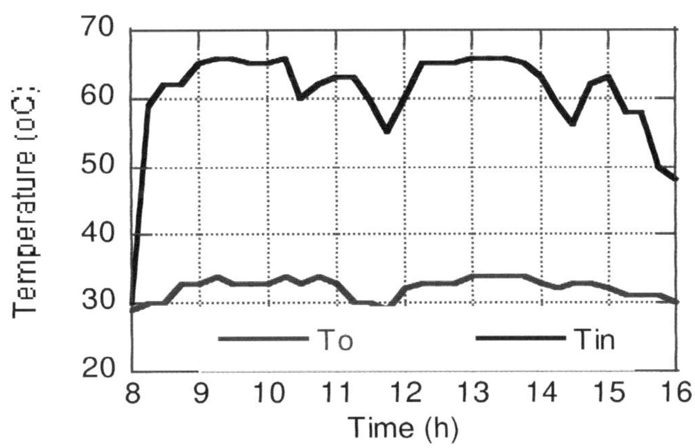

Figure 2. The trend of hot air temperature produced (Tin) from solar can collector

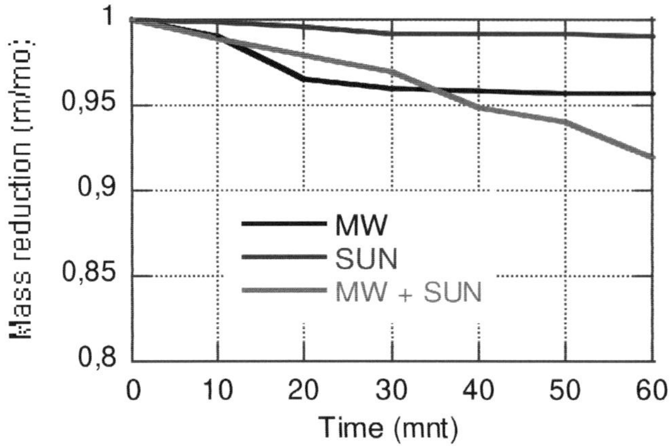

Figure 3. Reduction rate of the moisture in the grain products

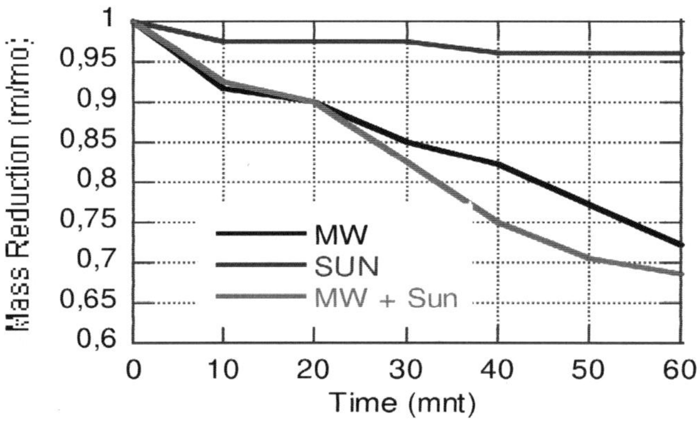

Figure 4. The reduction rate of the moisture in the potato slices product

Moisture diffusion is mainly mass transfer of water content from inside to the surface material followed by evaporation process to the air. The change of moisture content with drying time of the three different heating processes of grain products can be seen in Figure 3. Moisture content escaped from product is expressed in mass reduction during drying time. Mass reduction was calculated based on the product mass recorded at the time divided by initial mass of the products. The lower mass reduction, the better the moisture evaporated from the products. It is clear that mass reduction of the moisture content decreases continuously during time heating. Comparing of the three heating schemes, it is revealed that combination drying of microwave and hot air (MW + SUN) is very effective drying scheme. In 60 minutes drying, the mass reduction can reach 68% while others heating schemes can reach 96% for Microwave (MW) and 98% for natural open sun drying (SUN).
Same trend of the three heating characteristics for potato slices can be shown in Figure 4. The moisture content of potato was very high in the initial state of drying which resulted in higher absorptions of microwave power and higher drying rate. It is also clear from the figure that moisture content escape rapidly from the product using MW+ SUN drying scheme following with MW and SUN drying. By 60 minutes drying time, the moisture reduction can be reduced to 68% in MW+SUN, 72% in MW and 96% in SUN respectively.

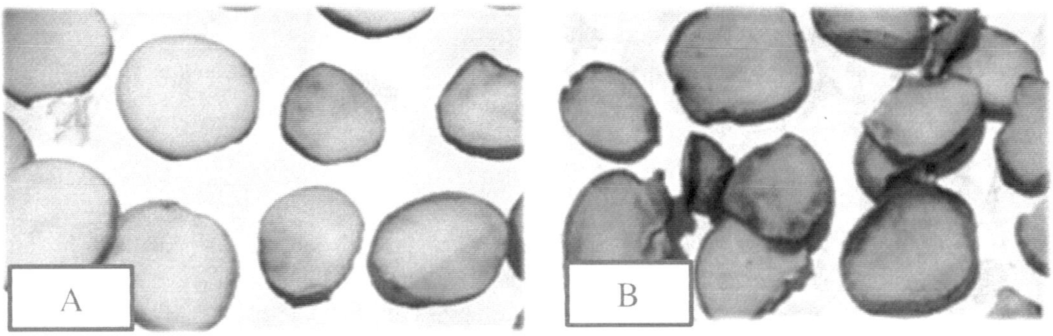

Figure 5. The images of drying products at the two different heating schemes. A) is for MW+SUN and B) is for SUN heating

The particular structural of the product defines the volume, size and shape of the sample. Whenever the water content escaped from material, the unbalance pressure between inner and outer spaces is happened. This phenomenon can generate contracting stresses that leads to material shrinkage, changing in shape and sometimes cracking on the surface of the product [15]. Figure 5 shows the images of potato slices for two different heating, MW+SUN and SUN. Heat generated of microwave inside material tends to increase the inside pressure of moisture and pump the water to the surface and later evaporates to the air. The figure shows a shrinkage on the surface of potatoes dried by the SUN system. This characteristics may indicates the higher temperature at the surface than at the inner part of product. Higher surface temperature can increase the moisture evaporation from the surface and shrink the surface of product. It is totally different for heating product by microwave that generates heat inside material. The balance pressure between surface and inner part can inhibit the shrinkage. From previous figure can be seen that moisture content of product heated by MW + SUN is lower than heated by SUN that affect the shape and size of products.

Power of microwave drier has significant effect on the mass transfer of the moisture from the product. Figure 6 and Figure 7 show the mass reduction rate of the grain and potato for different power input during time drying. It is revealed that drying time can be reduced by increasing the power input of microwave. Thus, the higher power input can increase the heating rate of the product. These figures also shows that heating rate can be affected by the shape and size of material exposed microwave energy. Penetration depths of microwave is affected by the frequency of electromagnetic wave generated from magnetron of microwave. Microwave penetrates uniformly throughout the volume of product being

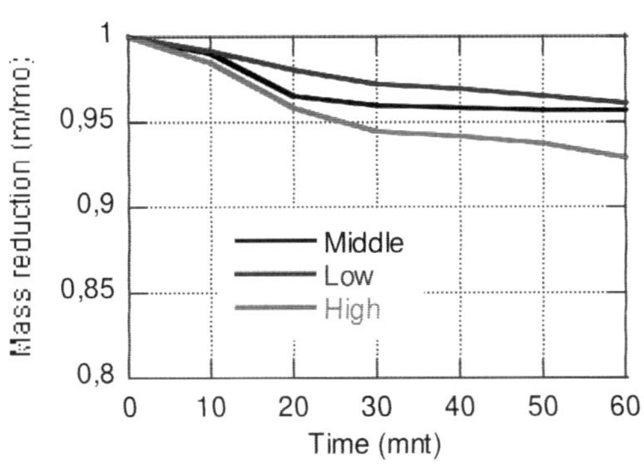

Figure 6. Input power effect on the reduction rate of the moisture in the grain product

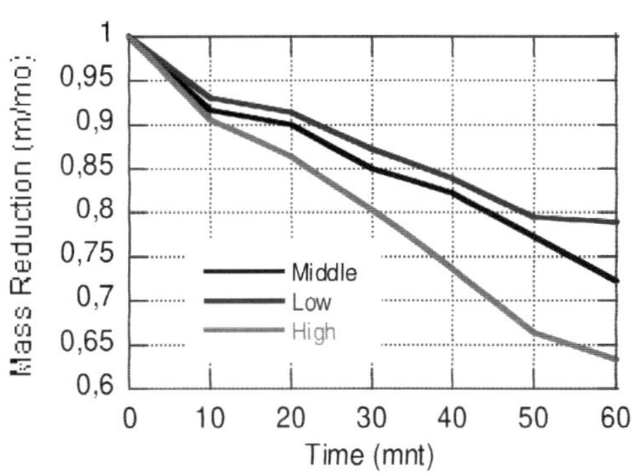

Figure 7. Input power effect on the reduction rate of the moisture in the potato products

heated and this is called volumetric heating. It is contras with the traditional thermal heating which relies on conduction and convective heating from surface to the inner parts of the products. Combination of the two schemes can be a smart heating system for drying system of product.

4. Conclusions

Volumetric heating effect of microwave proposes the effective drying system for the agricultural crops Combination of microwave and hot air heating schemes brings the new drying characteristics such as increasing the drying rate and improving the final quality. Power input can control the heating rate for microwave drying system and the shape of the product is the important thing to be mentioned.

References

[1] Monteiro, R.L., Carciopy B.A.M., Marsaioli A., & Laurindo J.B., 2015, How to Make A Microwave Vacuum Drier with Turntable, Journal of Food and Engineering, Vol.166 pp. 276–284

[2] Liberty J.T., Okonkwo W.I., & Ngabae S.A, 2014, *Solar Crop Drying - A Viable Tool for Agricultural Sustainability and Food Security, International Journal of Modern Engineering Research, Vol 4(6) pp.8-19*

[3] Raman V.V.S., Iniyan S., Goich R., 2012, A review of solar drying technology, Renewable and sustainable energy review, 16(5) pp,2652-2670

[4] Haghi A.K., and Amanifard, N., 2008, Analysis of Heat and Mass Transfer During Microwave Drying Of Food Products Brazilian Journal of Chemical Engineering Vol. 25, No. 03, pp. 491 – 501

[5] Oliveira M. E. C. and Franca A. S., 2002, Microwave heating of foodstuff, Journal of Food Engineering, 53, pp. 347–359

[6] Motevali A., Minaei S., Khostagaza M.H., 2011, Evaluation of energy consumption in different drying methods, Energy Conversion andn Management 52(2) pp.1192-1199

[7] Hossan, M.R., Byun D.Y. dan Dutta P., 2010, Analysis of Microwave Heating for Cylindrical Shaped Objects, International Journal of Heat and Mass Transfer, 53, pp. 5129-5138

[8].Chandrasekaran S., Ramanathan S., and Basak T., Microwave food processing- A Review. Food Research International 52, (2013), 243-261

[9] Pithcai,K., Birla, S.L., Jones D., and Thippareddi, H., 2012, Coupled Electromagnetic and Heat Transfer Model for Microwave Heating in Domestic Ovens, Journal of Food Engineering,112 ,pp100-111

[10] Oliveira M.E.C., Franca A.S., Microwave heating of foodstuff, Food Engineering 53, 2002, pp.347–359

[11] Mangalla, L.K., Enomoto H., Rianse U, Pasolon Y.B., Spray Formation of Ethanol Fuel in Local Contact Microwave Heating Injector (LMI) System, WSEAS Transaction on Heat and Mass Transfer, Vol. 10, pp.44-52

[12]Osepchuk JM., 2002, Microwave power applications. IEEE Transactions on Microwave Theory and Techniques Vol.50 (3), pp975–985

[13]Beaudry C, Raghavan G.S.V, Ratti C. and Rennie T.J., 2004, Effect of four drying methods on the quality of osmotically dehydrated cranberries. Drying Technology Vol 22(3), pp521–539

[14]Piotrowski D, Lenart A. and Wardzynski A., 2004, Influence of osmotic dehydration on microwave-convective drying of frozen strawberries. Journal of Food Engineering Vol.65 (4), pp519–525

[15]Koua B.K., Koffi P.M.E. and Ghaba P., 2017, Evolution of shrinkage, real density, porosity, heat and mass transfer coefficient during indirect solar drying of cocoa beans. Journal of the Saudi Society of Agricultural Sciences, available on http://dx.doi.org/10.1016/j.jssas.2017.

IOP Conf. Series: Materials Science and Engineering **797** (2020) 012018 doi:10.1088/1757-899X/797/1/012018

Composition Chemistry And Distribution to Quality of Essential Oil From Rhizome *E.elatior* and *E.calophyris* By Gas Chromatography-Mass Spectrometry (GC-MS)

Muhammad Aswan[1], Nur Arafah[2], I. Sahidin[3], Imran[4], Suryani[5], La Ode Ngkoimani[5]
[1]Departement of Chemistry, Post Gruaduated
[2]Pharmacy of Faculty
[3]Departement of Chemistery
[4]Departement of Enviromentally
[5]Departement of Geophysic
[12345]Universitas Haluoleo, 93232 Kendari, Sulawesi Tenggara, INDONESIA

*Corresponding author's e-mail: aswan.muhammad.biologi@gmail.com

Abstract. The composition chemistry and distribution of essential oil from the rhizome of two plants family Zingiberaceae is *Etlingera* species include *E.elatior* and *E.calophyris* from South East Sulawesi. The aim of this study is to determine the chemical composition of rhizomes two plants and compared the contain essential oils of the rhizome. The rhizome of the plants was extracted by hydrodistillation and analyzed was by Gas Chromatography-Mass Spectrometry (GC-MS). The essential oils from rhizome *E.elatior* and *E.calophyris* had a different compound among, with 25 compounds for *E.elatior* and 27 compounds identified for *E.calphyris*. The rhizome of *E.elatior* had the highest yield essential oils are dodecanal (36,09 %) and Dodecanoic acid (15, 54 %) and rhizome of *E.calophyris* is n-Decanal (34, 09 %) and Dodecanoic acid (14,5 %) into the mayor compound. While the compound minor for *E.elatior* had the highest 2-Undecanon (0,61%) and 2-Tridicanon (0,51 %) and *E.calophyris* is (-) Caryophyllene oxide (0,95 %) and 2-Undecanon (0,71%).

1. Introduction

Research on traditional medicinal plants in Indonesia currently shows a very significant speed. Natural resources that live in Indonesia are very useful for the development of medicinal plants. Which one from the ginger plant includes the Zingiberaceae family is the largest family of the Zingiberales order and one of the largest families of the Monocotyledonous class [1], which consists of 53 genera and more than 1200 species [2]. Zingiberaceae is conventionally in a different genus, where each genus usually consists of several species. Some examples of the genus are the genus *Curcuma, Kaempferia, Alpina,* and *Etlingera.* One of the largest genera of the Zingiberaceae family is the genus Etlingera plant [3], this genus includes more than 1,200 known species [2]. Plants from the genus *Etlingera* can be used by locals or commercially. In addition, in Sabah, the Malaysian part of the *Etlingera* genus plant can be used as spice ingredients such as young shoots, flowers and fruit. Also, in countries such

IOP Conf. Series: Materials Science and Engineering **797** (2020) 012018 doi:10.1088/1757-899X/797/1/012018

as Thailand, young fruits, stems and vegetable flowers [4]. In the Genera Etlingera rhizomes contain essential oils, in which the highest content typically found in the rhizomes [5]. Southeast Sulawesi Province generally has *Etlingera* genus plants. Based on the identification, there are three species of plant species is *E. elatior* [6], and E. calophyris. As a large community of Southeast Sulawesi utilizes as a seasoning for cooking or other traditional medicine. Therefore, the potential of *E. elatior* , and *E. calophyris,* plant which are less explored both from essential oil components, which can be found using GC-MS. The aim of this study was to determine the essential oil compounds of two Etlingera plant species (*E. elatior* , and *E. calophyris*) and compared the contain essential oils of the rhizome.

2. Method

2.1 Sample Preparation used Hydodestilation

Samples of the Etlingera (*E. elatior* and *E. calophrys*) plants on the rhizomes, which contained in a 2000 mL Erlenmeyer flask, flattened and distilled air extracted essential oils. and a hot plate for heating, the temperature on the hot plate is around 230 °C, and that will produce steam and air inside the distillation. Then hydrodistillation has waited for 4-5 hours and sampling is carried out for 3 or 4 times until essential oils are obtained. After getting the essential oil, the N-hexane solvent is dripped into the distillate.

2.2 Analyzed Essential Oils used GC-MS

Essential oils that have been produced from the processes contained in the spectrum data generated from the Gas Chromatography-Mass Spectrometry method. This measurement uses GC-MS Agilent 19091 S-433 series: 70eV model with column pillars (30m X 250µm) at moderate polarity. Helium mobile phase: with a pressure of 8.71 psi, with a current of 1.0m3 / minute and an average speed of 37 cm / s. The temperature initially was 60°C (for 2 minutes), then 60-120 °C (for 2 minutes) with a temperature of 3 °C / minute and 120-270 °C (for 2 minutes) with a programming temperature of 2 °C / min and an injection temperature of 250 °C. Data for *E. elatior* and *E. calophrys* were analyzed accurately.

3. Result

3.1 Yeild of Extraction

Table 1. The Yeild of Extraction

Sample	Wieght Sample	Essential oils	Visualization	Flavour
E.elatior	2,5 kg	10 mL	Yellow	Sharp, strong and distinctive
E. calophyris	2,5 kg	10 mL	Yellow	Sharp, strong and distinctive

3.2 Chemical Composition

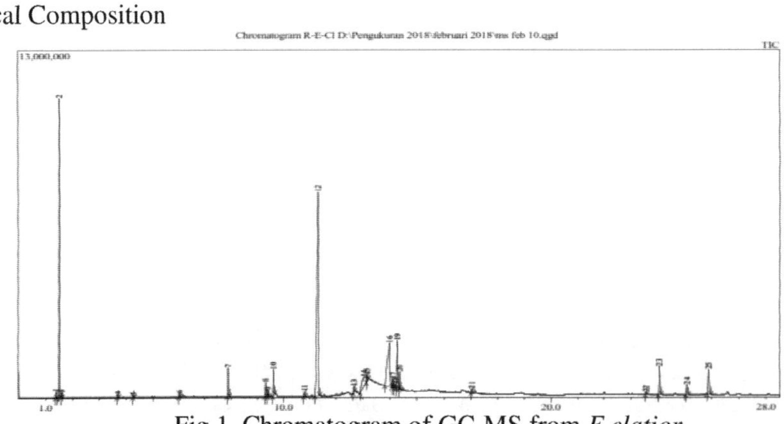

Fig 1. Chromatogram of GC-MS from *E.elatior*

Tabel 2. Chemical composition of *E.elatior*

No	Peak	Retention Time	% Area	Compound
				Mayor Compound
1	12	11,319	36,09	Dodecanal,
2	2	1,520	16,90	*n*-Hexana
3	16	14,002	15,54	Dodecanoid Acid
4	14	13,030	5,07	Cyclododecana
5	19	14,295	4,69	Dodecyl Acetat
6	23	24,062	3,09	1,54-Dibromo-tetrapentakonta
7	25	25,892	3,72	Dodesil dodekanoat
8	7	7,893	2,46	*n*-Decanal
9	10	9,606	2,31	Undecanal
10	20	14,348	1,59	Tetradecanan
11	24	25,892	1,36	1,54-Dibromo-tetrapentakonta
12	8	9,299	1,30	*α*, -Fenkil acetat
13	15	13,146	1,09	1-Chloroheptadecana
14	17	14,199	1,03	(-)-Caryophyllene Oxide
				Mayor Compound
No	Peak	Retention Time	% Area	Compound
15	9	9,606	0,61	2-Undecaanon
16	13	12,654	0,51	2-Tridecanon
17	11	10,804	0,45	α. –Copaena
18	18	14,199	0,39	11-Dodecanal
19	21	17,034	0,38	Dodesil acetat
20	1	1,391	0,31	Aceton
21	3	1,597	1,03	Methylcyclopentana
22	6	6,042	0,25	*n*-Undecane
23	4	3,682	0,22	α,-Pinene
24	22	23,553	0,22	Decanoid acid
25	5	6,042	0,15	2-β-Pinene

Source: Data of GC-MS Results Compound Components

The highest component of major compounds based on % area is dodecanoic acid compound with% area 37, 96% and the lowest is Tetradekanal with% area 1.79%. The highest component of minor compounds is based on % area, namely Dodecyl acetate with% area of 0.82% and the lowest is n-Dekana with% area of 0.08%. Which is grouped into several groups of compounds such as aldehyde (Tetradekanal (1.79%), dean (0.52%), undecanal (3.81%), dodecanal (11.43%), ketone (2-propanone (1, 99%), 2- Tridekanon (0.34%), 2-Undecanoon (0.61%) 3- (3,3-dimethylbutyl) cyclohexanone (0,14%)), esters (dodecanoic acid (37,49%), dodecyl acetet (3, 65%), tetradekanoic acid (2.48%)), and alcohol class (1-Hentetracontanol, (1.83; 0.26%)),

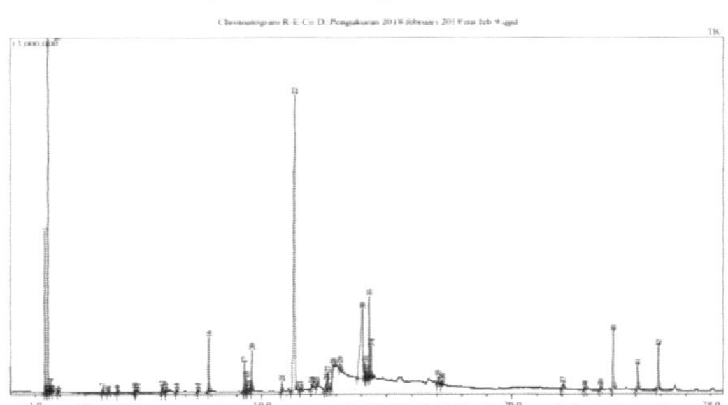

Fig 2. Chromatogram of GC-MS from *E.calophyris*

Tabel 3. Chemical composition of *E.calophyris*

No	Peak	Retention Time	% Area	Compound
				Mayor Compound
1	22	11.332	34.09	*n*-Decanal
2	30	14.043	14.5	Dodekanoat Acid
3	3	1.524	13.79	*n*-Hexana
4	1	1.394	5.49	Acetone
5	33	14,299	4,62	Dodesil acetat
6	40	24.065	3.78	1,54-dibromo- tetrapentakontana,
7	42	25.894	3.42	Dodesil dodecanoat
8	16	7.894	2.51	*n* –Decanal
9	20	9.605	1.97	Undecanal,
10	41	25.064	1.74	1,54-dibromo- tetrapentakontana
11	34	14.373	1.66	Tetradecanal,
12	27	12.660	1.43	2-Tridekanon,
13	17	9.300	1.40	Endobornil asetat
14	28	12.874	1.25	1-Tetradekanol
15	26	12,610	1,12	Cycloheptan
				Mayor Compound
No	Peak	Retention Time	% Area	Compound
16	31	14.126	0.95	(–)-Caryophyllene oxide
17	18	9.379	0.71	2-Undecanon
18	21	10.805	0.49	α-Copaena
19	37	22.015	0.41	1-Pentacontanol
20	32	14.202	0.40	11-Dodecen1-Al
21	35	17.034	0.38	Dodesil acetat
22	29	13.144	0.36	Kloro-6-heptadekena
23	4	1.601	0.33	Methylcyclopentana
24	19	9.505	0.30	α-Terpinena
25	7	3.683	0.25	α-Pinene
26	36	17.183	0.24	6-(2,6,6-Trimetil-1-Sikloheksana-4-Metil-(E)
27	39	23,552	0,24	*n*-Decanoat acid
28	12	6.043	0.22	*n*-Undecanan
29	24	12.005	0.21	β-Farnesena
30	9	4.266	0.18	β-Pinena
31	11	5.060	0.18	Eucaliptol (1,8-Sineol)
32	25	12.207	0.17	Allo-osimena
33	38	22.888	0.17	3,5,24-Trimetill- tetrakontana
34	2	1.475	0.14	2-Methylpentana
35	5	1.700	0.14	Cyclohexana
36	13	6.182	0.13	n-Nonanal
37	23	11.552	0.13	*trans*-Caryophylen
38	10	4.948	0.12	Benzena, metil(1-metiletil)
39	15	7,470	0,12	Bicylo(3,1,1)heptana,
40	8	3.891	0.10	Campena
41	14	6.619	0.10	α-Campolen aldehida
42	6	1.925	0.07	Metilsikloheksana

Source: Data of GC-MS Results Compound Components

The highest component of the major compound is based on% area, namely n-deanal compound with% area 34, 09% and the lowest is cycloheptane with% area 1.12%. The highest component of minor compounds is based on% area, namely (-) Caryophylen oxide with% area of 0.95% and the lowest is Methylcyclohexane with% area of 0.07%. Which is grouped into several groups of compounds such as aldehyde (n-decanal (34.09%; 2.51%), dean (0.52%), undecanal (1.97%), dodecanal (11.43%), tetradekanal (1.66%), 11 -dodecanal (0.40%), n-nonanal (0.13 %), ketone (2-propanon (5.49%), 2-Tridekanon (1.43%), 2-Undekanon (0.71%)), esters (dodecanoic acid (37, 49%; 14.5%), dodecano acetate (4.62%; 0.38%), n-decyl decanoate (0.24%)), alcohol class (1- Tertadekanol, (1.25%)), sesquerterpen class (β-Farnesena (0.21%, trans-Kariophyllen (0.13%)), monoterpenes (β- Pinene (0.18%). Eucalyptol (1.8-Sineol) (0.81%), allo-osimene (0.17%) campene (0.10%), α-Kampolen aldehyde (0.10%).

3.3 Compound Components Distribution

Table 4. Distribution of Compund Major and Minor

Retention Time	Compound	Molecule Formulation	Molecule Weigth	Rhizome	
				E. calophrys	*E.elatior*
11,313	Dodekanal,	$C_{12}H_{24}O$	186,33424	√	√
14,057	Asam dodekanoat	$C_{12}H_{24}O2$	200,31776	√	√
1,523	*n*-Heksana	C_6H_{14}	86,17536	√	√
1,394	2-Propanon	C_3H_6O	58,07914	√	-
14,296	Dodesil Asetat	$C_{14}H_{28}O_2$	228,37092	√	√
24,674	1,54-Dibromo-tetrapentakonta	$C_{54}H_{108}BR_2$	915,266	√	√
25,291	Dodesil dodekanoat	$C_{28}H_{48}O_2$	368,63672	√	√
7,8993	*n*-Dekanal	$C_{10}H_{20}O$	156,2652	√	√
9,607	Undekanal	$C_{11}H_{22}O$	170,29178	√	√
14,363	Tetradekanat	$C_{14}H_{28}O$	212,37152	√	√
9,299	*α*, -Fenkil asetat	$C_{12}H_{20}O_2$	196,286	-	√
14,199	(-)-Kariofilen oksida	$C_{15}H_{24}$	220,356	-	√
16,647	Asam tetradekanoat	$C_{14}H_{28}O_2$	228,37092	-	-
26,591	1-Hentetrakontanol	$C_{41}H_{48}O$	593,10506	-	-
12,66	2-Tridekanon	$C_{13}H_{26}O$	198,34494	√	-
9,3	Endobornil Asetat	$C_{12}H_{20}O_2$	196,286	√	-
12,874	1-Tetradekanol	$C_{14}H_{30}O$	214,3874	√	-
12,61	Sikloheptana	$C_{15}H_{24}$	204,35106	√	-

Source : Data Result of GC-MS

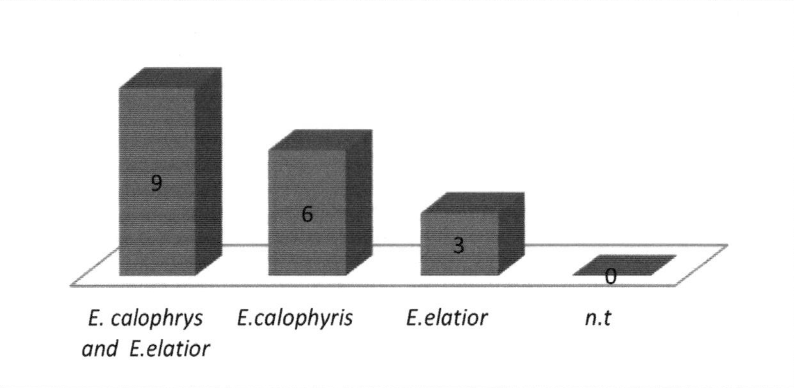

Fig 3. Total of Distribution Compund

4. Conclusion

The compound of the E. elatior rhizome contains 25 types of compounds and E.calophyris contains 42 types of compounds divided into components of major compounds and minor compounds.

5. Refences

[1] Aswani K., M. Sabu dan K. P. Smisha, 2013. Reproductive Biology Of Etlingera Elatior (Jack.) R. M. Sm. Ornamental Torch Ginger. *Internartional Journal of Plant, Animal dan Enviromental Sciense*.**3 (4)** : 75-80.

[2] Poulsen, A.D. 2006. *Etlingera of Borneo.* Sabah: Natural History Publication.

[3] Chiang, E.C.W., Yan, L.Y., Ali, N.A.M. (2010). Composition and Antibacterial Activity of Essential Oil from Leaves of Etlingera Species (Zingiberaceae). Int. J. Adv. Sci. Art. **1(2)**:1-12.

[4] Sirirugsa P. 1998. Thai Zingiberaceae: Species diversity and their uses. *Pure Appl Chem.* **70(11):**11–19.

[5] Chua, S. K., 2008. *Monitoring the Quality Of Essential Oil From Etlingera Sp.4 (Zingiberaceae) By Gas Chromatography-Mass Spectrometry (Gc-Ms).* Faculty of Chemical & Natural Resources Engineering. Universiti Malaysia Pahan.

[6] Chan EWC, Lim YY, and Wong SK.2011. Phytochemistry dan pharmacological properties of *Etlingera elatior*: A Review. *Pharmacognosy Journal*; **3(22)**:69-71.

IOP Conf. Series: Materials Science and Engineering **797** (2020) 012019 doi:10.1088/1757-899X/797/1/012019

Fuzzy Logic Methods to Identify Potential Area Mapping for Mangrove Forests in Kendari using Landsat Image

Ika Purwanti Ningrum[1], Mutmainnah Muchtar[1*], Rizal Adi Saputra[1], Adha Mashur Sajiah[1], Syamsul Razak Harati[2], Hasran Jaya[1]
Jurusan Teknik Informatika, Fakultas Teknik, Universitas Halu Oleo[1]
Fakultas Ilmu dan Teknologi Kebumian, Universitas Halu Oleo[2]

E-mail: ika.purwanti.n@uho.ac.id, *mutmainnah@uho.ac.id, rizaladisaputraa@uho.ac.id, adha.m.sajiah@uho.ac.id, syamsulrazakh@uho.ac.id

Abstract. Indonesia is an archipelago country with widest mangrove areas in the world. This potential is supported by length the coastline reaches + 81,000 km. As one of city in Indonesia, Kendari surrounded by the bay with temperatures reaching 22-31°C makes it potential to has mangrove forests. But the mangrove areas in Kendari decrease by 30% in recent year. This study aims to determine the level of mangrove fertility so we hope in future we could identify potential area Mapping for Mangrove in Kendari to increase more mangrove area. To analyze the quality of mangroves area through remote sensing data can be used Normalized Difference Vegetation Index (NDVI) method. By utilizing Landsat 7.0 ETM + image can make it easier to get information without touching the object of research. This experiment used fuzzy logic method (Fuzzy Mamdani) to get fertility rate of land soil. The result showed the highest fertility rate of mangrove area is 58,2034 which categorized as Very Good and the lowest score is 36,991 which categorized as Average.

1. Introduction

One of the most important parts of Indonesia's geographical condition as an archipelago is a fairly large coastal area with a coastline reaching 95,181 km [1]. This number makes Indonesia a country with the fourth longest coastline in the world, in harmony with its vast coastal area. Indonesia has tremendous potential for coastal natural resources with a diversity of ecosystems [2]. Indonesia is an archipelago country with the largest mangrove area in the world. This potential is supported by the length of the coastline which reaches ± 81,000 km and makes Indonesia a country with the second longest coastline in the world after Canada [3].

Kendari Region is also the capital city of Southeast Sulawesi Province, geographically located with coordinates of 3°54'30"- 4°3'11" S and 122°23'-122°39'E, and the location of the region stretching around Kendari bay [4]. While the average air temperature in Kendari city ranges between 22 °C - 31 °C [4] where the surface temperature is very important for the development of mangroves which acts as a parameter that is sensitive to the existence and the type of vegetation that covers the object of study, as well as air humidity and surface moisture. These parameters will affect temperature variations both temporal and spatial of a surface. Basically, the estimation of surface temperature using satellite imagery utilizes the concept of electromagnetic wave emission of objects that are unique based on surface temperature. Objects with different temperatures will emit maximum electromagnetic waves at different wavelength ranges [5].

Mangrove trees are no stranger to our lives. The benefits are so many, especially for the balance of the aquatic environment. So, it is only natural that Kendari city, Southeast Sulawesi, is participating in preserving the 'Green Belt' plant as an environmental defense because the area of mangrove forest in the coastal city of Kendari is shrinking. Data obtained from the Southeast Sulawesi Provincial Forestry Service stated that the area of mangrove forests in Kendari bay in 1960 was 542.58 ha, but experienced a drastic decline in 1995, leaving with only 69.85 ha. of total area [4].
The purpose of this paper is to get information about the level of mangrove fertility in Kendari City where mangroves play an important role in the stabilization of shoreline, can prevent erosion as a result of waves and also play a role in the addition of coastal land.
The method in this study uses the Mamdani method which is part of fuzzy logic. The Mamdani method itself, both input variables and output variables are divided into one or more fuzzy sets. The stages are data collection and data processing. Data retrieval consists of NDVI data, surface density and temperature. Data processing uses Fuzzy Logic decision making to determine the value and status of mangrove fertility levels.

2. Research Methods

The Mamdani method is often also known as the Max-Min Method. Some defuzzification methods on Mamdani's composition rules are [6]:

a. Centroid Method (Composite Moment). In this method, the crisp solution is obtained by taking the center point (z*) of the fuzzy region, which is generally formulated:

$$z^* = \frac{\sum_{j=1}^{n} z_j \, \mu(z_j)}{\sum_{j=1}^{n} \mu(z_j)} \tag{1}$$

b. Bisector Method. In this method, the crisp solution is obtained by taking values in the fuzzy domain that has a membership value of half of the total membership value in the fuzzy area, which is generally written:

z_p so that

$$\int_{R1}^{p} \mu(z) = dz \int_{p}^{Rn} \mu(z)dz \tag{2}$$

c. Mean of Maximum Method (MOM). In this method, the crisp solution is obtained by taking the average value of the domain that has the maximum membership value.
d. Largest of Maximum Method (LOM). In this method, the crisp solution is obtained by taking the largest value from the domain that has the maximum membership value.
e. Smallest of Maximum (SOM) method. In this method, the crisp solution is obtained by taking the smallest value from the domain that has the maximum membership value.

This study was done by collecting data from previous research. Data retrieval is needed as a reference for further analysis and calculation using fuzzy logic so the value of the mangrove fertility level in Kendari city can be known. In the fuzzy logic control system, there are several operational stages including:

2.1 Fuzzification

The fuzzification process is done by converting non-fuzzy variable data (numeric variables) into fuzzy variables (linguistic variables). Since fuzzy inference systems work with fuzzy rules and inputs, the first step is to change the firm input received, into a fuzzy input.

Table1. Fuzzy Set

Function	Variable Name	Range
Input	NDVI	0 - 1
	Density Level	0% - 100%
	Surface Temperature	0°C - 40°C
Output	Fertility Status	0 - 90

For each input variable, a fuzzification function is used to change the crisp input variable (the usual expressed in real numbers) becomes the value of the fuzzy approach.

The NDVI input variable has three forms of fuzzy set, namely: Low [0 0 0.18 0.32], Medium [0.3 0.4 0.47], and High [0.42 0.7 1 1]. The Density Level input variable has three forms of fuzzy sets, namely: Sparse [0 0 10 21.4], Intermediate [20 42 61.8], and Dense [60 80 100 100].

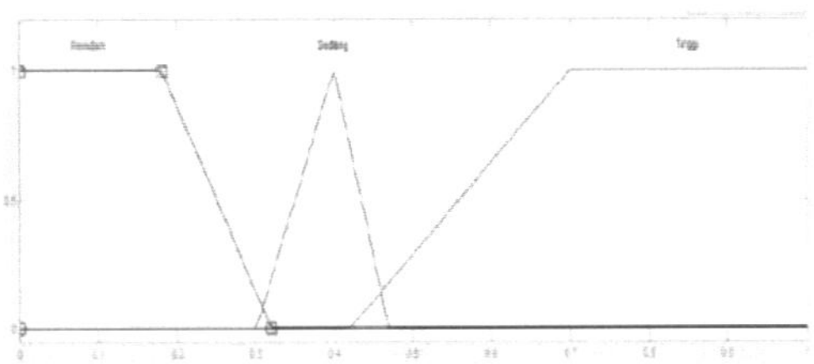

Figure 1. NDVI Variable

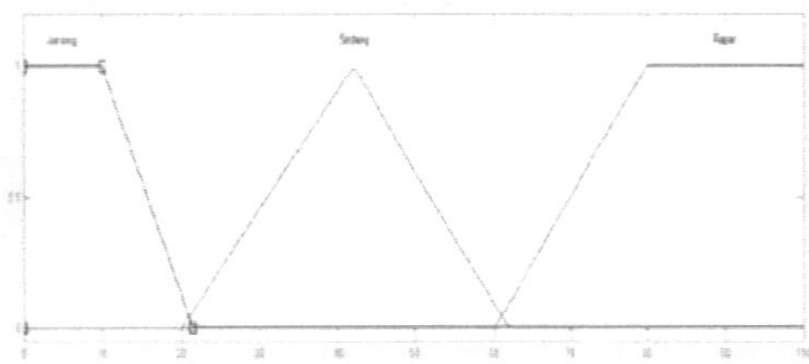

Figure 2. Density Level Variable

Surface temperature's input variables has three forms of fuzzy sets, namely: Cold [0 0 10 20], Warm [18 25 33], and Hot [32 35 40 40]. While the Fertility Status input variable has three forms of fuzzy sets, namely: Not Fertile [0 0 15 25], Medium Fertility [20 35 55], and High Fertility [50 70 90 90].

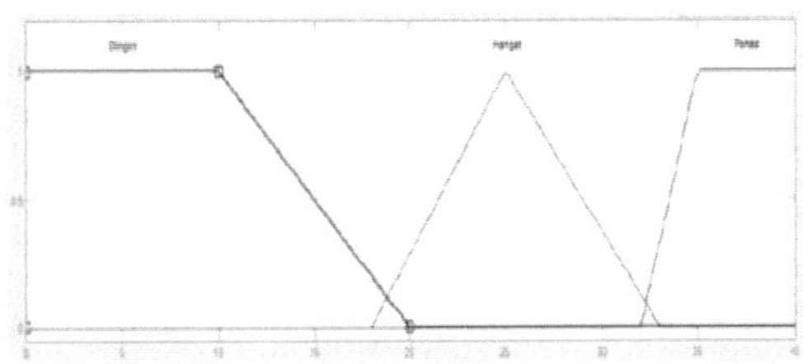

Figure 3. Surface Temperature Variable

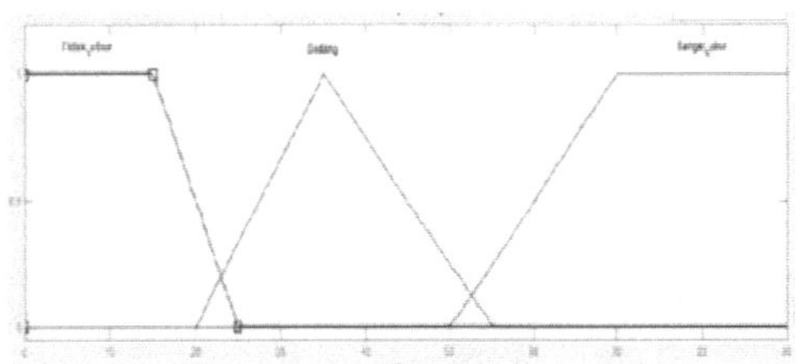

Figure 4. Fertility Status Variable

2.2 *Fuzzy Knowledge Base*

The knowledge base of a fuzzy inference system consists of a database and a rule base. Database is a set of membership functions from fuzzy sets that are related to linguistic values of variables involved in the system. The rule base is the set of fuzzy implications that apply as rules in the "*If ...and ...then...*" system. In this paper there are 27 rules used by the author to get the value of Mangrove fertility in Kendari City. The rules are:

Table 2. Knowledge Base

No	Rules
R1	If (NDVI is Low) and (Density Level is Sparse) and (Surface Temperature is Hot) then (Fertility Status is Unfertile)
R2	If (NDVI is Low) and (Density Level is Sparse) and (Surface Temperature is Warm) then (Fertility Status is Unfertile)

No	Rules
R3	If (NDVI is Low) and (Density Level is Sparse) and (Surface Temperature is Cold) then (Fertility Status is Unfertile)
R4	If (NDVI is Low) and (Densitiy Level is Intermediate) and (Surface Temperature is Cold) then (Fertility Status is Unfertile)
R5	If (NDVI is Low) and (Densitiy Level is Intermediate) and (Surface Temperature is Warm) then (Fertility Status is Medium Fertility)
R6	If (NDVI is Low) and (Densitiy Level is Intermediate) and (Surface Temperature is Hot) then (Fertility Status is Medium Fertility)
R7	If (NDVI is Low) and (Densitiy Level is Dense) and (Surface Temperature is Cold) then (Fertility Status is Unfertile)
R8	If (NDVI is Low) and (Densitiy Level is Dense) and (Surface Temperature is Warm) then (Fertility Status is Medium Fertility)
R9	If (NDVI is Low) and (Densitiy Level is Dense) and (Surface Temperature is Hot) then (Fertility Status is Medium Fertility)
R10	If (NDVI is Sedang) and (Densitiy Level is Sparse) and (Surface Temperature is Cold) then (Fertility Status is Unfertile)
R11	If (NDVI is Medium) and (Densitiy Level is Sparse) and (Surface Temperature is Warm) then (Fertility Status is Medium Fertility)
R12	If (NDVI is Medium) and (Densitiy Level is Sparse) and (Surface Temperature is Hot) then (Fertility Status is Medium Fertility)
R13	If (NDVI is Medium) and (Densitiy Level is Intermediate) and (Surface Temperature is Cold) then (Fertility Status is Unfertile)
R14	If (NDVI is Medium) and (Densitiy Level is Intermediate) and (Surface Temperature is Warm) then (Fertility Status is Medium Fertility)
R15	If (NDVI is Medium) and (Densitiy Level is Intermediate) and (Surface Temperature is Hot) then (Fertility Status is Medium Fertility)
R16	If (NDVI is Medium) and (Densitiy Level is Dense) and (Surface Temperature is Cold) then (Fertility Status is Unfertile)
R17	If (NDVI is Medium) and (Densitiy Level is Dense) and (Surface Temperature is Warm) then (Fertility Status is Medium Fertility)
R18	If (NDVI is Medium) and (Densitiy Level is Dense) and (Surface Temperature is Hot) then (Fertility Status is Medium Fertility)
R19	If (NDVI is High) and (Densitiy Level is Sparse) and (Surface Temperature is Cold) then (Fertility Status is Unfertile)
R20	If (NDVI is High) and (Densitiy Level is Sparse) and (Surface Temperature is Warm) then (Fertility Status is Medium Fertility)
R21	If (NDVI is High) and (Densitiy Level is Intermediate) and (Surface Temperature is Cold) then (Fertility Status is Unfertile)
R22	If (NDVI is High) and (Densitiy Level is Intermediate) and (Surface Temperature is Warm) then (Fertility Status is Medium Fertility)
R23	If (NDVI is High) and (Densitiy Level is Intermediate) and (Surface Temperature is Hot) then (Fertility Status is Medium Fertility)
R24	If (NDVI is High) and (Densitiy Level is Dense) and (Surface Temperature is Warm) then (Fertility Status is Medium Fertility)
R25	If (NDVI is High) and (Densitiy Level is Dense) and (Surface Temperature is Warm) then (Fertility Status is High Fertility)
R26	If (NDVI is High) and (Densitiy Level is Dense) and (Surface Temperature is Hot) then (Fertility Status is Medium Fertility)
R27	If (NDVI is High) and (Density Level is Dense) and (Surface Temperature is Hot) then (Fertility Status is High Fertility)

2.3 Defuzzification

Defuzzification is the process of mapping fuzzy sets to a strict set (crisp). This process is the opposite of the fuzzification process.

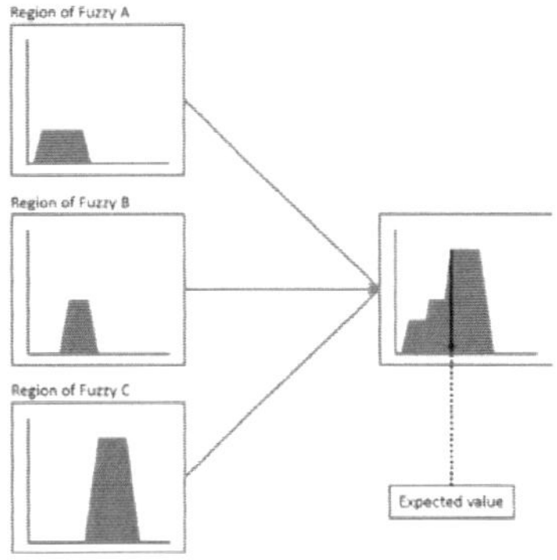

Figure 5. Defuzzication Process with Mamdani Method

3. Results and Discussion

To analyse the quality of mangroves through remote sensing data, vegetation index method normalized NDVI (Normalized Difference Vegetation Index) can be used. This calculation method will divide the mangrove health level based on the canopy density value, which are:

1) <=20% (Very Sparse), the range of NDVI values> 0.01 to 0.18
2) 21,40% (Sparse), the range of NDVI values is 0.18 to 0.32
3) 41,60% (Medium), the range of NDVI values is 0.32 to 0.42
4) 61,80% (Heavy), the range of NDVI values from 0.42 to 0.47
5) >=80% (Very Heavy), the range of NDVI values > = 0.47

Determination of mangrove fertility status using the Mamdani method of Fuzyy Logic. Based on the regional data of Kendari city in 2014 (NDVI values, Density and Temperature), the results obtained for the calculation for fertility status of mangroves are as shown in Table 3 below.

Table 3. Mangrove Fertility Level Status

No	NDVI Value	Density (%)	Temperature (°C)	Fertility Value	Fertility Status
1	0.256	26.0818	21.7372	37.1731	Medium Fertility
2	0.430	66.3110	21.8944	41.4167	Medium Fertility
3	0.874	68.0142	22.3661	56.951	High Fertility
4	0.345	48.4718	21.7372	36.991	Medium Fertility
5	0.808	80.2783	21.7372	57.6743	High Fertility
6	0.202	11.0731	21.5800	11.0481	Unfertile
7	0.519	80.2910	21.8944	56.714	High Fertility
8	0.510	79.1245	21.7372	56.5603	High Fertility
9	0.824	77.7830	22.3661	58.2034	High Fertility
10	0.414	63.2599	21.7372	37.2844	Medium Fertility

Based on these data, mangroves that have the highest fertility value is 58.2031 with their Mangrove Fertility Status is in "High Fertility", this is because the environment in which mangroves grow is very suitable for mangroves to live. The area is dominated by mangroves because it has not been touched by development for human settlements or fishponds. The place was also protected by the government. Besides that, fresh water supply, and groundwater salinity are the most influential factors on mangrove growth in this region. Its location close to the sea (salt water) that also influences its growth. With the influence of sea water, if mangroves are not able to adapt to existing environmental conditions it can cause growth ability to be slightly inhibited.

Whereas mangroves that have lowest fertility value are 11.0481 with the Fertility Status mangroves "Unfertile" which is on the coast of Kendari city. The main cause besides low's density is due to its ability to adapt to existing environmental conditions. The utilization pattern is one of the factors that greatly affect the presence of mangroves in this place. With the increasingly high rate of exploitation, the mangroves here are very vulnerable to damage because many are used as a material for consumption by the community. Its use as firewood and other household necessities by the surrounding population is a major factor in the reduction of mangroves population in this place.

For the condition of density of mangroves tree in the Bay of Kendari as a whole is still categorized as "Dense" where the value of density reaches 80.2910 because the location of mangrove growth in this place gets sufficient water supply. In addition, because mangroves in this place grow along beaches that are protected from large wave activity and strong tidal currents, the substrate needed for mangrove growth is very suitable so that this area is highly recommended for Mangrove cultivation. Given the type of substrate in the form of sandy mud and peat soil, this type requires special adaptation, namely in the roots. Also due to regeneration factors that are often limited.

4. Conclusion

Based on the experiment results it can be seen that the level of fertility of mangroves is very high. The highest fertility value is 58,2034. The recommended area for Mangrove cultivation is the Bay of Kendari city because the density of mangroves in the area reaches 80.2910. The role of the community and government is needed to conserve mangrove ecosystems as an effort to maintain the existence of mangrove ecosystems.

REFERENCES

[1] Wijaya, N. I., Damayanti, I. R., Patwati, E., & Wismayanti, S. Perubahan Luas Ekosistem Mangrove Di Kawasan Pantai Timur Surabaya. *Prosiding Seminar Nasional Kelautan dan Perikanan III 2017*, Universitas Trunojoyo Madura, 7 September 2017.

[2] MUKHTAR. (2009). Garis Pantai Indonesia Terpanjang Keempat di Dunia. http://www.kkp.go.id/index.php/arsip/c/1048/Garis-Pantai-Indonesia-Terpanjang-Keempat-di-Dunia/?category_id=. (Access online: 29 Mei 2018).

[3] Forestian, O. K. T. A. M. A. (2011). Estimasi Biomassa dan Kerapatan Vegetasi Mangrove Menggunakan Data Landsat ETM+. Fakultas Kehutanan Institut Pertanian Bogor (1-62).

[4] Laremba, S. (2014). Sebaran dan Kerapatan Mangrove di Teluk Kota Kendari Sulawesi Tenggara. Makassar: FIKP UNHAS.

[5] Wulandari, D. (2013). Identifikasi Sebaran Dan Tingkat Kesuburan Mangrove Melalui Pemantauan Indeks Vegetasi Dari Satelit Landsat 7.0 ETM+ Menggunakan Fuzzy Logic (Studi Kasus Pantai Pesisir Timur Surabaya/Pamurbaya). *Jurnal Neutrino*, 60-67.

[6] Sivanandam, S. N., Sumathi, S., & Deepa, S. N. (2007). Introduction to fuzzy logic using MATLAB (Vol. 1). Berlin: Springer.

IOP Conf. Series: Materials Science and Engineering **797** (2020) 012020 doi:10.1088/1757-899X/797/1/012020

Floating Houses Technology as Alternative Living on The Water

Tri Endangsih[1], **Ikaputra**[2]
[1] Program Study of Architecture, Faculty of Engineering, Budi Luhur University, Jakarta, INDONESIA
[2] Department of Architecture and Planning, Faculty of Engineering, Gadjah Mada University, Yogyakarta, INDONESIA

E-mail: tri.endangsih@budiluhur.ac.id[1], ikaputra@ugm.ac.id[2]

Abstract. Increasing population has resulted in increased housing needs, but currently the available land for housing is decreasing. In addition, climate change has resulted in rising sea levels. With these three problems, it is necessary to develop a place to live with limited land and utilize the potential of Indonesian waters. The efforts that have been made by the current government by building settlements vertically, but have not been able to meet the needs of the residence. One type of house that can be developed by utilizing the potential of the waters is floating houses. A floating house is a residential building that is built on water by applying floating structures, materials, and shapes to suit the construction site. The floating house will function properly with proper planning, especially in the selection of technology at the bottom, floats, and the stability of the building above the water. Besides technology, other things that need to be taken into consideration are the selection of materials and the selection of the type of structure that will be applied. This technological innovation is expected to be able to meet the needs of residences that can utilize the territorial waters.

1. BACKGROUND

Experts argue that increasing concentrations of greenhouse gases in the atmosphere can cause climate change to warm up dramatically. Impact The global warming of the Coastal / City Region is a decrease in land area due to sea level rise and excessive evaporation of sea water which results in flooding which reduces the land area. Most cities in Indonesia are located on the coast, with characteristics of high growth rates. Changes in population density due to increasing population will increase the demand for land for settlement needs [1] and [2].

Reduced land area due to flooding as a result of global warming, so that building technology is needed that easily adjusts the height position when there is a flood, with minimal risk that can cause material loss to residents, namely the development of floating houses[3], [4], and [5]. Technological advances have led to various innovations in the construction of houses and public facilities on water. Currently the innovation of floating houses is recommended to be the right solution to meet housing needs.

2. PURPOSE OF RESEARCH

The purpose of this research is to get floating house technology that is able to accommodate more occupants with a structure that meets building safety and stability standards. Research on floating house technology is certainly related to the selection of structures and materials that will be applied to the building.

3. RESEARCH METHOD

This research is a theoretical study that can be developed in a floating house. The method used is to dialogue the theory to apply floating house technology that can meet more occupant capacity with

IOP Conf. Series: Materials Science and Engineering **797** (2020) 012020 doi:10.1088/1757-899X/797/1/012020

quality that meets the occupancy requirements and the structure of floating houses that are able to support more residential capacity so that it functions better. The theory used to support this writing is a combined theory of several experts according to the field of study. The choice of theory in this writing by considering the relationship that is relevant to the problem of floating home technology as an alternative to residential houses on water.

4. RESULT AND DISCUSSION

4.1. Definition of Floating houses

The floating house as a reflection of the waterfront architecture has many different terms and terms based on its location and structural system, some of the concepts put forward, among others, by the Chaichongsrak in Denpaiboon "define it as a shop house that can float in water, move and can be used for life and sleeping floating houses is a building that floats in water, can move and can be used as a place to live, while Denpaiboon has its own definition that is a floating house with another name raft house is a traditional architectural style building that uses a raft structure and is located in water [6] , and (Iskandar & Yuliandani, 2001) Floating buildings use aquatic areas such as lakes, seas, beaches, rivers, and their parts and are not on land or land with floating type and amphibian structures [8]. or a raft house, it can be concluded that a floating house or raft house is a residential building that is built on water by applying floating structures, materials, and shapes that correspond to the construction site. Based on the explanation above, it can be explained with the following picture:

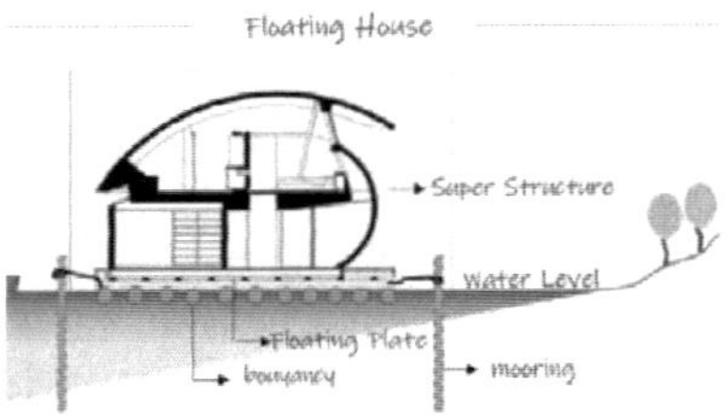

Figure 1. Terminology of Floating Houses
Source: Results of analysis, 2018

The floating house is above the water with the position of the building following the water level. It needs the principle of balance so that the building is not tilted, and needs a fastening pole so that the house does not move following the flow of water.

4.2. The Principle of Floating

Understanding Archimedes Law (flotation law) is when an object is partially or completely immersed in water, then the object will get a compressive force that points upwards which is equal to the weight of the water that is moved by the part of the immersed object. The period of the object does not change, the reduction in the weight of the object is caused by the upward force of the water that is moved by the part of the object that is in the water (force of buoyancy), with the direction of work the style is directed upwards; while the line of work is in line with the work line of gravity.

Balance will be achieved if the amount of buoyancy force is equal to the weight of the water transferred by the part of the object that is in the water or if the object floats with half the volume of the image [9] and [10]. The principle of balance applies to floating houses, the condition of the buoy is caused by the balance of enforcement of vertical forces, and will experience movement or shaking by lateral forces of attraction due to pressure of river water flow or flooding. Water surface is dynamic, whose conditions are strongly influenced by the flow of the generator which causes horizontal and vertical thrust.

4.3. Building Stability Based On Building Expenses

The workload on floating buildings is divided into dead and live loads. Dead load is a burden that will not change (very much) from time to time and the burden of life is a burden that changes in different circumstances. Dead load above the floating structure includes cargo from buildings, facilities and infrastructure. More weight above the floating structure results in a greater load. This static load needs to be well distributed to prevent rotation and tilt of the float structure. The dead weight of the platform and upper structure depends on the design of the building. In addition to dead load, the platform is also subject to a live load. Life burden is a burden that changes in different situations. Some examples of living loads in floating houses are storage loads, traffic loads, population loads, wave loads, wind loads, water pressure and buoyancy[11].

In planning low-rise floating buildings, it is important to know that the total area available for floating buildings is estimated to be 50% of the floating structure. In planning high-rise floating buildings, it is important to know about the number of occupants who will live in one flat / apartment, the number of floors, floor to floor height, and the total height of the structure of the floating structure will be tilted if the eccentric strength is large (dead weight of the building), so the building is placed in center of the floating structure.

The more central the buoyancy shifts and shifts with a certain rotation, the higher the center shift, so the more stable the buoyancy structure is. Rectangular float structure is more stable than other shapes such as cylinders or triangular shapes. This is because when the float element rotates, the rectangular shape replaces most of the water[12]. The ratio of width to depth largely determines the shape of the float element.

4.4. Floating Houses Structure

Floating house, is the concept of a residence or stop by using floating structure media. The concept of a floating structure or often called a 'Floating Structure' is used as a substitute for land in the construction of a building. Besides being an alternative prearrangement of the area besides reclamation, because the structure is able to float on water [13]. The structure of a traditional floating house consists of a bottom structure of round wood, a bamboo raft and on the upper part is a traditional low-rise stage house. The use of the lower structure of the raft and low mast models aims to anticipate catastrophic winds so that the house is not easily thrown by the wind and carried away by the flow of water. In addition, the use of a single fastening pole that makes the house float rotates up to 360 degrees every day, because it follows the wind direction[14]. In some regions of Indonesia, which have typologies of floating houses with lower structures of wood, bamboo rafts, and used drums, and the main structure of wooden houses are very vulnerable to high winds and water currents. [15] dan [4].

4.4.1. Floating Structure

In the era of the emergence of floating houses in other countries the structure of floating houses in the form of wooden boats, but with advances in technology and increasingly limited building materials made from wood, in some countries, for example the Netherlands, New Zealand and other countries use a combination of structural steel and reinforced concrete structures. Just reinforced, and specifically in the Netherlands at this time many floating polystyrene buoyancy structures have been developed into hollow concrete blocks [5]. This is considered safer and economical compared to the use of old designs that use steel and concrete structures. An alternative development of Polystyrene foam float material in the Netherlands because architects think that the use of steel and concrete as floats can damage the environment.

This floating system is based on a polystyrene foam core (EPS, Expended Poly Styrene) with a surrounding concrete shell. EPS is a lightweight material that provides exceptional buoyancy and can be used in shallower waters. This floating system development contributes to the formation of large floating residences in the cities of Seattle and Vancouver[11].

The floating structure called the floating plate is where the building mass is attached. Part of this floating plate is a cover layer, frame plate and float. This floating plate shape will affect the buoy configuration used. Configure the float illustration in the following picture:

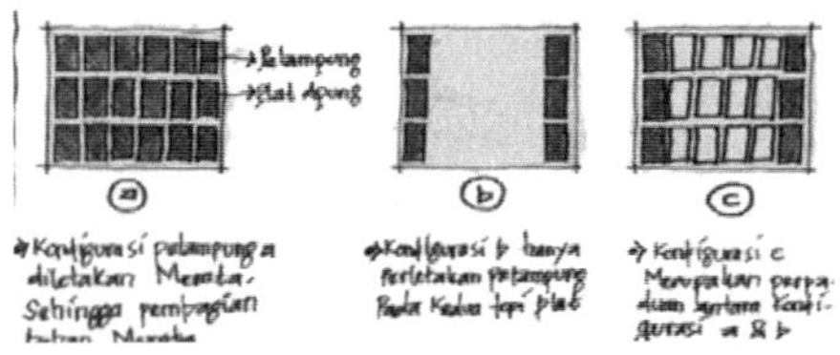

Figure 2. Configuring The Buoy
Source: Results of Analysis, 2018

The floating structures developed include; pontoon, eps foam, drums, etc.[16] and [8]. Examples the development of the structure of floating houses in Indonesia, nowadays wood and metal drums have been replaced by HDPE Floating Cube made of HDPE plastic which has proven to be strong and reliable 20 years without damage.

4.4.2. Mooring System

Each floating structure requires a mooring system that is sufficiently rigid and strong to limit the movement of the structure to external forces from wind, currents, waves or others such as the movement of ice if there is a winter. There are many belay systems in the form of temporary / temporary or permanent systems. Distribution of types of belay can also be in the form of mooring from inside or outside. Basically, the type of mooring there are several variations as follows[18]:

1. One mooring system with truss (*Attached Mooring System).*
   Basically, the mooring is done by tethering a special part of mooring line to the floating structure. Additions can be in the form of turrets placed on the inside (mounted inside end) or outside (mounted with the addition of a structure at the end). This system is caused by waves. This mooring system uses truss that is tied to one side of the building. This mooring system is more suitable for use on large-scale floating buildings.

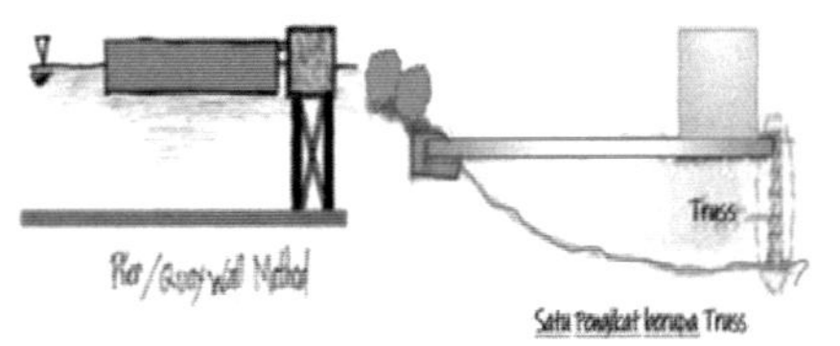

Figure 3. Attached Mooring System
Source: Results Of Analysis, 2018

2. The mooring system with piles
   This mooring system uses piles to keep the building in position. This system uses the principle of stake. Whereas to maintain the position of the building, the configuration of the supporting pole is also a consideration in this mooring system.

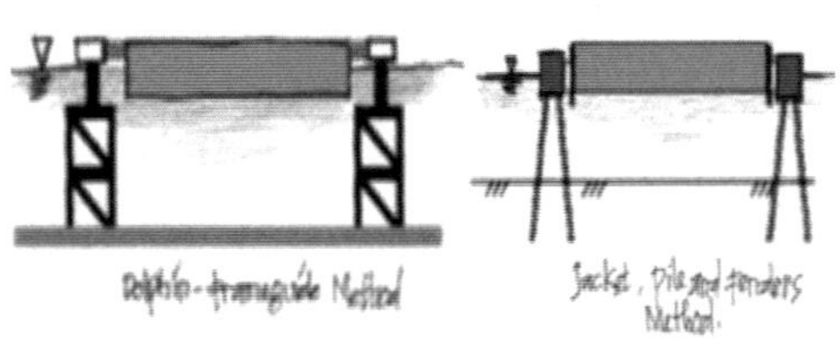

Figure 4. Pile Mooring System
Source: Results Of Analysis, 2018

3. The mooring system with a cable (Spread Mooring System)
   This type anchors the floating structure with a fixed heading direction. In this configuration is not needed swivel component. This configuration is only suitable for a location that is relatively quiet and has a change in the direction of loading that tends to be constant / not large. In general, locations such as the Malaka Strait or the Java Sea Coast may be suitable for this configuration.

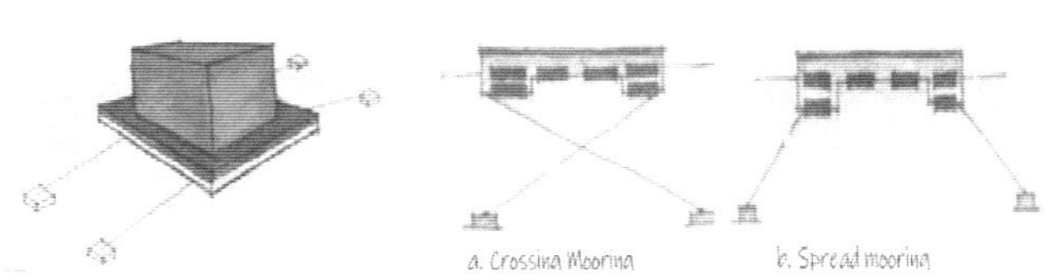

Figure 5. Spread Mooring System and Type Of Cable Configuration
Source: Results Of Analysis, 2018

There are two types of cable configurations in this mooring system, including crosses and crosses. a configuration a is more suitable for use in environments with calm water conditions because the cable only functions to maintain the position of the building.

4.5. Construction System

Construction systems in small-scale floating buildings generally use lightweight construction systems. The use of building materials is well considered to be able to withstand environmental conditions and not damage the environment. Whereas, for the floating building construction system, the mass of the construction process is carried out in the factory by building modules and assembled on the site.

4.6. Floating Houses Technology Application

Based on the theory regarding the stability of the floating house structure system, the factors that influence the selection of the structure type of floating houses used for residence include; structure that is able to provide stable conditions for waves and wind forces that work around it, relatively low prices, high durability / low maintenance, low weight, high buoyancy capacity, and easy to build shape.

For stability of buildings other than building dimensions, another factor that must be considered is the higher the building, the greater the dimension of the floating plate. In addition, the most suitable configuration of building plans is a long rectangular shape. That's because when the float element rotates, the rectangular shape replaces most of the water. The mooring system used is a wire mooring system because this system has more economic value than other systems and ease of construction. While the flexibility of the building is calculated based on the type of building lock that can be in the form of a mooring system with the installation of stakes with a sea flex anchoring system that has the highest flexibility. In addition, this system has been certified and has been used in many buildings on the water.

Figure 6. The Mooring System and Illustration of Floating Plate Configuration
Source: [19] and Results of Analysis, 2018

The float plate used is a square floating plate with a parallel float configuration on the edge. The square shape is chosen because this form has ease in the construction process, especially the mass of floating buildings is a mass with a large number. While the float configuration on the edge of the plate can save construction costs.

4.7. Construction System and Material

In the selection of this material, materials that are environmentally friendly are sought so as not to cause damage to the environment and surrounding habitats. Material selection based on parts of the building structure is explained as follows;

a. A binding floating structure based on air conditions can use cables and anchors.
b. Basic structure / foundation as buoy using concrete materials, composite panels of Rexwall, EPF (Expanded Polystyrene Foam), Platform, Steel, HDPE Floating Cube and concrete pontoons. The selection of this material is carried out on the ability to withstand structural loads. Middle structure,

building frame made of wood, steel and concrete blocks. Wall materials from lightweight concrete, wooden planks, plywood, and glass.

c. Upper structure, roof truss of wood, mild steel and for cover made of concrete, and Zincalum. Selection of materials for basic structures and materials, resistant to local climate, able to provide comfort for residents and easy maintenance.

5. CONCLUSIONS
The conclusion of the theoretical dialect of this floating house is as follows:

1. Looking at the development of science and technology in the field of architecture, Floating building is one solution to problems in meeting housing needs.
2. The use of appropriate technology and structures is the main thing that must be considered in designing floating buildings.
3. The things that need to be considered in the planning of a floating house are the use of components of the shape of a floating building to be symmetrical and using constructions with light weight, floating shapes and plate materials do not damage the environment, mooring the system must pay attention to changes in water on the site, and inter circulation building so as to facilitate users.

ACKNOWLEDGEMENTS
Thank you to Mr. Ikaputra as the lecturer in Architecture Theory for a Doctoral Program Architecture Science at Gadjah Mada University, for his learning and input in preparing waterfront city dialectics with the theme of floating house technology.

REFERENCES
[1] W. Hantoro, 'Effects of Sea and Beach Characteristics on the Development of Coastal Cities', 2002.
[2] National Population and Family Planning Agency (BKKBN), Bumi Increasingly Increased Earth Population', Jakarta, 2011.
[3] Koen Olthuis, *Building on Water to Combat Urban Congestion and Climate Change*. 2007.
[4] Puspitasari, Kadri, Indartoyo, and L. Kusumawati, 'Microclimate and architectural tectonic: vernacular floating house resilience in Seberang Ulu 1, Palembang', in *The 4th International Seminar on Sustainable Urban Development*, 2017, pp. 1–6.
[5] A. Ambica and K. Venkatraman, 'Floating Architecture : A Design on Hydrophilic Floating House for Fluctuating Water Level', vol. 8, no. November, 2015.
[6] C. Denpaiboon, M. Tohiguchi, H. Matsuda, and S. Hashimoto, 'Typology and Life Style Analysis of the Raft House (Ruan Pae) in Riverine Settlements in Thailand', vol. 1, pp. 173–180, 2000.
[7] S. Syamsiar and Irhamna, Banjar Traditional Architecture of South Kalimantan. Banjarmasin: Association of Indonesian Architects of South Kalimantan, 2001.
[8] Wijanarka, 'Maritime Architecture: Water Based Architecture, Solution to development in the Indonesian Archipelago', *Workshop Continuing Profesional Development (CPD)* Architectural Expert Organized by the Directorate General of Construction of the Ministry *PUPR*, Jakarta, 2016.
[9] R. Riski, 'Archimedes' Law (Law of Floating)', 2009.
[10] 212BC Archimedes of Syracuse, 'Archimedes principle', 2008.
[11] K.K.M. Ko, 'Realising a floating city Appendix', Delft University of Technology, Netherlands, 2015.
[12] Winkelen, 'How high can you float', Delft University of Technology, 2007.
[13] Frick.H, Ecological Architecture. The concept of ecological architecture in tropical climates, greening of ecological cities and cities, and renewable energy. Yogyakarta: Kanisius, 2006.
[14] N. Naing, 'Creating a Better Living Condition at the Floating House in Lake Tempe', 2008.
[15] N. dan Naing and H. Halim, 'Structure System of Floating House at Tempe Lake in South Sulawesi', *J.* Settlement., vol. Vol. 8 No., 2013.
[16] M. Khamdevi, 'Floating Architecture Approach', *IPLBI*, 2014.
[17] R. Didik, 'Floating Houses', 2018.
[18] C. M. W. E. Watanabe and T. U. A. T. Moan, 'Very Large Floating Structures: Applications, Analysis And Design', Singapore, 2004.
[19] Marina Housing, 'Marina Housing Floating Concept and Projects', 2009.

IOP Conf. Series: Materials Science and Engineering **797** (2020) 012021 doi:10.1088/1757-899X/797/1/012021

Design Concept Long Span Bridge with Floating Foundation

Abdul Kadir[1*], **Baso Mursidi**
Department of Civil Engineering, Universitas Haluoleo, 93232 Kendari, Sulawesi Tenggara, INDONESIA

*Corresponding author's e-mail: kadir12340@yahoo.com

Abstract. Indonesia is a maritime country where the sea area is larger than the land area and consists of a group of islands stretching from west to east. Inter-territorial accsebility can become faster and smoother by structur long span bridges. In the deep sea and tidy ground conditions allow use of pile foundatio, however the floating foundation is an option that can be taken. The purpose of this paper is to propose the concept of a long-span bridge floating foundation that is capable and stable to support the load or forces that work.

1. Introduction

Indonesia is a archipelago country that consists of a group of islands stretching from west to east. Inter-territorial accsebilitty can become faster and smoother by structure long span bridges. In the deep sea and water where the bottom is too soft for pillars, the floating foundations is an option that can be taken.

Floating bridge has been known for 4000 years, generally used for temporary military purposes and materials made of wood. Now the floating bridge can be built by used of concrete, steel or a combination of concrete and steel. Floating bridge has been used in many countries such as; USA, Canada, Norway, Englan, Japan, Maldives. Built-in bridges are dominated by short span.

Long bridge spans such as arch bridge, cable stayed, suspension and extradosed relatively rare. The use of long spans floating foundation that has a high clearance will facilitate the movement of marine or river transportation. The first long-span buoy bridge is the Yemenai Japanese bridge that was completed in 2000 [1]. Now a study and analysis of long span bridge floating foundation is being developed in European and Norwegian.

2. Floating foundation bridges concept

The basic concept of floating bridges is simple: the piers or foundations are replaced with floating elements floating freely or with mooring lines (Figure 1). The foundation bridge floating bridge is inspired by the natural law of the lifting force of water (buoyancy force). Vertical loads, (live and dead loads) are held by a pontoon system consisting of single columns or multiple column combinations. Horizontal loads are retained by the mooring system. The effectiveness of the work and the function of floating bridge is influenced by the interaction of several aspects namely: 1) enviroment and loading, 2) stability, 3) station keeping, 4) motion response, 5) structural integrty.

3. The Elements of long span floating foundation bridges

The main element of the bridge floating foundation is the pontoon and mooring system. The pontoon consists of two forms: a continuous pontoon and a separate pontoon. On a suitable long span bridge of pontoons is a separate pontoon that directly supports the upper structure. A separate pontoon can be a single pontoon or some cells arranged horizontally or vertically. A single cell pontoonsection can be; a) circle, (b) base case (D <W <2D), (c) oblong (W> 2D), (d) oval (D> W). Horizontal combined pontoons are composed of two cells where the lower cell has a larger diameter. Vertically combined cells can be made with a unified arrangement without a gap or with horizontal connecting rods.

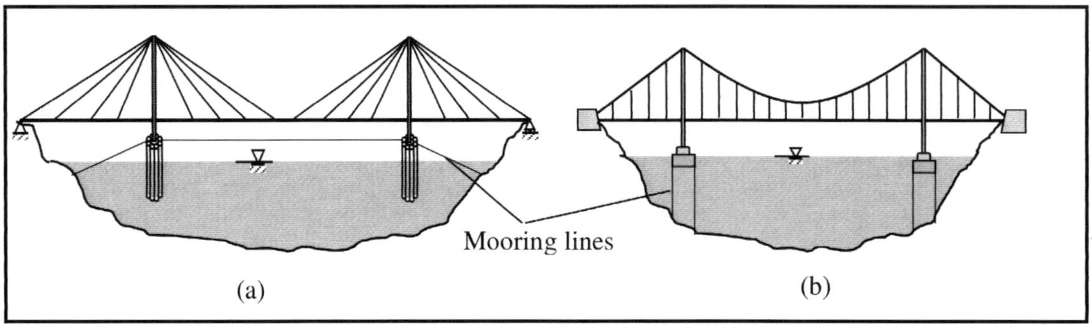

Figure 1. Floating foudation long span bridges, (a) Cable stayed, (b) Suspension bridge

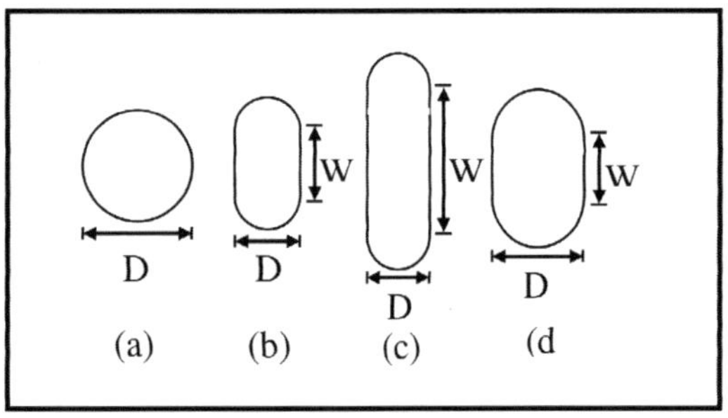

Figure 2. Shaps of one cell pontoon, (a) Circular, (b) basecase, (c) oblong, (d) oval

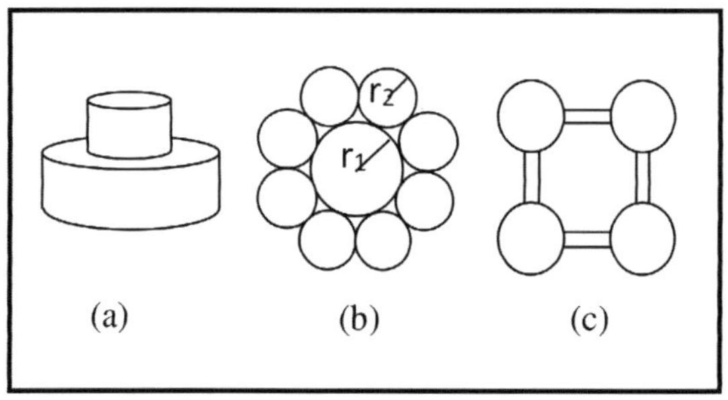

Figure 3. The shaps of unified cell pontoon, (a) horizontal unified cell, (b) vertical unified cell, (c) verical unified cell with rod conection

Mooring system is used to reduce the movement of the bridge. There are several methods or mooring systems used is; 1) The dolphin-guide frame, 2) Pier/quay wall, 3) Sliding piles, 4) Cable/chain and anchor (catenary), 5) Tension leg method (taut and vertikal tension leg).
On long floating bridge, mooring type that can be used is cable type, chain with achor, and tension leg. Chain provides greater stiffness than cable / wire. More vertical tension legs can produce better stability than leg or catenary links, but will require a larger cost of anchoring.

4. Load and load combination

Floating foundation must be able to support the loads and combinations of the load, to the ultimate, service, fatique, accidental and postdamage conditions. By their feature the load acting on floating foundations is classified as in Table 1.

The combination of loads on the floating structure according to [2] and [3] summarized by Chandrasekaran, 2015[4] is shown in Table 2.

Table 1. Loading Type Category Descriptions

Load Type	Description
Permanent (load) loads (PL)	Weight of structure bridge and any permanent equipment that does not change during the life of the structure.(structural weight, dry equipment, ballast, and hydrostatic pressure)
Live loads (LL)	Loads that can be changed, moved or removed during the life of the structure. (vessel, personnel, and liquid)
Environmental loads (EL)	Loads on the structure due to the action of wind, wave, current, tide, earthquake, tsunami, or ice
Deformations loads (DL)	Loads on the structure due to out-of-level supports and subsidence
Accidental load (AL)	Loads on the structure due to dropped object, ship impact, blast, and fire.

Table 2. Load factor [3]

Limit State	Load categories				
	PL	LL	DL	EL	AL
ULS (normal)	1,3	1,3	1,0	0,7	0,0
ULS (extremed)	1,0	1,0	1,0	1,3	0,0
SLS	1,0	1,0	1,0	1,0	0,0
FLS	0,0	0,0	0,0	1,0	0,0
PLS (accidental)	1,0	1,0	1,0	1,0	1,0
PLS (post-damage)	1,0	1,0	1,0	1,0	0,0

5. Movement

5.1 Stablity

The stability of a floating foundation is related to the rotational movement of a floating. The floating is stable if the rotational motion can return to its initial equilibrium position. The inverting force in the buoyant structure can be generated by: floating moment, ballast and mooring.

The stability of a floatings system consists of static equilibrium and dynamic equilibrium. If the floating structure shakes with a fixed angle of rotation then two possible moments are the reversal moment (P.b) and the falling moment (P'.b'). The relation moment and the failing moment are denoted by Figure 4. Figure 4 (a) shows that the floating structure will be stable if the height of the Meta center is positive or greater than the gravity center (G). The greater the value of meta center (M) then the degree of stability is higher. The dynamic stability of the floating structure can be evaluated by figure 5[1]. According to [1] dynamic stability will occur if the area area (BCD + OBDE)> 1.4 x area (OBDE + OAB) or righting moment is greater than 1.4 times heeling moment.

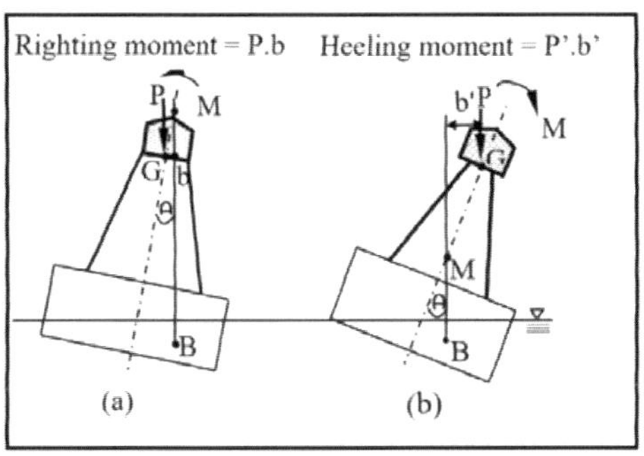

Figure 4. Floating structure stability, (a) stable, (b) unstable

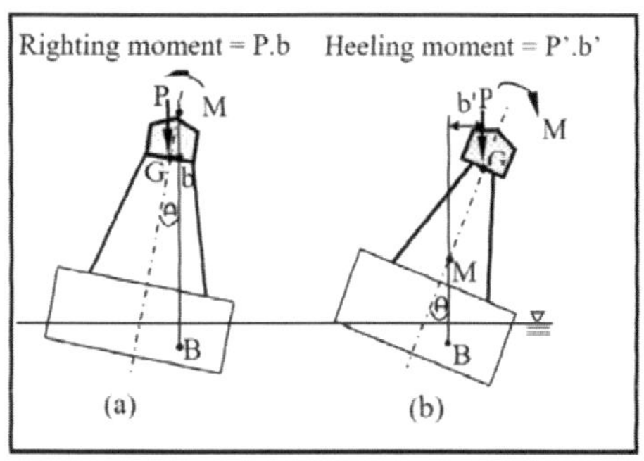

Figure 5. Dynamic stability evaluation [1]

IOP Conf. Series: Materials Science and Engineering **797** (2020) 012021 doi:10.1088/1757-899X/797/1/012021

5.2 Deflection and displacement

Floating bridge is very sensitive to movement. Floating bridges should be designed to meet the criteria of deflection and movement boundaries to provide the comfort and safety of bridge users. The bridge must be safe under normal conditions (1-year storm) and strong enough/not to collapse under extreme conditions (100-year storm).

Table 3. Tabel Deflextion and motion criteria governing the design (1-year storm) [7] and [8])

Loading condition	Type of deflection or motion	Maximum deflection	Maximum motion
Trafic load	vertical	L/800	
	Rotation (roll)		$\pm 1,0^{o}$
Environmental load	Vertical (heave)	$\pm 0,3$ m	0,5 m/s^2
	Lateral (sway)	$\pm 0,3$ m	0,5 m/s^2
	Rotation (roll)	$\pm 1,5^{o}$	0,05 rad/s^2

5.3 Dynamic movement respon of floating foundation

The floating structure movements consist of six degrees of freedom, three translational movements and three rotational movements. The determinant factor in the analysis of floating structural movement is period, damping and stiffness. The general equation of the floating structure movement is expressed by;

$$[M + A]\ddot{X} + [B_1 + B_2]\dot{X} + [C + c]X = F(t) \quad (2)$$

where

$X.\dot{X}\ddot{X}$ are generalized displacement, velocity and acceleration at each degree of freedom, M is mass–inertia matrix of the structure, B_1 is structural damping coefficient, B_2 is hydrodynamic damping coefficient, C is structural stiffness matrix c is hydrostatic stiffness (hydrostatic restoring forces), F(t) is forces acting on the structure.

The structural damping is the damping related to the internal friction in the material of the structure. In the analysis, the damping structural can be expressed by classical Rayleigh proportional to a coupling between the velocity, and the strain velocity at each node-point.

Hydrodynamical damping is due to the waters interaction with the submerged pontoons, which is combined from geometrical damping and drag damping. The geometrical damping is also called wave or diffraction damping is caused by structural vibrations that generate waves in the fluid around the structure. The drag damping or viscous damping is due to viscous effects and vortex shedding in the water around the structure.

According to [9], [10] there are two analytical methods used in the dynamic analysis of floating structures:

1. Frequency domain analysis (Probobilistic analysis)
2. Time domain analysis (Deterministic analysis)

Frequency domain analysis is a simulated event at a certain time with a predetermined frequency interval. Frequency domain can be used to estimate the random wave response including movement and acceleration of floating structure. The advantage of this method is more saving time calculation. Disadvantages in this method ie for transient response and nonlinear response require non-linear equations to be changed in linear or nonliner form eigenmodes the integration over a wide range of frequencies [11] and [12].

With the Fourier transform, the equations of motion can be expressed in the frequency termsω;

$$\{-\omega^2(M + A) + i\omega(B_1 + B_2) + (C + c)\} = \{F(\omega)\} \quad (3)$$

For systems without attenuation the above equations can be simplified in the eigenvalue problem.

The floating structure response can be analyzed using the Operator Amplitude Response (RAO) function which converts the external load in the form of a wave in response to a structure. The relationship between RAO and spectral structural response for unit amplitude units is;

$$S_u(\omega) = \left(RAO(\omega)\right)^2 S(\omega) \tag{4}$$

where

$$RAO = \frac{F(\omega)}{[-\omega^2(M+A) + i\omega(B_1+B_2) + (C+c)]} \tag{5}$$

Time domain analysis is the completion of dynamic movement of structures based on time function. The approach adopted in this method uses time integration procedure and will result in time history response based on time function x (t). Quite accurate to analyze wind sensitive bridges and nonlinear behavior such as long span bridges [13] and [14]. The dominant time gain is that all non-liner models (system matrix and external load) can be modeled correctly but take longer time.

6. Modeling floating foundation

The finite element model of a floating foundation consists of blocks, springs, additional masses and damping. Ponton is modeled as a very rigid fictitious beam. Linear or non linear springs are used to model the horizontal and vertical stiffness of the mooring or the tether. Hydrodynamic Properties consist of hydrodynamic added damping and hydrodynamic added mass. Hydrodynamic added damping can be liner or nonliner.

7. Design procedures long spans briges with floating foundation.

Procedure design of long span floating foundation can be described as the flowchart in Figure 6.

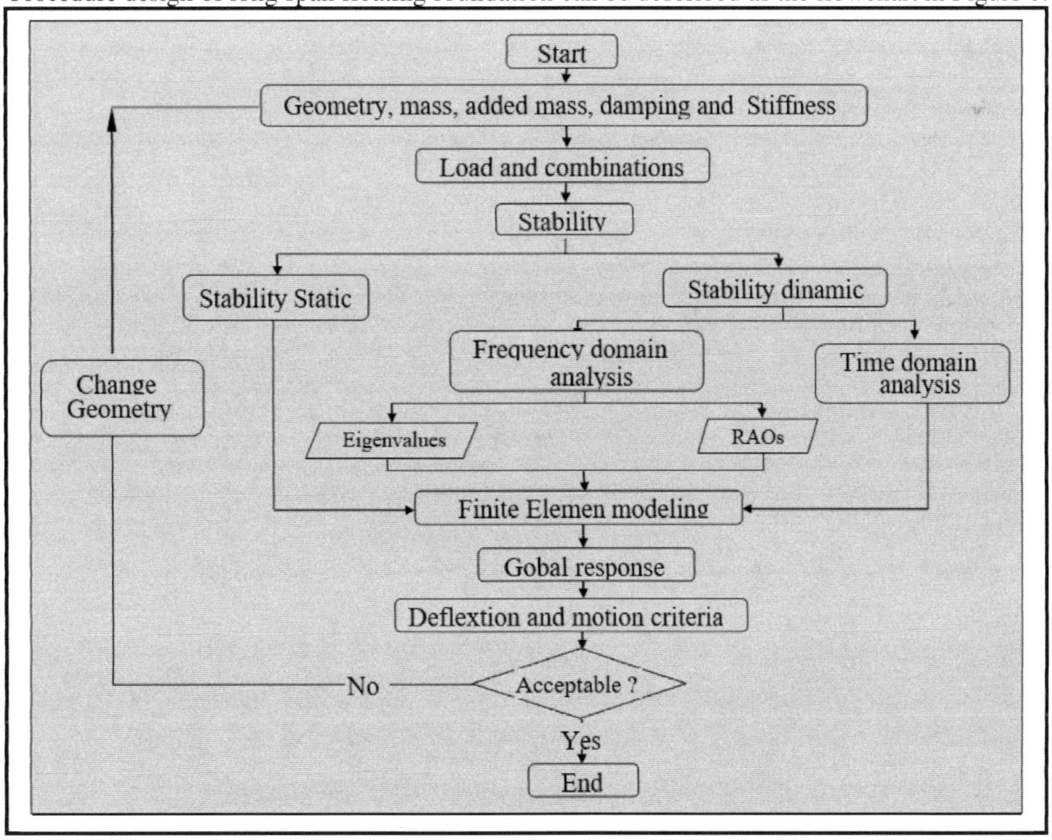

Figure 6. Steps and procedural of design the long span floating bridges

IOP Conf. Series: Materials Science and Engineering **797** (2020) 012021 doi:10.1088/1757-899X/797/1/012021

8. Discussion

Floating foundation and mooring system must be designed to be able to carry the load and its combination without losing its stability and without experiencing shaking and displacement beyond the tolerable limits. Modeling floating foundation structures are carried out to simulate and evaluate the global behavior of structures.

References

[1] Wang C M, Wang B T 2015 *Large Floating Structures* Ocean Engineering & Oceanography, Springer

[2] International Organization for Standards 2006 *Floating offshore structures—monohulls, semisubmersibles and spars*, ISO 19904-1.

[3] NORSOK Standard C-001 2012 *Integrity of Offshore Structures* Edittion 8

[4] Chandrasekaran S 2015 *Dynamic Analysis and Design of Offshore Structures* Ocean Engineering & Oceanography, Vol. 9, Springer, India.

[5] Chakrabarti S. K 1987 *Hydrodynamics of offshore structures* WIT Press

[6] Dagli B Y , Yigit M E, and Gokkus U 2017 *Behaviour of Large Cylindrical Offshore Structures Subjected to Wave Loads* TEM Journal Vol 6 p.550-557.

[7] Aas Jacobsen 2016 *Curved Bridge – Navigation Channel in South* Cowi As Oslo

[8] L win MM 2000 *Floating Bridges* Bridge Engineering Handbook. Ed. Wai-Fah Chen and Lian Duan Boca Raton: CRC Press

[9] API 2010 *Planning, Designing, and Constructing Tension Leg Platform* API Recommended Practice 2 T, 3rd edition.

[10] Det Norske Veritas 2017 *Global performance analysis of deep water floating Structures* DNVGL-RP-F205

[11] Watanabe E, Utsunomiya T, Wang C 2004 *Hydroelastic analysis of pontoon-type VLFS: a literature survey* Eng Struct. 26:245– 256.

[12] Salvatori L, Borri C 2007 *Frequency-and time-domain methods for the numerical modeling of full-bridge aeroelasticity* Comput Struct. 85:675–687.

[13] Santos J, Miyata T, Yamada H 1993 *Gust response of a long span bridge by the time domain approach* Proceedings of the Third Asia-Pacific Symposium on Wind Engineering, Hong Kong.

[14] Cao Y, Xiang H, Zhou Y 2000 *Simulation of stochastic wind velocity field on long-span bridges* J Eng Mech. 126:1–6.

IOP Conf. Series: Materials Science and Engineering **797** (2020) 012022 doi:10.1088/1757-899X/797/1/012022

Effect of Compaction Pressure on Quality of Activated Charcoal Briquette Made from Sago Stem Midrib Material

Lina Lestari[1], V I Variani[1], Muh. Zamrun Firihu[1], S Raharjo[2], Ismail Saleh[1], Nining Aprilla[1]

[1)]Department of Physics, Faculty of Mathematics and Natural Sciences, Universitas Haluoleo, 93232 Kendari, Sulawesi Tenggara, INDONESIA

[2)]Department of Chemistry, Faculty of Mathematics and Natural Sciences, Universitas Haluoleo, 93232 Kendari, Sulawesi Tenggara, INDONESIA

*Corresponding author's e-mail: linalestari68@yahoo.com

Abstract. Fabrication and characterization of activated charcoal briquette made from sago stem midrib have been performed. The sago stem midrib firstly dried before carbonized using kiln drum method. Subsequently, the sago stem midrib charcoal was activated using electric furnace with 650oC of temperature for 5 minutes. Then, the activated briquette mashed and sieved by 80-mesh strainer before moulded into a compaction pressure. The pressure that applied were 34.66 kg/cm^2, 69,32 kg/cm^2 and 103,98 kg/cm^2. Ignition time, burning rate, and burning test were performed to charaterize the briquette quality. The result shown that the higher the pressure applied, the lower ignition time, burning rate and the longer of burning time of sago stem midrib briquette, compared to the unactivated sago stem midrib briquette.

1. Introduction

By the increasing of human population in the world, necessity of energy also increase. The main energy source for nowadays is taken from the fossil energy, such as oil. Because of its non-renewable characteristic and highly utilization, the oil reserves are getting runs out. If there are no new oil reserve that could be found, especially in Indonesia, it is estimated that the oil will completely disappear in 15 years latter [1]. An alternative energy source that could be developed to substitute the oil energy is biomass [2-4]. Biomass popularly used as alternative fuel [5-6]. Another form of biomass energy is briquette that is produced from organic materials. There are some advantages of biomass briquetting such as no sulfur and fly ash in its burning proccess, easy to ignite and could solve the residual disposal problem [7]. There were many briquettes that have been produced from organic materials, such as cocoa shell [8-10], palm shell [11], durian peel [12], banana waste [13], rice husk and corn cob [14].

Sulawesi Tenggara province is one of the biggest sago producing regions in Indonesia, besides Papua as the biggest producing regions in Indonesia [15]. Sago flour is produced from sago stem, while the sago stem midrib will become unuseful materials. By the abundant availability of sago trees, the existence of sago stem midrib waste is also huge and could affect the environment. By the reason, in this research, sago stem midrib is used as main material to produce the activated charcoal briquette and the characteristic of the briquette, such as igniton time, burning rate and burning test, will be studied by varying the compaction pressure on its fabricating proccess.

2. Material and methods

Material used to fabricated the briquette is sago stem midrib that is taken from Konda region, Konawe Selatan regency, Sulawesi Tenggara province. Before carbonated, the sago stem midrib wastes were dried firstly. Then, the sago stem midrib was carbonized using kiln drum method [8]. The sago stem midrib charcoal from carbonization process was sieved using 80-mesh strainer to yield charcoal powder. The charcoal powder is then activated using electric furnace with 650°C of temperature for 5 minutes [8]. Before adding the adhesive agent, the charcoal was cooled using desicator. Sago powder is used as the adhesive agent with mass ratio 9:1 [8]. After mixing with the adhesive agent, the charcoal powder was compressed with different compaction pressure: 34.66 kg/cm^2, 69,32 kg/cm^2 and 103,98 kg/cm^2 [8]. The characterization of the briquttes include the ignition time, burning rate and burning test. Temperature is measured using infrared thermometer. Briquette fabrication and characterization are shown in Figure 1.

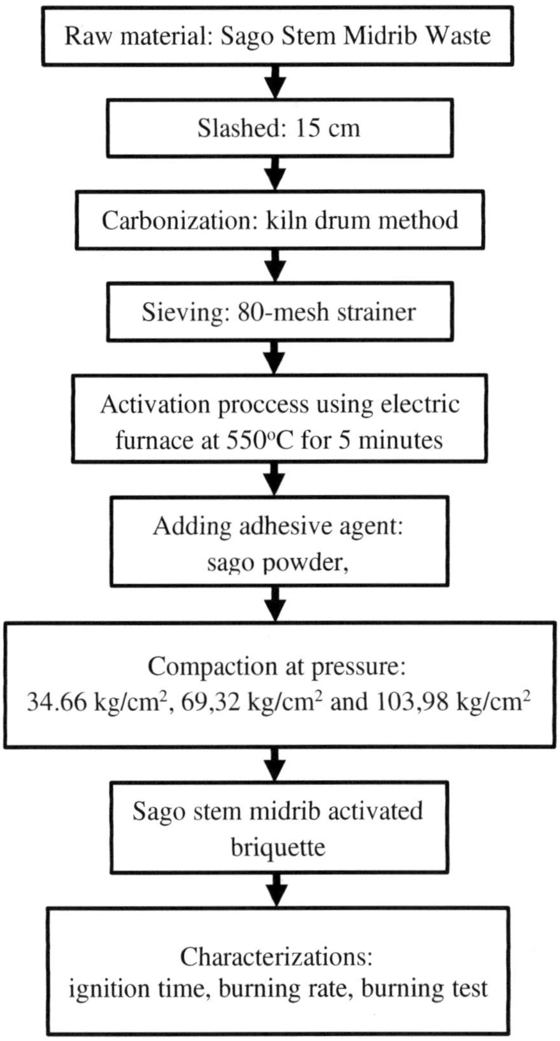

Figure 1. Flow chart of sago stem midrib briquette fabrication and characterization proccess

3 Result and discussion

3.1 Ignition time

Ignition time represents the time that is needed to burns the briquette. Ignition time was observed when the briquette starts to burn. Figure 2 shown the the ignition time of sago stem midrib briquette with different compaction pressure. The blue line shows the briquette that was not activated, while the red one represents the activated briquette. It can be shown that the higher pressure that applied, the shorter of ignition time, both for the unactivated and activated briquette. Applied high pressure causes the decreasement of porosity so the ignition time become faster [8]. The shortest ignition time is reached on the activated briquette condition, which is 0.093 minutes of time or 5.58 seconds on 103,98 kg/cm^2 of pressure.

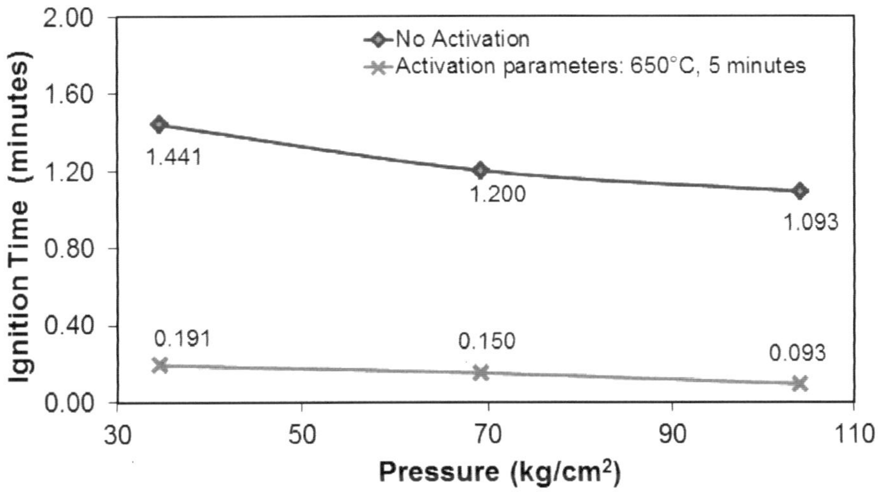

Figure 2. Ignition time of sago stem midrib briquette with different compaction pressure

3.2 Burning rate

Burning rate represents the time that is needed to consumes 1 gram of briquette when it starts to blame until become ash. Figure 3 shown the burning rate of sago stem midrib briquette with different compaction pressure. The blue line shown the unactivated condition, while the red one shown the activated briquette with 650°C of temperature for 5 minutes. It can be shown that the increase of compacting pressure causes the burning rate decrease. It can be caused by the decrease of porosity inside the briquette at higher compaction pressure [16]. Decrease of porosity caused the reduce of air cavities inside the briquette. Mallika et al [16] reports that the reduce of air cavities inside the briquette will curb the mass and heat transfer in combustion process and consequently the lower compaction pressure will produce briquette with higher burning rate .

IOP Conf. Series: Materials Science and Engineering **797** (2020) 012022 doi:10.1088/1757-899X/797/1/012022

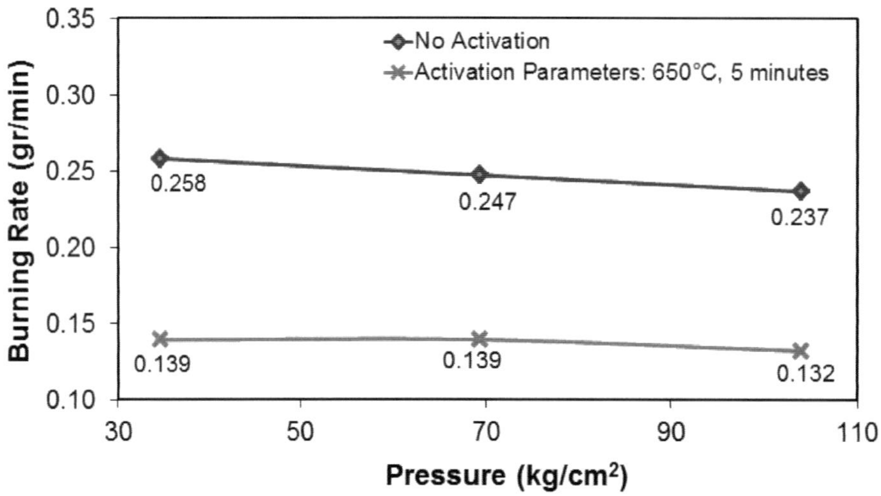

Figure 3. Burning rate of sago stem midrib briquette with different compaction pressure

3.3 Burning test

Burning test was held to determine the maximum temperature in briquette burning proccess. The temperature measurement was done in every 1 minute using infrared thermometer. Figure 4 shown the burning test for the sago stem midrib briquette without activation, while the Figure 5 shown the burning test result for the activated sago stem midrib briquette. It can be seen that the graph in Figure 4 is narrower than the the activated briquette graph. This means that the activated briquette spends more times to burning than the unactivated briquette. The unactivated briquette stops to burn in 17-22 of minutes, while the activated briquette stops to burn in 35-38 of minutes.

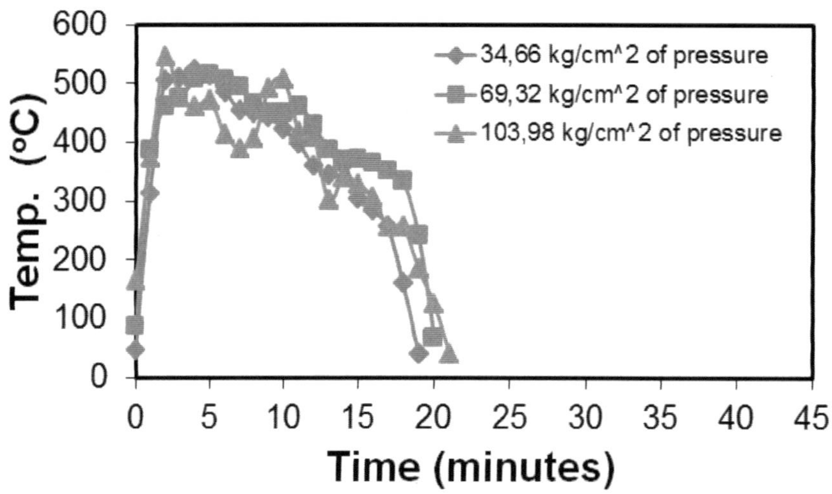

Figure 4. Burning test of unactivated sago stem midrib briquette with different compaction pressure

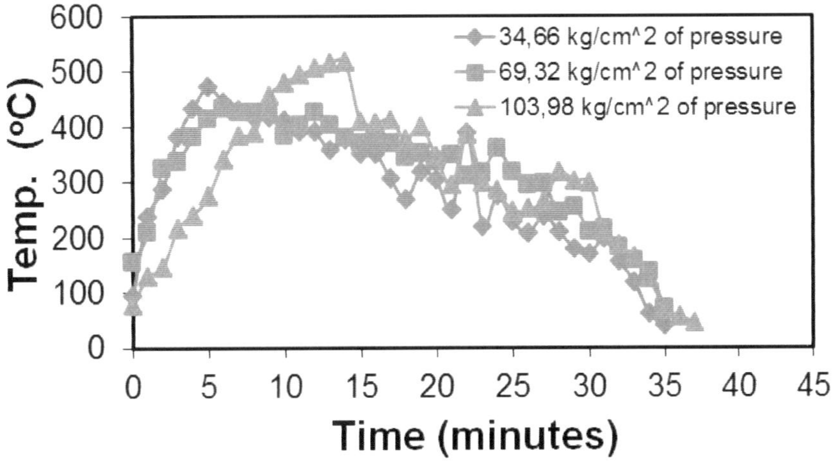

Figure 5. Burning test of activated sago stem midrib briquette with different compaction pressure

4. Conclusion

Fabrication of activated charcoal briquette made from sago stem midrib has been successfully done. After carbonizing the sago stem midrib, the charcoal is then mashed and sieved using 80-mesh strainer. Then the charcoal grain is activated using the electric furnace with 650°C of temperature for 5 minutes. Sago powder is mixed to the activated charcoal to adhere the material. Subsequently, the mixed material was compacted with different compaction pressure: 34.66 kg/cm^2, 69,32 kg/cm^2 and 103,98 kg/cm^2. The result shown that, the higher compaction pressure, the shorter the igniton time, the lower burning rate, and the longer the burning test, compared with the unactivated sago stem midrib briquette.

References

[1] *ekonomi.kompas.com*, accessed on August 31th, 2018
[2] Geoffrey P H and Ross V M M 2018 *Applied Energy* **224** 136-146
[3] Shruti A B and P B Kalburgi 2016 *Procedia Environmental Sciences* **35** 555-562
[4] Perez-Sarinana B Y, Saldana-Trinidad S, S E L Fernando, P J Sebastian, D Eapen 2014 *Energy Procedia* **57** 950-956
[5] Kusmiyati and Heru S 2015 *Procedia Environmental Sciences* **23** 199-206
[6] S Suryaningsih, O Nurhilal, Y Yullah, C Mulyana 2017 *IOP Conf. Series: Earth and Environmental Sciences* **65** doi:10.1088/1755-1315/65/1/012012
[7] Shaikh A A, Anita K, and Ketan M 2014 *International Journal of Inovations in Engineering and Technology* (IJIET) **3** 4
[8] Lina L, Erzam S H and Risna 2017 *Jurnal Aplikasi Fisika* **13** 2 [In Indonesian]
[9] N Syamsiro and H Saptoadi 2007 *Seminar Nasional Teknologi*, ISSN: 1978-9777 [In Indonesian]
[10] I Nyoman S, Lina L, Muhammad Z F, Yuni A K, Gusti E S, Yustin B, L Arfad, Purwo A S, and Endang S 2017 *Jurnal Aplikasi Fisika* **13** 1 [In Indonesian]
[11] Djoko P, Retno U P and Santy D S 2015 *Journal Research of Forest Product Industry* **7** 2 [In Indonesian]
[12] Wahidin N, Nurfa A and Martana 2014 *Energy Procedia* **47** 295-302

IOP Conf. Series: Materials Science and Engineering **797** (2020) 012022 doi:10.1088/1757-899X/797/1/012022

[13] Erna Rusliana M. Saleh 2010 *Seminar Rekayasa Kimia dan Proses* ISSN: 1411-4216 [In Indonesian]

[14] Arry Y N, Yuda C H, and W Hasanah 2016 *Agriculture and Agricultural Science Procedia* **9** 386-395

[15] *www.viva.co.id/berita/bisnis*, accessed on August 31th, 2018

[16] Mallika T, Thancanok P, Kasidet P, Pisit M, and Prasong W 2015 *Energy Procedia* **79** 890-895

Design of Prototype System for Monitoring Air Quality for Smart City Implementation

Yuni Aryani Koedoes[1], Samuel Jie[1], Muh. Nadzirin Anshari Nur[1], Bunyamin[1], Alwan Astari[1]
[1]Department of Electrical, Faculty of Engineering, Universitas Haluoleo, 93232 Kendari, Sulawesi Tenggara, INDONESIA

*Corresponding author's e-mail: yuniarafkendari@gmail.com

Abstract. Air pollution can be caused by natural sources and human activity. Now many say the polluted air around us. Carbon monoxide (CO) and carbon dioxide (CO2) is a parameter for determining the quality of clean air and polluted. If levels of these two substances excessive inhaled by humans, can be detrimental to health. By using the Arduino Mega 2560, air quality monitoring system in this study using LM35 sensor to determine the temperature of the air, MQ-7 sensor to determine the levels of CO and sensor MQ-135 to determine the levels of CO2 that is in the room. Sensor MQ-7 will be equipped with red and green LED indicator as the level of air quality is polluted or not polluted. If the levels of CO < 5 ppm, the green LED turns on and vice versa CO > 5 ppm, the red LED lights up. Likewise, CO2 levels <350 ppm, the green LED turns on and vice versa CO> 350 ppm, the red LED lights up. Research and simulation results show that the system works well. In the application of smart city, one of the important parameters is the monitoring system, in this research will be designed air quality monitoring system for the application of smart city.

1. Introduction

High levels of air pollution through a good smoke pollution from vehicles, factories as well as the impact of forest fires has increased especially in the big cities, along with the problem yet so real efforts to prevention through early warning against air pollution. Air pollution can be caused by both natural sources originated from human activities. Type-the type of air pollutants is based on the raw quality of air inhaled daily, according to the Government Regulation Number 41 year 1999, which include: sulfur dioxide (SO2), carbon monoxide (CO), nitrogen dioxide (NO2), oxidants (O3), carbon and hydrogen. (HC), PM 10, PM 2.5, TSP (dust), Pb (Lead), dustfall [1]

Health effects from air pollution can get into the body through the respiratory system. Away the penetration of substances polluters into the body depends on the type of contaminant. Larger particulates can be stuck in the upper respiratory tract, while small-sized particulate and gas can reach the lungs. From the lungs, polluters are absorbed by circulatory system health effects the most common are RESPIRATORY (acute respiratory tract infections), including, asthma, bronchitis, and other breathing disorders. Impact on crop plants that are grown in areas with high levels of air pollution can be disrupted its growth and prone to disease, such as chlorosis, necrosis, black spots[2].

Carbon monoxide (CO) and carbon dioxide (CO2) is the parameter to determine the air quality clean and polluted. If the levels of these substances are excessively inhaled 2 by humans then can impair health. Carbon monoxide (CO) is produced by the combustion process is not perfect from the carbon-based material, such as wood, coal, fuel oil and other organic matter carbon dioxide resulting from the combustion process for example burning plastic, paper and also smoke cigarettes as well as on the larger case is forest fires[2]

Ambient air quality of a region is determined by the power the area's nature as well as support a number of sources of pollution or contamination loads from existing resources in the area. In some developed countries have implemented system monitoring the air, such as South korea and Taiwan use applications with IoT collect data environment that is massive as well as identified a number of particles/hazardous substances [3] Substances emitted by the source of the contaminant into the air and could affect air quality among other things the gas Nitrogen oxides (NOx), Sulfur dioxide (SO2), dust as well as Lead (Pb) in the dust.

Table 1. Clean air and Polluted Air according to WHO (Mukono H.J, 2005)[6]

Parameters	Clean Air	Polluted Air
Particle Materials	0,01 – 0,02 mg/m3	0,07 – 0,7 mg/m3
SO_2	0,003 – 0,02 ppm	0,02 – 2 ppm
CO	< 1 ppm	5 – 200 ppm
NO_2	0,003 – 0,02 ppm	0,02 – 0,1 ppm
CO_2	310 – 330 ppm	350 – 700 ppm
Hydrocarbons	< 1 ppm	1 – 20 ppm

Often occurs in the middle of the community particularly the area vulnerable to air pollution especially in the big cities will implement smart city may not necessarily be directly perceived society takes long enough that the air in the polluted, it These can be detected when many societies have been hit by the effects of the pollution for example respiratory disease, the limited views and more. With the conditions and problems of the air quality monitoring system is needed to anticipate the impacts of air pollution, on the concept of smart city one of the aspects that can be applied is system monitoring, by it that researchers are designing a prototype system for monitoring air quality for the application of smart city.

2. Methods and Literature

Until recently, research and development of monitoring system for air quality continues to be done, such as research conducted with the title "tool Monitors air pollution From carbon monoxide Gas (CO) at Room AT89S51 Microcontroller-based". The study designed a system of monitoring the levels of pollutant gases using gas sensor TGS 2442 are sensitive to carbon monoxide gas and transcription results will show up in the LCD the difference with the research that the researcher will be minutely present on the sensors are used i.e. only using the TGS 2442 whereas researchers MQ-7 and not just the levels of CO are measured but CO2 levels and temperature of the room [4].

Other research is system monitor air quality with the title "the making of Gauges the level of air pollution using AT89S51 Microcontroller-based sensor TGS 2600" research consists of a microcontroller circuit sensor TGS AT89S51, 2600, Op-Amp and LCDS as appearance. This tool works when detecting the levels of carbon monoxide gas is passed through the sensor, its output voltage to be converted by the Op-Amp to be processed into the microcontroller and LCD in order to make it easier to display readings [5].

Based on some research that has been done before, then on the research will be carried out development to add some sensors for more versatility. Development is done using the on-board Arduino Mega 2560 to simplify the system. This is possible because the Arduino Mega board has been integrated with an ATmega168 microcontroller and there have been pin-a pin that contains all required by the microcontroller. In addition, the Arduino Mega 2560 also have special applications with programming language has been simplified, making it easier to do the development of the system.

3. Result and Discussion

3.1 System Design

The System Monitors the air quality automatic monitoring system is composed using the LM35, MQ-7 as well as MQ-135 and 4x20 LCD display by utilizing the Arduino Mega 2560, sensors and LCD 4x20. Block diagram of the system listed in Figure 1.

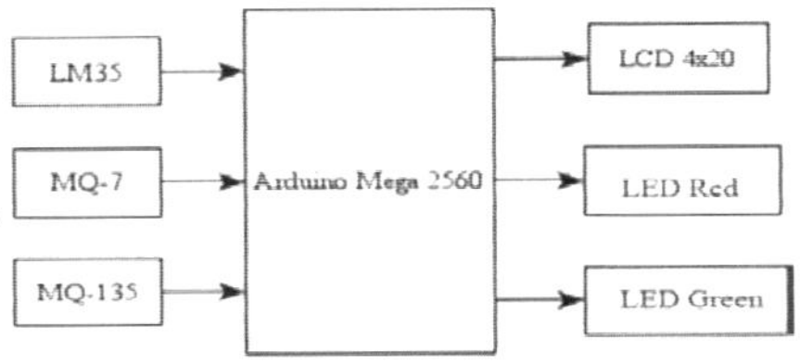

Figure 1. Block Diagram of the system of air quality Monitoring

Figure 1 describes an overview of the design of the air quality information systems are designed. The system consists of several parts which are connected in such a way so as to form a single unity of the systematic work function. The main part of the system is the Arduino Mega 2560 that serves as the central control system. Input on the Arduino consists of LM35 temperature sensor probes as the MQ-7 as the quality of the monitor co., and sensor MQ-135 as the quality monitoring of CO2 in the room.

The output of the Arduino Mega i.e. LCD 4x20. LCD 4x20 function to display information the room temperature and air quality levels of CO and CO2 in the form of a value from the sensor reading of results. LCD 4x20 is part of the output of the Arduino which is automatically read and the monitoring of indoor air quality. The ability of the system to monitor air quality is highly dependent on the performance of the sensor. The function of the automatic control system that is owned by the air quality monitoring function which allows we can monitor air quality with air so that we feel safe and good if bad can be avoided and anticipated.

3.2 System Implementation

In this section will be discussed regarding the design of a series of port, results of implementation tools, testing and analysis of the test results. The purpose of the test is done to find out the extent to which performance results of design that has been discussed, as well as to know the success rate of each specification system. Testing conducted include testing against the sensor-sensor is used, LEDs as air quality information, and display on the LCD.

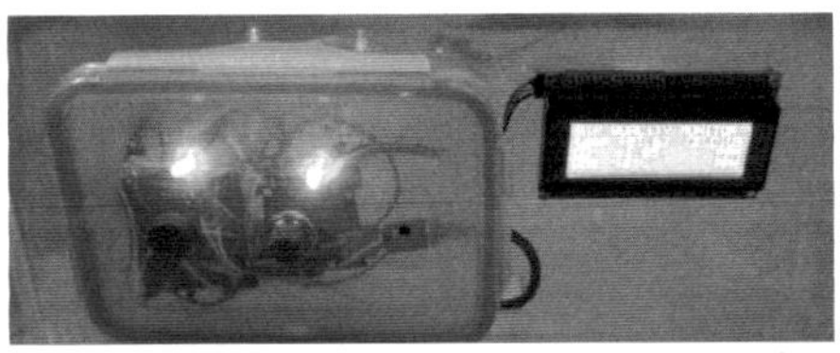

Figure 2. Implementation of Air Quality Monitoring System

3.3 Design of the circuit at the Port the Arduino Mega 2560

The design of the overall system series on the port the Arduino Mega 2560. In this system there are several port circuit subsystem, ranging from sensor-sensor is used, on display on the LED and LCD 4x20.

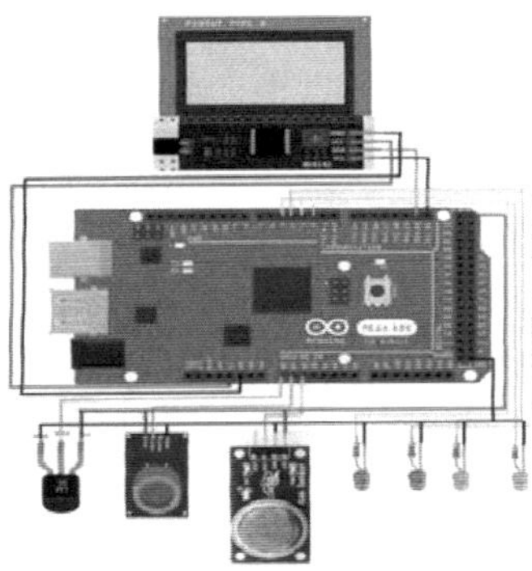

Figure 3. Designing the overall system series on the port the Arduino Mega 2560

IOP Conf. Series: Materials Science and Engineering **797** (2020) 012023 doi:10.1088/1757-899X/797/1/012023

3.4 Designing on Sensor LM35, MQ and MQ-135

This is a design of the sensor LM35 series, MQ and MQ-7-135 on the Arduino Mega 2560 and equipped its configuration table.

Table 2. Configuration pin sensor LM35, MQ-7, MQ-135 with the Arduino Mega 2560

Pin configuration of the sensor with Arduino Mega 2560		
Sensor	Pin Sensor	Pin Arduino Mega 2560
LM35	Vss	5 V
	Vout	Pin A0 (Analog Input)
	GND	GND
MQ-7	Vcc	5 V
	Aout	Pin A1 (Analog Input)
	Dout	-
	GND	GND
MQ-135	Vcc	5 V
	Aout	Pin A2 (Analog Input)
	Dout	-
	GND	GND

3.5 Design on the LED

Here is the design of a series of LED as an indicator of air quality polluted and unpolluted on Arduino Mega 2560 and equipped its configuration table.

Table 3 LED Configuration with the Arduino Mega 2560

LED configuration with the Arduino Mega 2560		
LED	**LED**	**Pin Arduino Mega 2560**
LED 1 (Green)	Positif - Resistor	Pin 2 (PWM)
	Negatif	GND
LED 2 (Red)	Positif - Resistor	Pin 3 (PWM)
	Negatif	GND
LED (Green)	Positif - Resistor	Pin 4 (PWM)
	Negatif	GND
LED 2 (Red)	Positif - Resistor	Pin 5 (PWM)
	Negatif	GND

3.6 Design on the LCD 4x20

This is the design of a series of LED as an indicator of air quality polluted and unpolluted on Arduino Mega 2560 and equipped its configuration table.

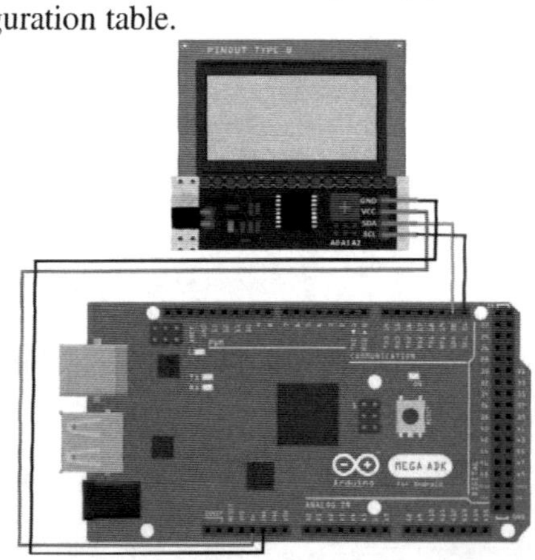

Figure 4. system arrangements of LCD Design 4x20 on port the Arduino Mega 2560

IOP Conf. Series: Materials Science and Engineering **797** (2020) 012023 doi:10.1088/1757-899X/797/1/012023

4. System Testing

4.1 Testing system at Normal air

On testing was performed against the normal air, not polluted by smoke.

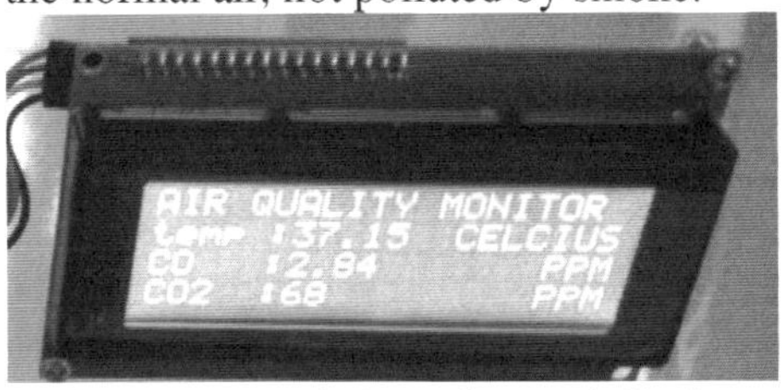

Figure 5. Testing on normal air testing and display on the LCD

From the results of testing the normal air, it appears that the system is running well according to the instructions.

4.2 System testing using MQ-7 sensor

In this test, researchers will simulate indoor air quality with exhaust emissions on the motor. Exhaust emissions on the motor are carbon monoxide substances whose levels can be detected by the MQ-7 sensor.

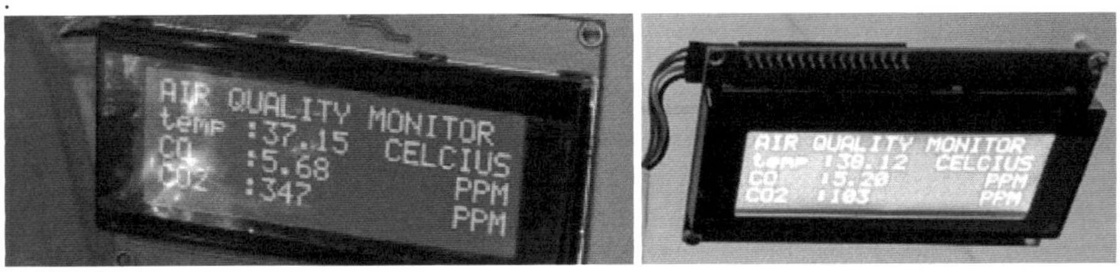

Figure 6. Testing of MQ-7 sensors on exhaust emissions and wood burning smoke

From the test results above it can be seen that CO levels in exhaust emissions are 5.68 ppm and the indicator LED on the MQ-7 sensor is red and indicates that the air quality is contaminated. In this test, researchers will simulate the quality of air in a room with carbon. Wood is one of the carbon substances. Smoke from burning wood is a carbon monoxide substance whose levels can be detected by the MQ-7 sensor. From the test results, it can be seen that CO levels in wood burning smoke are 5.20 ppm and the indicator LED on the MQ-7 sensor is red and indicates that the air quality is polluted.

4.3 System testing using MQ-135 sensor

In this test, researchers will simulate indoor air quality with cigarette smoke. Cigarette smoke is one of the carbon dioxide substances whose levels can be detected by the MQ-135 sensor.

Figure 7. Testing of MQ-135 sensors in cigarette smoke and paper burning smoke

From the test results above it can be seen that the CO2 level in cigarette smoke smoke is 235 ppm and the indicator LED on the MQ-135 sensor is green and indicates that the air quality is not polluted. In this test, researchers will simulate indoor air quality with paper burning smoke. Paper burning smoke is one of the carbon dioxide substances whose levels can be detected by the MQ-135 sensor,

From the test results, it can be seen that the CO2 level in the paper burning smoke is 206 ppm and the indicator LED on the MQ-135 sensor is green and indicates that the air quality is not polluted.

4.4 Testing the system using an LM35 sensor

4.4.1. Testing 1

In this test, researchers will simulate indoor air temperature with fire as a heat temperature that can be detected by an LM35 sensor. By simulating with fire the researcher will know the sensor works.

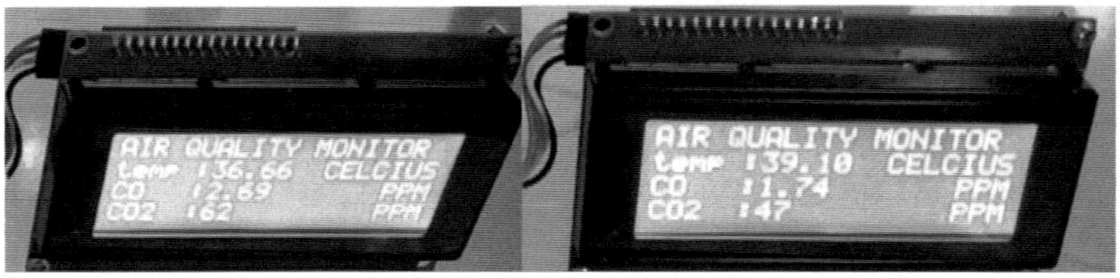

Figure 8. Testing of LM35 sensors after being tested with fire and display on the LCD

From the simulation results, it can be seen that after the LM35 sensor after being simulated with fire, the temperature on the LCD increases from 36.66℃ to 39.10℃. This indicates that the LM35 sensor works because the temperature has increased.

5. Conclusion

From the results of testing and analysis system of the author's conclusion is then Informed air quality can be monitored directly by human beings through LED 4x20 LCD screen and as the charge indicators will, humans can know the levels of CO, CO2 and temperature indoors so can benefit additionally Information level air quality polluted and unpolluted knowable through the red and green indicator LEDS, from the results of the simulation levels of CO exhaust emissions amounted to 5.68 ppm and smoke wood-burning of 5.20 ppm, so that indicates that both the contaminated air, CO2 levels of simulated results with cigarette smoke and paper gets results that can function properly, from the results of testing of prototype can run well and for implementation at smart city needed further development especially on the device online as Internet of Things, and from the results of this research, the application of this prototype more appropriate in use at home with the smart home concept is applied to the system before the smart city

References

[1] Parameter Pencemar Udara dan Dampaknya Terhadap Kesehatan, http://www.depkes.go.id/downloads/Udara.pdf, access date 11 Maret 2018.

[2] Tomie Hermawan Soekamto, David Perdanakusuma. Intoksikasi Karbon Monoksida, Jurnal, SMF Ilmu Beda Plastik Universitas Airlangga, access date 28 Agustus 2018.

[3] Monitor Kualitas Udara dengan IoThttp://www.smartcityindo.com/2018/01/smart-city-monitor-kualitas-udara.html access date, 29 september 2018

[4] Farli Rizki. Alat Pemantau Polusi Udara Dari Gas Karbon monoksida (CO) pada Ruangan Berbasis Mikrokontroler AT89S51 [Skripsi]. UPN Jatim. 2011.

[5] Vadlya Maarif, Nuzul Iman Fadlilah. Pembuatan Alat Pengukur Tingkat Polusi Udara Berbasis Mikrokontroler AT89S51 menggunakan sensor TGS 2600 [Seminar Nasional ke-9]. Amik Bina Sara Informatika Purwokerto.

[6] Buletin Who Mukono, 2005.

[7] Arduino. Introduction To Arduino {Online}. Italy: Arduino https://www.Arduino.Cc/En/Main/Arduinoboardmega2560, access date 29 Juli 2016

[8] Roberts, M. Mc, 2009, Arduino Started Kit Manual, Earthshine Electronics.

[9] Nur M. Alat Ukur Suhu Otomatis Berbasis Mikrokontroler ATMega 8535 dengan Penampil LCD [Tugas Akhir]. Medan: Universitas Sumatera Utara; 2012.

IOP Conf. Series: Materials Science and Engineering **797** (2020) 012024 doi:10.1088/1757-899X/797/1/012024

Shear Strength Analysis of Reduced Beam Section (RBS) on Castellated Beam

Nini Hasriyani Aswad[1]* Tachrir[2], Herman Parung[3], A. Arwin Amiruddin[3]
[1]Department of Civil Engineering, Universitas Haluoleo, 93232 Kendari, Sulawesi Tenggara, INDONESIA
[2]Department of Electrical Engineering, Universitas Haluoleo, 93232 Kendari, Sulawesi Tenggara, INDONESIA
[3]Department of Civil Engineering, Universitas Hasanuddin, Jl. Poros Malino km.6, Kampus Teknik Gowa, Kab. Gowa, Sulawesi Selatan, INDONESIA

*Corresponding author's e-mail: niniaswad@gmail.com

Abstract. Shear strength analysis of the reduced beam section (RBS) on the castellated beam with exterior connection aimed to analyze the structural elements to determine the shear force acting due to cyclic loading. The study was developed from the theory of moments due to shear forces based on the American Institute of Steel's Prequalified Connection Construction for Special and Intermediate Moment Frames for Seismic Applications (AISC 358-10). This study used an experiment of RBS on a castellated beam with an exterior connection model subjected to cyclic loading. The result shows that the shear strength occurred after the bending failure of the beam.

1. Introduction

Seismic design for steel moment frame is aimed to make structure elastic during small to medium seismic events. In large earthquakes, the structure must be strong and secure from collapse. Investigation of the 1994 Northridge earthquake and the earthquake in Kobe show that the damage occurred to connections to the frame structure during the earthquake. [1] reported that the crack model on the Northridge Earthquake began at the point of intersection between the weld access hole and the beam bottom flange. In Kobe Earthquake, [2] described the fractures occurred on metal welding, heat-affected zones, base metals, and diaphragm plates.

Many experiments and analysis were done to improve the performance of the connection. One popular method is the reduced beam section (RBS). In connection with the RBS method, the beam flange is trimmed to ensure the flexural mechanism strategy in the desired RBS on the beam. Meanwhile, [3] explains that beam with RBS experienced failure caused by bending on the beam. In the study, the top of the beam had a maximum tensile while the bottom of them had compressive stress. Moreover, the shear stress on the beam web was larger than on the beam flange.

[4] examined the performance of frame beam connection with RBS against fires by numerical methods. The results show that the strength of axial forces and deflections was generated in beam connections with the RBS method. The first yield in the RBS region was well exposed to fires locally and global, which shows the bending failure on the beam.

In a study conducted by [5], the connection shows a reliable ductility to maintain the desired inelastic deformation with no brittle fracture. In the test, the three specimens showed damage to the beam flanges while the RBS area did not have faults on the column-beam welded connection. Thus, the behavior of the castellated beam without RBS with an opening distance of 15 cm from the column face is strong enough as it can deform and withstand a load of 12.5 kN. Meanwhile, with a load of 10.9 kN the power decreased by 12.8%. With ductility of 4.1, the castellated beam has the ability to withstand cyclic loading [6].

This paper used experiments of the reduced beam section. The resulting shear analysis of the RBS on the castellated showed sufficient shear capability, where the shear force occurred after the bending force.

2. Reduced beam section (RBS)

Reduced beam section (RBS) is a modification of the beam cross-section by trimming the flange at a specified distance from the column face. The reduction was made in a way that all melting and plastic joint occurred in the RBS. Moreover, the trimmed area also could reduce the moment on the column and control the inelastic deformation on the column. According to AISC 358-10 subchapter 5.8 [7], the procedure of designing RBS is as follows.

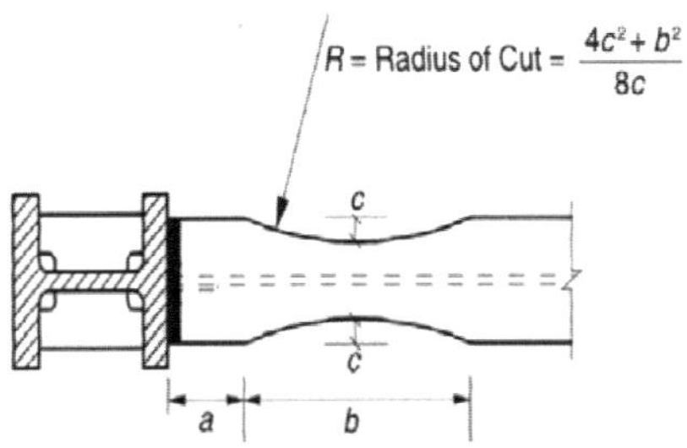

Figure 1. RBS design on beam flange.

Determination of RBS geometry (Figure 1) is limited by:

$$0.5\ bf \leq a \leq 0.75\ bf \tag{1}$$

$$0.65\ d \leq b \leq 0.85\ d \tag{2}$$

$$0.1\ bf \leq c \leq 0.25\ bf \tag{3}$$

with:
bf = width of beam flange
d = depth of the beam

3. Calculation of shear force in the center of RBS (V_{RBS})

The shear force was determined by calculating the effect of yield moment on the center of RBS (M_{RBS}) and the shear effect of a combination of gravity (w) and earthquake force. The loading combination used was 1.2D + f_1L + 0.2S.

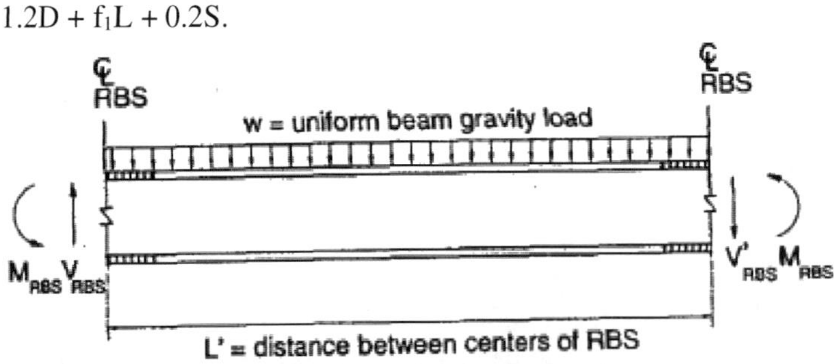

Figure 2. The free body used to calculate V_{RBS}.

The maximum moment that may occur on the column face can be calculated from the free body diagram of the moment in the RBS center as can be seen in Figure 2. Based on the diagram, the moments on the column face are as follows:

$$Mf = Mpr + V_{RBS}Sh \tag{4}$$

With:
Mf = the maximum moment that may occur on the column face (N-mm)
M_{pr} = the plastic moment that may occur (N-mm)
V_{RBS} = the maximum shear force of two shear forces in the center of RBS at each end of the beam (N)
Sh = a + b/2 (mm)

The plastic moment on the beam based on the expected yield strength was calculated as follows.

$$Mpe = ZbRyFy \tag{5}$$

with:
Mpe = plastic moment based on the expected yield strength (N-mm)
Zb = modulus of the plastic cross-section (mm^3)
Ry = ratio of the expected yield strength to the minimum yield strength
Fy = minimum yield strength (MPa)

The Mf must be lower than Φd Mpe. If not, then the value of c was increased of the values of a and b were decreased.

$$Mf \leq \Phi d\ Mp \tag{6}$$

with:
Mf = maximum moment that may occur on the column face (N-mm)
Φd = reduction factor of the maximum ductility
Mpe = plastic moment based on the expected yield strength (N-mm)

Meanwhile, the ultimate shear force of the beam was calculated using the following equation.

$$Vu = \frac{2\ Mpr}{L'} + V_{grav} \tag{7}$$

With:
Vu = the ultimate shear force of the beam
L' = distance between the center points of RBS (mm)
V_{grav} = the beam shear force from 1.2D + 1.6L + 0.2S

The thickness limit of column flange was calculated using the following equation.

$$t \geq \frac{(d_z + w_z)}{2} \geq\geq \tag{8}$$

with:
t = thickness of column flange (mm)
d_z = thickness of the panel (mm)
w_z = width of the panel (mm)

The ratio of column moment to the beam was calculated using the following equation.

$$\Sigma\, Mpb \approx \Sigma\, (Mpr + Mv) \tag{9}$$

with:

IOP Conf. Series: Materials Science and Engineering **797** (2020) 012024 doi:10.1088/1757-899X/797/1/012024

M_{pr} = plastic moment that may occur (N-mm)
M_V = V_{RBS} (a +b/2 +d_C/2)
V_{RBS} = the maximum shear force of two shear forces in the center of RBS at each end of the beam (N)
d_C = column height (mm)

4. Experimental design

The welding procedure was done with a system that was considered semi-rigid on the plate and connection with bolts and endplates as shown in Figure 3. The exterior joint was subjected to the additional cyclic loading as specified. It was set at the end of the beam with the displacement of the roller joint. The test was adopted to support the lateral loads near the actuator, which aimed to prevent lateral deformation of the beam and damage to the actuator. Global deformation of the specimen was measured during the test to determine the component of the beam end deformation that was contributed to the column.

Figure 3. Setup of the experiment.

5. Results

The identified parameters were the plastic moment, local buckling, and yield stress. These parameters were used to analyze the strain and stress distribution resulting in a plastic limit on the beam with RBS. Based on the loading and straining data, the first yield point of the steel material (f_y) was obtained at 350 MPa, divided by 200,000 steel elasticity to obtain the value of 0.0017. The first yield point occurred at the strain of 0.0017 or as shown in Figure 4, the first yield point occurred at 1700 μ. The yield point was connected with perpendicular lines drawn from the P load.

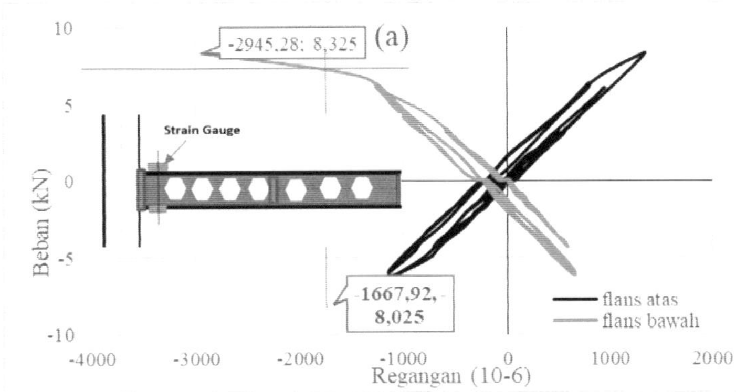

Figure 4. Relation between load and yield point on the beam flange.

As can be seen in Figure 4, the yield point occurred at the strain of 1700.10^{-6} at the load of 9,015 kN, deflection of 32.86 mm at the push-positive direction and the load of -9.015 kN, deflection of 32.82 mm at the pull-negative direction. This shows that the first yield point occurred in the upper flange with the strain value reaching 1692 μ, then moved to the lower flange at the load of -9.015 kN. The results show that the damage occurred in the column face due to the pure bending of the beam, and did not indicate failure at the connection of the beam and column [6].

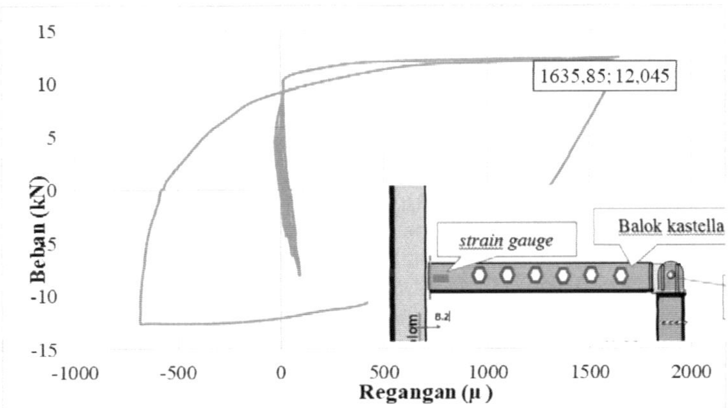

Figure 5. Relation between load and yield point on the beam web.

Four models of castellated beams with various type of RBS and one model of the castellated beam without RBS was conducted in this study. Based on the result, the difference of the maximum load and maximum displacement was not significant for RBS than without RBS, Therefore, RBS technique on castellated beam has the ability to create plastic moment far from the column face without significant strength loss [8].

As can be seen in Figure 5, the yield point on the web occurred on the shear area, with the strain yield point occurred on the load of 12.04 kN on the push-positive direction. This caused an increased load that led to shear failure on the beam. Figure 6 shows the peak load at 12.045 12,045 reached deflection of 97.5 mm that caused shear failure and buckling on the beam web.

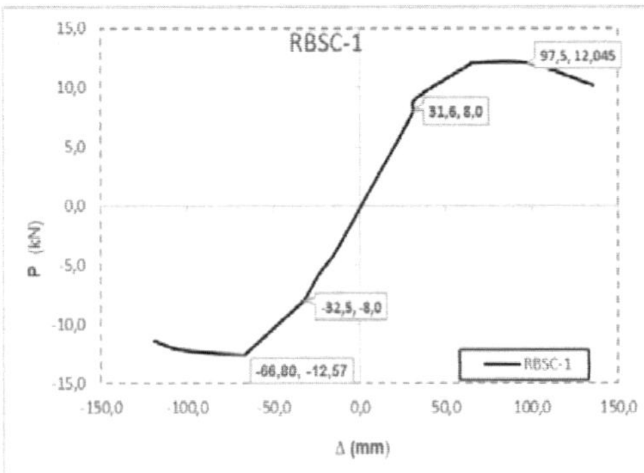

Figure 6. Relation between load and deflection.

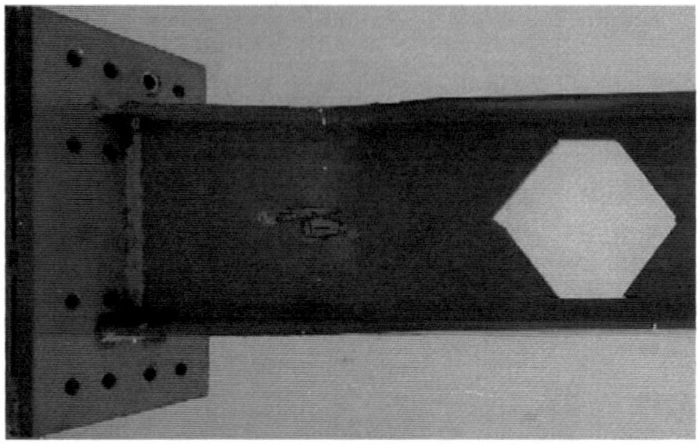

Figure 7. Damage caused by shear force on the beam.

The results show that the initial failure was caused by the bending on the beam, then it spread to the beam web due to the shear force as displayed in Figure 7. The failure in the RBS area was caused by buckling due to the shear force. This is similar with the study conducted by [3], where beam with RBS experienced failure caused by buckling on the beam with the shear stress on the beam web was larger than on the beam flange.

6. Conclusion

The use of reduced beam section on the beam flange could suppress the effect of cyclic loading; thus, the damage did not occur on the connection of beam and column flange. The damage only occurred on the column face. The experiment results show that shear failure occurred on the column face of RBS area. This means that the castellated beam had a buckling on beam web due to the cyclic loading, which caused a damage to the RBS area.

References

[1] Miller D K 1998 Lessons learned from the Northridge earthquake *Eng. Str.* **20** 249–60
[2] Nakashima M, Inoue K and Tada M 1998 Classification of damage to steel buildings observed in the 1995 Hyogoken-Nanbu earthquake *Eng. Str.* **20** 271–81.
[3] Sungkono K K D 2017 Stress analysis of RBS steel moment connection (AISC 358-2005) *e-journal Teknik sipil dan arsitektur Fakultas Teknik UTP Surakarta* **21**
[4] Mehrdad M and Hussam M 2014 Performance of steel moment resisting frame with RBS connection under fire loading *J. Eng. Str.* **75** 126–38
[5] Chen C C, Lee J M and Lin M C 2003 Behaviour of steel moment connections with a single flange rib *J. Eng. Str.* **25** 1419–28
[6] Aswad N H, Parung H, Irmawaty R and Amiruddin A A 2017 Evaluasi posisi jarak bukaan balok kastella terhadap daktilitas *KONTEKS 11* SK91–99.
[7] American Institute of Steel Construction 2010 Prequalified connections for special and intermediate moment frames for seismic applications. *ANSI/AISC 358-16 Standard*
[8] Aswad N H, Parung H, Irmawaty R and Amiruddin A A 2017 The effect of reduced beam section on castellated beam. Proceedings of the 6th International of Euro Asia Civil Engineering Forum

IOP Conf. Series: Materials Science and Engineering **797** (2020) 012025 doi:10.1088/1757-899X/797/1/012025

Dynamic Behavior of Human Tympanic Membrane Perforation Using Finite Element Method

Hidayat[1], **Sudarsono**[2], **Rozaini Othman**[3]
[1]Mechanical Engineering Department, Politeknik Negeri Samarinda, INDONESIA
[2]Mechanical Engineering Department, Universitas Haluoleo, 93232 Kendari, Sulawesi Tenggara, INDONESIA
[3]Faculty of Mechanical Engineering, Unversiti Teknologi MARA Pulau Pinang, MALAYSIA

[*]Corresponding author's e-mail: hidayat@polnes.ac.id

Abstract. Human tympanic membrane also known as an eardrum is a thin membrane that separtes the outer ear and middle ear. The tympanic membrance perforation can cause longer lasting hearing loss. The purpose of this research is to simulate the dynamic behavior of tympanic membrane perforation using finite element method. In this paper, two types analysis model, namely normal tympanic membrane and tympanic membrane perforation were developed. Firstly, the geometrics model of human tympanic membrane was generated by CAD Software (Solidworks) using physical properties reported by other researchers. Then, eigenvalue analysis was carried out to simulate the dynamic behavior of two types of human tympanic membrane. As for the boundary conditions of human tympanic membrane, torsional springs were used for the rotational motion around one axis of the local coordinate frame on the boundary of tympanic membrane. Finally, the simulation of dynamic behavior of two types of tympanic membrane were compared.

1. Introduction

The human ear system is divided into three parts, namely outer ear, middle and inner ear. The middle ear system consists of a tympanic membrane, ossicles and some ligaments. As for the ossicles, these parts are three tiny bones linked each other connecting the tympanic membrane and cochlea (inner ear). The three tiny bones are malleus, incus and stapes but collectively called ossicles. The ossicles have functioned as a mechanical system to transmit the sound through from outer ear to the inner ear. A person unable to hear sound in one or both ears called hearing loss. There are three types of hearing loss, namely conductive, sensorineural, and congenital hearing loss. Many cases of conductive hearing loss happened due to problems in the tympanic membrane or ossicles — a hole in the thin membrane which separates the outer ear and middle ear called tympanic membrane perforation. The size of the hole and where is the hole in the eardrum can cause many problems. A tympanic membrane perforation or perforated eardrum such as a tympanic membrane having a hole accompanies liquid discharge from middle ear through the ear canal.

Many researchers on human middle ear system used finite element analysis to simulate the dynamic behavior of the human tympanic membrane. Finite element analysis is a powerful tool to simulate the tympanic membrane vibrations. The main benefits of finite element analysis in this research are to determine natural frequencies and vibration mode of tympanic membrane perforation which cannot be obtained by using traditional analysis method.

Wada et al. measured Young's modulus, thickness and damping ratio of the human tympanic membrane [1]. Cat eardrum was firstly investigated by using the finite element analysis on the curved conical eardrum by W. R. I. Funnell and C. A. Laszlo [2]. D. D. Greef et al. performed the dynamics analysis on a new anatomically-accurate model consists of the tympanic membrane and malleus using the finite element analysis [3]. Mehta, R.P et al. predicted the possibility of conductive hearing loss in tympanic membrane perforations [4]. Voss, S.E et al. developed a quantitative model of the human middle ear with eardrum perforation [5]. Saliba, I et al. carried out the differences of perforation sizes of the human eardrum and evaluated the effect of perforation on hearing levels and frequencies [6]. The previous research was carried out the eigenvalue analysis of tympanic membrane perforation with size of hole is 2.5 cm [7].

In this research, the dynamic behavior of the human tympanic membrane perforation was carried out. The finite element analysis software was used to carry out dynamics analyses of the human middle ear system with eardrum perforation.

Firstly, the geometric mode of the human tympanic membrane was generated by using CAD software. Then, geometric model as an IGS file was imported to the finite element analysis software. Furthermore, the eigen-value analysis of human eardrum perforations was carried out to obtain the natural frequencies and the vibration modes.

2. Finite Element Model of Human Tympanic Mebrane

Figure 1 shows the finite element model of human tympanic membrane perforation. In this research, the human tympanic membrane perforation with the Young's modulus and Poisson's ratio were 33 MPa and 0.3, respectively. As for the mass density, the value of 1.3×10^3 kg/m3 was used. The human tympanic membrane was generated by using CAD Software. One of the case human being have conductive hearing loss caused by the tympanic membrane perforation. The hole in tympanic membrane is around 50% of the membrane surface. As for the boundary condition of the tympanic membrane, the rotational springs elements were used around the tympanic membrane. The values 3×10^{-5} Nmm/rad was used as the rotational spring constant in order to the finite element model in this research become similar to the real human tympanic membrane.[8]

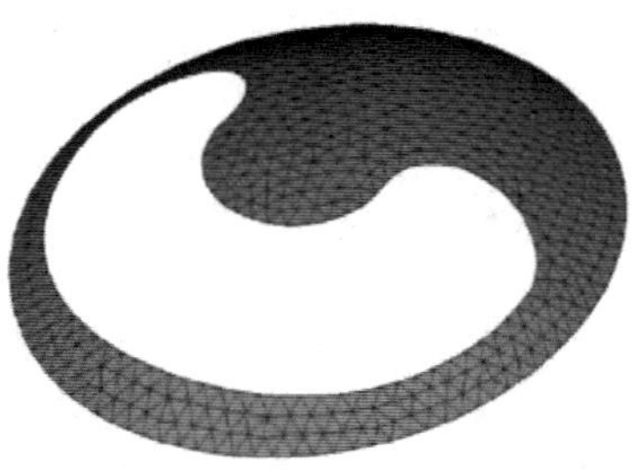

Figure1. Finite Element Model of Human Tympanic Membrane Perforation

3. Eigenvalue Analysis

In the dynamic analysis by using finite element method, the eigenvalue analysis was carried out to obtain the natural frequencies and the vibration modes of the finite element model without the damping. The equation of motion of the finite element model in the matrix-vector form is written by

$$\boldsymbol{M}\ddot{\boldsymbol{x}} + \boldsymbol{K}\boldsymbol{x} = \boldsymbol{f} \quad (1)$$

Where M, K, x and f are the mass matrix, the stiffness matrix, the displacement and the external force, respectively. Then M and K are symetric matrices of order n. The x and f are the vectors of n elements.

The natural frequencies and vibration modes can be obtained from the equation of motion by using a homogeneous equation. If there is no damping force, the homogeneous equation becomes

$$\boldsymbol{M}\ddot{\boldsymbol{x}} + \boldsymbol{K}\boldsymbol{x} = \boldsymbol{0} \tag{2}$$

In Eq. (2), x is assumed by using the fundamental solution, $e^{j\omega t}$ as follows.

$$\boldsymbol{x} = \boldsymbol{x}_o e^{j\omega t} \tag{3}$$

Where $\boldsymbol{x}_o$ and ω are the amplitude and the angular frequency of the finite element model, respectively. The velocity and acceleration vectors in Eq. (2) can be given by the first and second derivatives of the Eq. (3) shown as follows.

$$\dot{\boldsymbol{x}} = j\omega \boldsymbol{x}_o e^{j\omega t} \tag{4}$$

$$\ddot{\boldsymbol{x}} = -\omega^2 \boldsymbol{x}_o e^{j\omega t} \tag{5}$$

Then, by substituting Eqs. (3) and (5) into the Eq. (2), the following equation can be obtained.

$$-\boldsymbol{\omega}^2 \boldsymbol{M}\, \boldsymbol{x}_o e^{j\boldsymbol{\omega} t} + \boldsymbol{K}\, \boldsymbol{x}_o e^{j\boldsymbol{\omega} t} = \boldsymbol{0} \tag{6}$$

After simplifying Eq. (6), it becomes

$$\left[-\omega^2 \boldsymbol{M} + \boldsymbol{K}\right] \boldsymbol{x}_o e^{j\omega t} = \boldsymbol{0} \tag{7}$$

In Eq. (7), the angular frequency is not equal to zero ($\omega \neq 0$), then Eq. (7) should become

$$\left[-\boldsymbol{\omega}^2 \boldsymbol{M} + \boldsymbol{K}\right] \boldsymbol{x}_o = \boldsymbol{0} \tag{8}$$

or

$$\boldsymbol{\omega}^2 \boldsymbol{M}\, \boldsymbol{x}_o = \boldsymbol{K}\, \boldsymbol{x}_o \tag{9}$$

The Eq. (9) can be solved as generalized eigenvalue problem given by

$$\Omega_i^2 \boldsymbol{M}\, \boldsymbol{\phi}_i = \boldsymbol{K}\, \boldsymbol{\phi}_i \tag{10}$$

where Ω_i^2, Ω_i and $\boldsymbol{\phi}_i$ denote the *i*-th eigenvalue, natural angular frequency and eigenvector or vibration mode, respectively.

The *i*-th natural frequency can be written as follows.

$$f_i = \frac{\Omega_i}{2\pi}, \quad (i = 1, 2, 3\ldots, n) \tag{11}$$

Figure 2 shows the vibration modes of the normal and perforated eardrum. The natural frequency of tympanic membrane perforation is lower than normal eardrum. The effect of mass on tympanic membrane perforation reduce the natural frequency. The vibration modes of the perforated eardrum appeared around the hole of the eardrum.

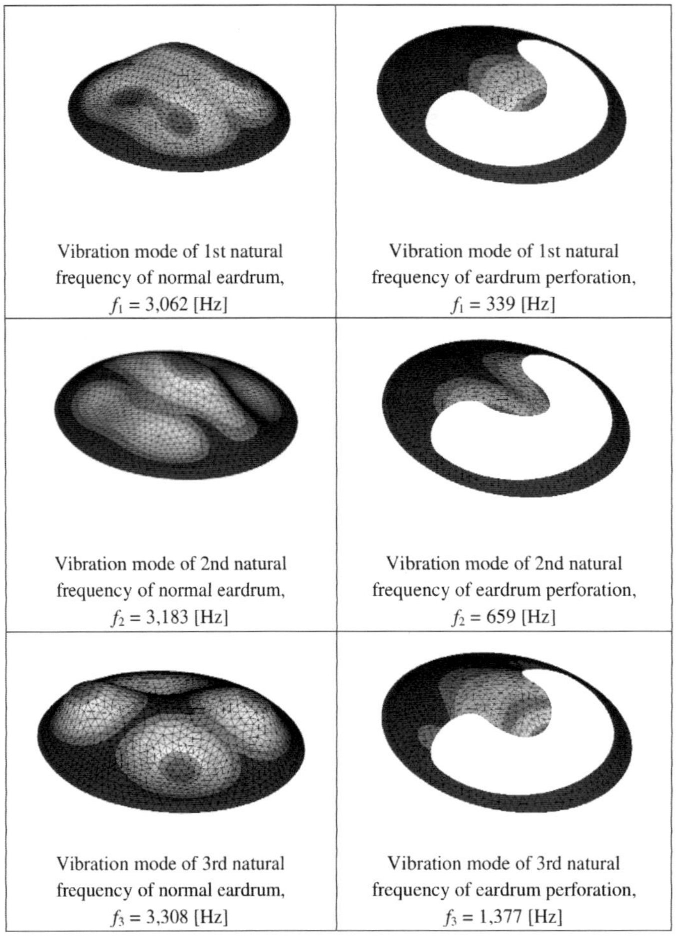

Fig. 2 Vibration modes of the normal and perforated eardrum

4. Conclusion

The summary of the results is shown below.

1. The finite element model of human tympanic membrane perforation had been developed by using CAD software.
2. The eigenvalue analysis of human tympanic membrane perforation had been performed to obtain natural frequencies and vibration modes.
3. The vibration modes of first, second and third natural frequency of human tympanic membrane perforation appear around the hole.

REFERENCES

[1] Wada H, et al, *Analysis of Dynamic Characteristics of Eardrum (Young's modulus, thickness and damping ratio of human eardrum)*, Japan Society of Mechanical Engineers, Vol. 56 No. 526, 1990, pp. 87-90.

[2] Funnel, W Robert J, et al. Modelling of the cat eardrum as a thin shell using finite element method, J. Acoust. Soc. Am. 63,1978, pp. 1461-1467.

[3] Graf Daniel De, et al, Viscoelatic properties of the human tympanic membrane studied with stroboscopic holography, *Hearing Research Journal* 312, 2014, pp. 69-80.

[4] Mehta, R.P., Rosowski, J.J., Voss, S.E., O'Neil, E. and Merchant, S.N. (2006) Determinants of Hearing Loss in Perforations of the Tympanic Membrane. Otology & Neurotology. vol. 27, pp. 136-143.

[5] Voss, S.E., Rosowsko, J.J., Merchant, S.N. and Pekae, W.T. (2001) Middle Ear Function With Tympanic Membrane Perforation. Acoustical Society of America. vol. 110, pp. 1445-1452.

[6] Saliba, I., Abela, A. and Arcang, P. (2011) Tympanic Membrane Perforation : Size, Site and Hearing Evaluation. *International Journal of Pediatric Otorhinolaryngology*. Vol. 75, pp. 527-531.

[7] Hidayat, Hi & Okamoto, S & Hoon Lee, J & Hato, N & Yamada, H & Takagi, D. (2017). Proper Material Properties and Dimensions of Sliced Materials Used in Myringoplasty of Humans.

[8] Hidayat, Okamoto, S., Lee, J.H., Matsuura, K., Hato, N., Yamada, H., & Takagi, D. (2016). ECBA-16 Dynamics Analyses of Human Middle Ear System Using Finite Element Method.

Effect of Current Density on Hardness of Low Carbon Steel Electroplated by Copper, Nickel and Copper-Nickel

Sudarsono[1*], Aminur[1], Irma Nurjannah[2], Hidayat[3], Rozaini Othman[4]
[1]Mechanical Engineering Department, Universitas Halu Oleo, INDONESIA
[2]Architecture Engineering Department, Universitas Halu Oleo, INDONESIA
[3]Mechanical Engineering Department, Politeknik Negeri Samarinda, INDONESIA
[4]Faculty of Mechanical Engineering, Universiti Teknologi MARAPulau Pinang, MALAYSIA

[*]Corresponding author's e-mail: sudarsono@uho.ac.id

Abstract. Coating process of metal surface undergoes rapid development, especially in electroplating method. This study aims to investigate effect of current density on hardness of low carbon steel plates electroplated by copper, nickel and copper-nickel. Low carbon steel plate coated with copper, nickel and copper-nickel using electroplating process with current variation (6, 9, 12 A for copper coating), (2.5, 3.5, 5 for nickel coating) and (2.5, 3.5, 5, 6, 9, 12 A for copper-nickel coating). Electroplating time for the three processes is 20 minutes in 5-liter solution at temperature range of 55-60°C. The electroplated steel plates are characterized with micro vickers hardness test. The results of this study indicated that the highest hardness is 290 kg/mm2 for nickel coating at current density of 5 A. Contrary, the lowest hardness is 111 kg/mm2 in copper coating at current density of 12 A, while the hardness for copper-nickel coating is 252 kg/mm2 at current densities of 12 A and 5 A.

1. Introduction

In recent decades, metal plating has evolved from an art to an exact science. There are many reasons to coate objects. These reasons include corrosion control, resistance to wear, hardness and decorative purposes. This development is seen as responsible for the ever-increasing number and widening types of application of this branch of practical science and engineering. There are some method coatings for metals range such as electrochemical and electroplating. Technology of electroplating has been known for a long time, and it has been used for preparation of coating in models of complex form (galvanoplatics), and for thin coating on subjects for the purpose of protecting metals mostly from corrosion (galvanization) as mentioned [1]. Electroplating process uses electric current and electrochemical reaction to make metal coatings. Coatings may be formed by copper, nickel, chromium, zinc, tin, cadmium, silver, gold and others. Electroplating is one of the most difficult surface adjustment technologies for pre-treatment products. It is clear, because the transmission of metal ion from solution on the surface of cathode and its incorporation to the crystal grid require perfect contact of phases. Selection of technology depends on the type of electroplating products and bath, electroplating technique, condition of surface and technical equipment of electroplating plant [2]. Nickel is one of the most important metals applied by electro-deposition [3]. Nickel plating for low carbon steel is widely used in industry application due to its benefit in improving corrosion fatigue life, increasing plating thickness to provide good protection against corrosion. Copper electroplating is widely used in the electronic industry for fabrication of electronic devices. It is particularly explored for fabrication of printed circuit boards and semiconductors. High throwing power copper electrolytes are becoming increasingly important due to electronic industry requirements for manufacturing of high aspect ratio circuit boards. Effect of plating parameters such as current density, electrolyte pH, electrolyte concentration, plating medium temperature, plating time, material size and plating bath volume on the mechanical properties must be investigated to obtain optimum parameters. Current density management is important for two main reasons. Firstly, the current density controls the rate of deposition. Secondly, the current density must be controlled within the correct operating range in order to obtain sound deposits having uniform appearance and free from burning or treeing. The purpose of this project is to study the effect of current density of on hardness of low carbon steel electroplated by copper, nickel and copper-nickel. This will enhance the engineer to have a formal knowledge on when and how to apply a particular concentration for a particular purpose such as in corrosion control, decorative purpose, hardness and wear resistance.

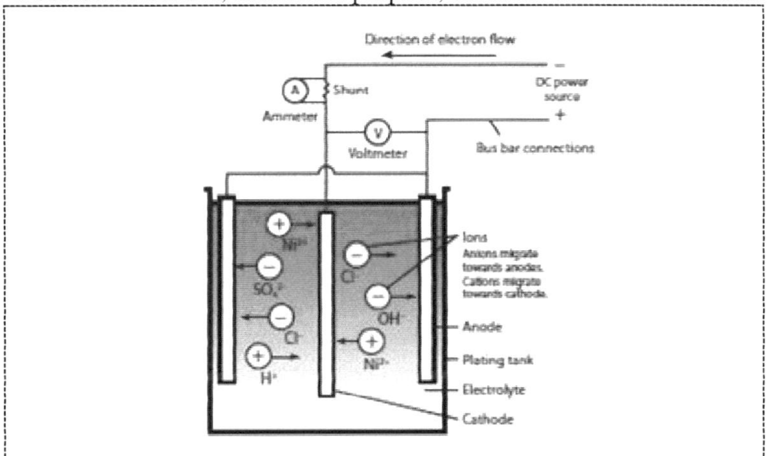

Figure 1. Schematics of electrical circuit for electroplating

2. Process of Electroplating

The electroplating process essentially involves passing an electric current between two electrodes immersed in an electrolyte. The positively charged electrode is anode while the negatively charged electrode is the cathode. The electrolyte contains electrically charged particles or ions. When an electrical potential or voltage is applied between the electrodes these ions migrate towards the electrode with the opposite charge-positively charged ions to the cathode and negatively charged ions to the anode. This results in the transfer of electrons, that is a current flow, between the electrodes-thus completing the electrical circuit [5]. Copper is soft, tenacious and not too oxidized to air. Because of its nature, being electropositive, copper is likely being gravitated to less-conductive metals, like iron. Copper plating is easily conducted, and so is its liquid controllable. Copper is appropriate to be utilized as the base coating prior to the following plating [4]. Nickel plating is used to prevent corrosion or to heighten luxury. Nickel is resistant to heat, not corrosive and preventable to salty water, freshwater and alkali. Nickel can be decayed by nitric acid and somewhat corrosive by cline acid and sulfuric acid. Nickel also has rigidity and medium endurance, ideal tenacity, fine conduction and good thermal. Generally, in electroplating, the anodes are composed of the metal being plated. These are referred to as 'soluble' anodes. During electroplating, the positively charged metal ions discharge at the cathode (the component to be plated) depositing metal on the surface. The component being plated therefore receives a coating of metal. The reverse effect occurs at the anode and, with a

IOP Conf. Series: Materials Science and Engineering **797** (2020) 012026 doi:10.1088/1757-899X/797/1/012026

soluble anode, metal ions are formed through dissolution. The electrical energy is supplied by a DC power source such as a rectifier. The basic electrical circuit is depicted in figure 1.

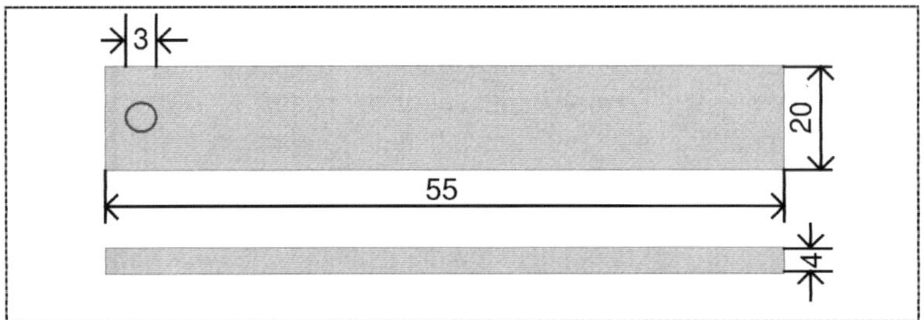

Figure 2. Dimension of substrate (mm)

3. Experimental Procedure

The substrates were low carbon steel plate with a hardness of 107 HVN. The palte dimensions were 55mm×20mm×4mm as shown in figure 2. Chemical composition of the substrate is shown in table 1. Before deposition, the substrates were mechanically polished surface roughness using sandpapers. Three electroplated coating processes were performed namely Cu, Ni, Cr-Ni coating. Cu coatings were electrodeposited with a bath containing copper cyanide (45 g/L), sodium cyanide (64 g/L), rochelle salt 30 (g/L) and coustic soda (20 g/L). The deposition current density were varied in three conditions (6, 9 and 12A) in the temperature range of 55-60 ◦C in 20 s at pH of 3.6. Ni coating were elekctrodeposited by using a bath containing nickel sulfate (250 g/L), nickel chloride (50 g/L), borid acid (45 g/L) and brightener 1(g/L). The deposition current density were varied in three conditions (2.5, 3.5, and 5A) in the range temperature 55-60 °C in 20 s at pH of 4,7. Cr-Ni coatings were firstly electroplated with Cu coating bath in 10 s. Then it were continued with Ni coating bath in 10 s. Three kinds of substrates were treated by combining of Cu-Ni coating, i.e (6A Cu - 2.5A Ni, (9A Cu – 3.5A) and (12A Cu – 5A Ni).

Table 1. Chemical composition of low carbon steel substrate.

Elements	Composition (%)
C	0.18
Si	0.08
P	0.10
Mn	0.62
Ni	0.75
Mo	0.082
V	0.005
Ti	0.002
Al	0.006
Mg	0.004
Sn	0.0008
Fe	97.60

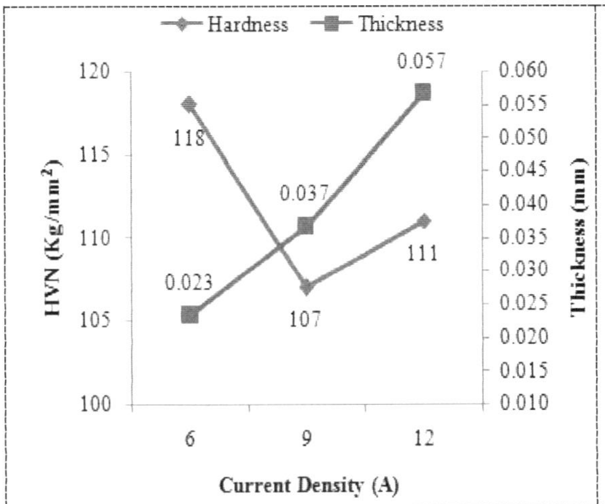

Figure 3. Hardness and layer thickness in copper electroplating on low carbon steel substrate with varied current density

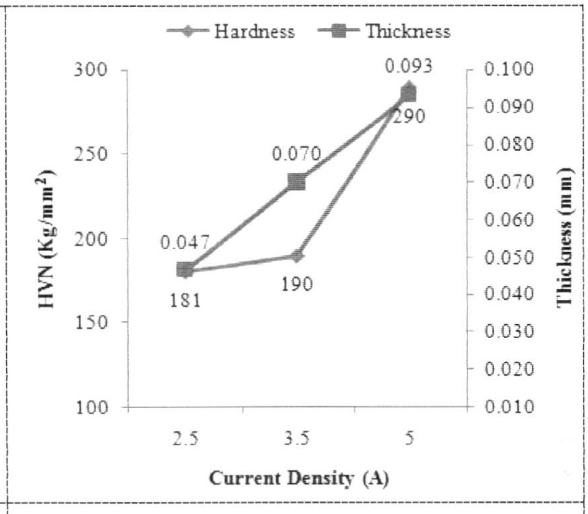

Figure 4. Hardness and layer thickness in nickel electroplating on low carbon steel substrate with varied current density

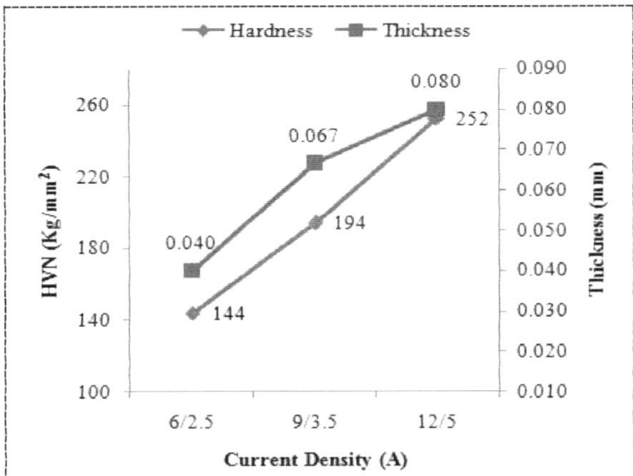

Figure 5. Hardness and layer thickness in copper electroplating on low carbon steel substrate with varied current density

4. Results and Discussion

Effect of Current Density

The hardness and layer thickness of the substrate in copper plating technique with different current density were shown in figure 2. The hardness of the substrate at current density of 6, 9 and 12A were 118, 107 and 111 HVN, respectively, while the thickness of the substrate at current density of 6, 9 and 12A were 0.023, 0.037 and 0.057, respectively. The hardest sample was obtained in the thickness of 0.023mm with current density of 6A . The results indicated that the hardness of substrate increases with increasing of current density. In contrast, the thickness of substrate increases with increasing of current density. The hardness and layer thickness of the substrate in nickel plating technique with different current density were shown in figure 3. The hardness of the substrate at current density of 2.5, 3.5 and 5A were 118, 190 and 290 HVN, respectively, while the thickness of the substrate at current density of 6, 9 and 12A were 0.023, 0.037 and 0.057, respectively. The hardest sample was obtained in the thickness of 0.093mm with current density of 5A. The results indicated that the hardness of substrate increases with increasing of current density. Moreover, the thickness of substrate also increases with of current density increasing. The hardness and layer thickness of the substrate in copper-nickel plating technique with different current density were shown in figure 4. The hardness of the substrate at three combination of current density of (6A Cu - 2.5A Ni, (9A Cu – 3.5A) and (12A Cu – 5A Ni) were 144, 194, and 252 respectively. Moreover, the thickness of the substrate for three combinations of current density i.e. (6A Cu - 2.5A Ni, (9A Cu – 3.5A) and (12A Cu – 5A Ni) were 144, 194, and 252 respectively. The hardest substrate was obtained in the thickness of 0.080mm with current density combination of 12A Cu -5A Ni.

5. Conclusions

The foremost task of this study was to investigate the effect of current density of copper, nickel and copper-nickel electroplating of low carbon steel alloy. According to the obtained results, the subsequent conclusions can be summarized: the hardness of electroplating of low carbon substrate was affected by current density and types of electroplating process. The highest hardness in copper, nickel dan copper-nickel electroplating was 118 kg/mm2 (6A), 290 kg/mm2 (5A), and 252 kg/mm2 (12/5A, respectively. The substrate hardness in nickel electroplating is higher than that in copper electroplating.

References

[1] Selhi A and Nikhil S 2014 *A Study of Electroplating Process Through Experiment and Simulation* (India: 5th International & 26th All India Manufacturing Technology, Design and Research Conference AIMTDR).

[2] Agboola O S, Sadiku E R, Ojo O I, Akanji O L, and Biotidara O F 2011 Influence of Operation Parameters on Metal Deposition in Bright Nickel-Plating Process (Portugaliae: Electrochimica, Acta) pp 91-100.

[3] Budman E and Sizelove R R 1998 *Zinc Alloy Plating Metal Finishing* Vol 98 pp. 334-339.

[4] Nikolova M and Watkowski J 2013 *Innovative High Throw Copper Electroplating Process For Metallization of PCB* (Waterbury, USA: Mac Demid Enthone).

[5] Rose I 2014 *Nickel Plating Handbook* (Belgium, Brussels).

IOP Conf. Series: Materials Science and Engineering **797** (2020) 012027 doi:10.1088/1757-899X/797/1/012027

Analysis of Islamic Center Planning with Ecological Architecture Approach in Kendari City

Adelila Haksika Fisika Ratna[1], **Irma Nurjannah**[1*]
[1]Department of Architecture, Faculty of Engineering, Universitas Haluoleo, 93232 Kendari, Sulawesi Tenggara, INDONESIA

*Corresponding author's e-mail: nurjannah.irma@gmail.com

Abstract. Indonesian religion plays an important role in people's lives, based on Indonesian ideology, Pancasila. Kendari City is one of the cities in Indonesia, which majorities are Muslim. Year by year population of Muslim in Kendari City have increased, but the increase is not directly proportional with the existing means of worship, for example like Islamic Center. Basic from status, Islamic Center in Kendari City consists of two types, they are Islamic Private Center and Government Islamic Center. The Government Islamic Centre, located in Abdullah Silondae street, but the building has been destroyed and turn into Head Office of PT Bank Sultra. Therefore, it is necessary to plan a new Islamic Center building for the people in Kendari. The Islamic Center planning was designed by applying the Ecological Architecture approach. The method used is descriptive analysis carried out by collecting data, then analyzed based on the theory of ecological architecture. The results of the analysis are micro and macro site processing which has minimal impact with the environment, and the application of a rectangle as a form of building and the use of sloping roofs to cover buildings. This form can maximize the lighting and the building's effect on the thermal comfort of the building user.

1. Introduction

Religion in Indonesia plays an important role in live, this is stated in the ideology of the Indonesian nation, in the first precepts of Pancasila, "God's Almighty". Based on population, Indonesia is the country with the most Muslim population in the world. Kendari City is one of the cities in Indonesia. As in other regions in Indonesia, the majority of Kendari City adheres to Islam with 267,064 people, spread in ten sub-districts, Abeli District with 22,257 people, Baruga District with 28,004 people, Kadia District with 24,181 people, Kambu District with 13,584 , Kendari Subdistrict as many as 24,378 people, District of West Kendari as many as 35,226 people, District Mandonga as many as 35,429 people, District Poasia as many as 35,429 inhabitants, District Puuwatu as many as 23,711 inhabitants and District Wua-wua as many as 24,931 inhabitants [1]. The number of mosque and musholla in Kendari City are 380 buildings, consisting of 336 Mosques and 44 musholla, 30 Protestant Churches, 2 Catholic Churches, 1 pura and 2 monasteries [2]. As Islam population, Muslim and Muslimah are bound by the rules and worship that must be carried out, such Islamic law, prayer, fasting, zakat and pilgrimage. Generally, these activities are carried out at the mosque and Islamic Center. Islamic Center is a forum for community activities based on Islam. Islamic sense as a religion or in a broader sense as a way of life. Those activities in it include the value of worship as well as social values [3].

The status of the Islamic Center is a semi-official (half-official) organization that is managed by the Ministry of Religion's Office (Depag) assisted by social preaching institutions and local religious education. Judging from the status of the Islamic Center, Islamic Centers in Kendari City consist of two types, namely Islamic Private Centers and Government Islamic Centers. Included in the Private Islamic Center are the Islamic Center of Muadz Bin Jabal and the Campus of the Islamic Center of Muadz Bin Jabal, which belongs to the Islamic Center of the Government, which is the Islamic Center located on Abdullah Silondae Street. Consisting of a management office, the office of BAZNAS (National Zakat Agency) and a hall that can be used to hold religious activities, the shortcomings of the Islamic Center located on Abdullah Silondae Road are not equipped with worship facilities, quoted from the online newspaper m.antarasultra.com Islamic This Center will be dismantled and used as the Head Office of PT Bank Sultra.

Based on the description above, it is necessary to plan a new Islamic Center that is more feasible for the people of Kendari City. The Islamic Center area as the center of Islamic civilization should be in direct contact with the daily activities of the community, both worship activities and community activities. In addition to providing religious facilities in the city of Kendari, the Islamic Center's regional planning also wants to respond to the issue of global warming that is sweeping various countries around the world. Wulfram Ervianto in his book, Save the Earth Through Green Construction, said that Indonesia is included in the top 15 CO_2 emitting countries (1.35% of the total) and ranked 5th of the top ten contributors to greenhouse gases in the world (2,045.30 mega tons of greenhouse gas emissions). As one of the cities in Indonesia, it should be if Kendari City participates in reducing global warming by using an environmentally friendly approach in every development plan. One approach that is environmentally oriented is ecological architecture.

Ecology is defined as the knowledge of the interrelationships between living things and their environment [4]. From the ecological understanding that has been described, the attention to architecture as a technical science has been transferred to humanitarian architecture which takes into account harmony with nature and the human interests of its inhabitants. The construction of a house or place of residence as a necessity for human life in a reciprocal relationship with its natural environment is called ecological architecture or eco-architecture [5]. Islamic Center Regional Planning with Ecological Architecture approach is expected to be able to meet the needs of facilities and infrastructure in the field of religion for the people of Kendari City without destroying the nature of Kendari City, besides that the application of Ecological Architecture is also in accordance with the understanding that Islam is a religion

2. Literature Review

2.1. Review of the Islamic Center

Islamic Center is a religious institution as well as a forum for Islamic activities, which includes religious values and social values. The function of the Islamic Center as a center for fostering and developing Islamic religion and culture is as follows:

- Center for shelter, preparation, formulation of results and ideas on the development of religious life and Islamic culture.
- Center for the implementation of non-formal education training programs.
- Center for research and development of religious life and Islamic culture.
- Center for broadcasting religion and Islamic culture.
- Coordination center, synchronizing coaching activities and developing Islamic da'wah.
- Information center, public communication in general and the Muslim community in particular.

IOP Conf. Series: Materials Science and Engineering **797** (2020) 012027 doi:10.1088/1757-899X/797/1/012027

The nature and institutional status of the Islamic Center according to the Islamic Center project implementation manual throughout Indonesia in 1976, is participatory coordinative in the sense that its handling and management are inter-departmental coordination at the central and regional levels of all local Regional Office and Religion Office, as well as participation in the meaning of the entire community was mobilized to carry out this project, both direct participation funds and religious social funds and personnel to complete this project. The organizational status of the Islamic Center is a semi-official (semi-official) organization in accordance with its purpose and function to mobilize community participation to build. In general, there are several facilities that will be accommodated in this Islamic Center Area, including: public facility, semi-public facilities, private facilities, supporting facilities

2.2. Review of the Ecological Architecture Approaces

The term ecology was first put forward by Ernst Haeckel, a zoologist in 1869 as a science of interaction between all types of living things and their environment. The meaning of the Greek word oikos is household or way of residence, and logos is scientific or scientific. Ecology can be defined as the study of the interrelationships between living things and their environment [6]. On the basis of the knowledge of the ecological foundations that have been described, the attention to architecture as a technical science has been transferred to humanitarian architecture that takes into account harmony with nature and the human interests of its inhabitants. The construction of a house or place of residence as a necessity for human life in a reciprocal relationship with its natural environment is called ecological architecture or eco-architecture [5]. The design context of the Ecological Architecture, seen in terms of micro (buildings inside the land itself) and macro (buildings against the surroundings) [7]. While ecological planning are as follows: *Balance with nature;* Understanding balance with nature contains the unity of living things (including humans) with the surrounding nature holistically. Great influence from nature and tropical climate on the surrounding environment, consisting of: sunlight and building orientation, building protection against the sun, lighting, wind and air space, temperature and heat protection, rainfall and humidity, structure and construction (floor, wall, and roof). *Making nature a planning pattern;* Nature as a holistic pattern of eco-architectural planning can then be concluded with the following requirements [6]: adjustments to the local natural environment, saves natural energy sources that cannot be renewed and save energy use, maintain environmental sources (air, soil, water), maintain and improve natural circulation, reducing dependence on systems of energy centers (electricity, water) and waste (waste water, waste), and workplaces and settlements.

3. Discussion

3.1. Project location

The chosen location is the District of Poasia because the District of Poasia fulfills the criteria that are taken into consideration in determining the location, including: allocation of locations in accordance with government regulations and directives, included in the city's strategic area and city development center, has many empty site alternatives, and a city utility network is available. Allotment (Trade and Services); Site Area (5.3 Ha); KDB (30%); GSB (10 m); GSS (50 m). The Islamic Center area planning location with the Ecological Architecture approach in Kendari City is on Jalan Madusila, Andounohu Village, Poasia District. The limits are as follows: North (Highway and Empang); South (Empty Land); East (Empty land); West (Citraland Housing).

3.2. Site Processing Concept

The land on the site is a swamp land that has been buried, with a slope of less than 5%. The slope of the site is to the west, so that rainwater on the site flows toward the river.

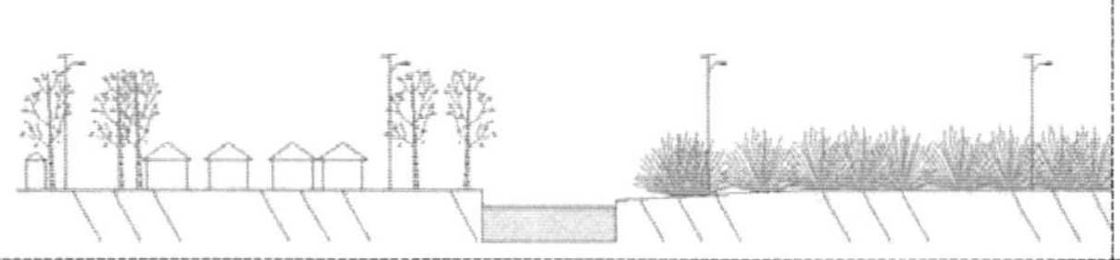

Figure 1. The slope of the site located near the river.

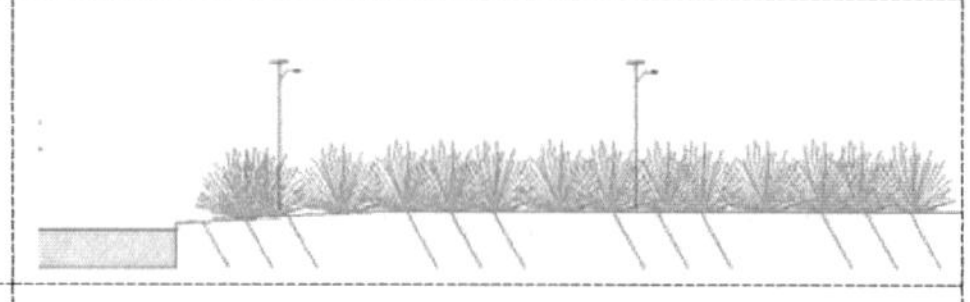

Figure 2. Vegetation in the site.

Due to ecological reasons, the general purpose of site development must be to minimize disruption to existing land forms and special characters. Then the land contour is maintained as it is. The vegetation found in the site is grass, and some shrubs. Vegetation in the form of weeds will not be maintained, while those in the form of trees will be maintained as long as they are not in the area to be built. To replace the removed vegetation, planting and structuring new vegetation is carried out. Sun orientation on the site influences the shape and orientation of buildings and natural lighting and the placement of openings in buildings.

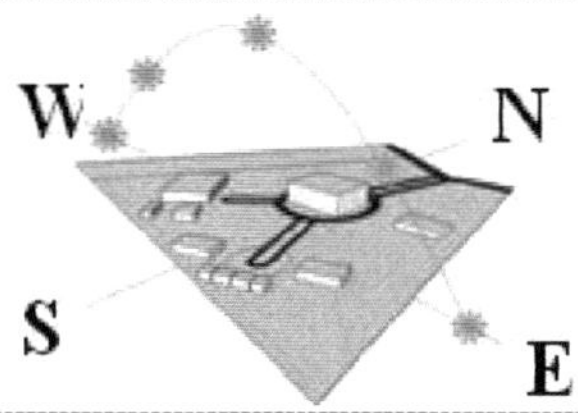

Figure 3. Sun orientation

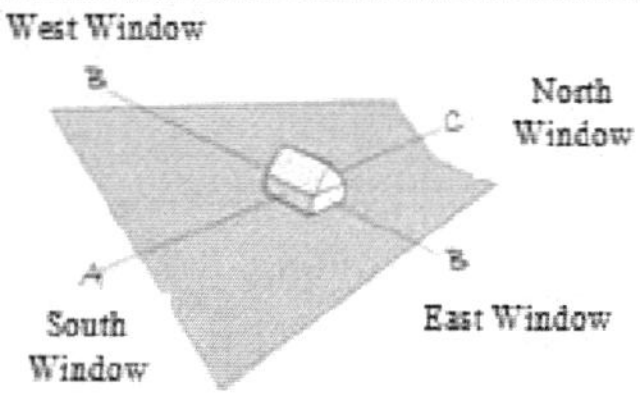

Figure 4. Windows Places

The recommended shape and orientation for buildings in the tropical climate is to extend from east to west with as little east and west as possible in order to reduce solar heating in buildings (Figure 5). The south-facing window is an ideal natural lighting source when horizontal shade can control excessive radiation and glare. Windows facing east and west require shade facilities to avoid dazzling morning and evening suns. The north-facing window enters soft, diffused light. The wind blows from all directions past the footprint. Wind flow is used as a natural environment for building by making openings in buildings from various directions. The falling rain on the site affects the model and construction of the roof structure, in addition to the impact due to the construction makes the discharge of rainwater that must be removed from the site become more. The roof model used is a sloping roof that can flow water easily. To drain rainwater out the site is carried out ground pavement and drainage planning.

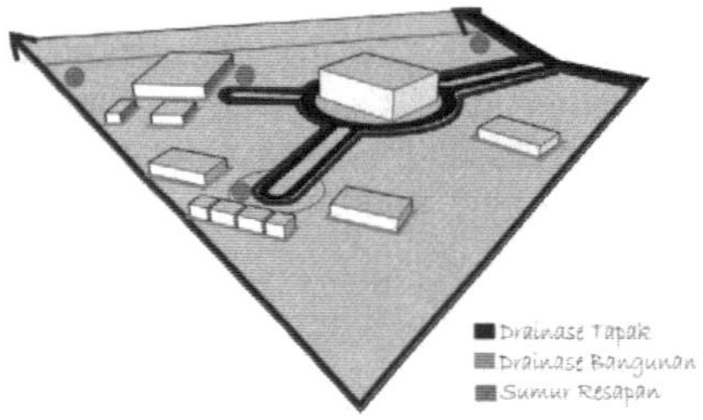

Figure 5. Solution to drain rainwater

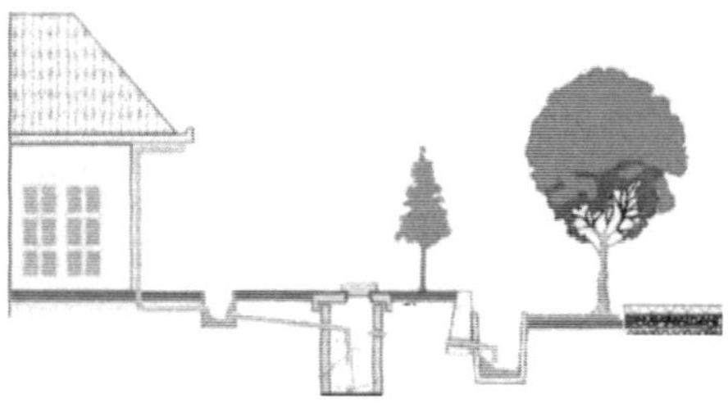

Figure 6. Infiltration system

Meanwhile, to anticipate excessive rainwater discharge, which has the potential to cause flooding, infiltration wells are made. Noise around is:

- High noise comes from vehicle traffic north of the site
- moderate noise comes from residential activities (residential citraland) in the west of the site
- Low noise comes from empty land in the south and east of the site

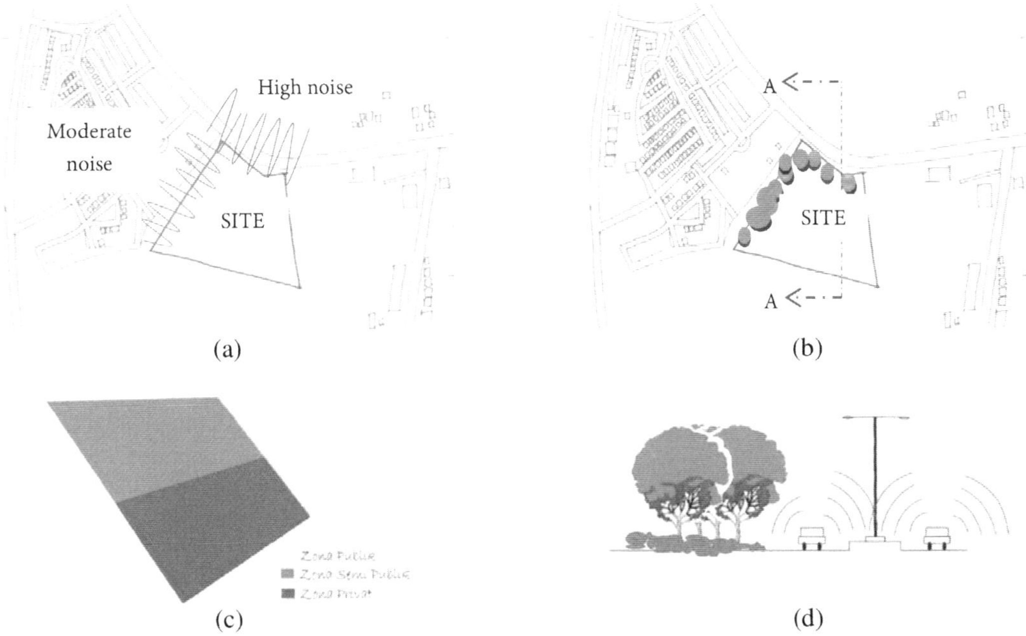

Figure 7. (a) Noise (b) To reduce noise into the site, vegetation is planted (c) Site Zones (d)section A-A The vegetation used is tanjung tree, because the tanjung tree has thick and tight canopy so that it can absorb noise maximally.

Site can be accessed via Madusila street using public transportation and private transportation.

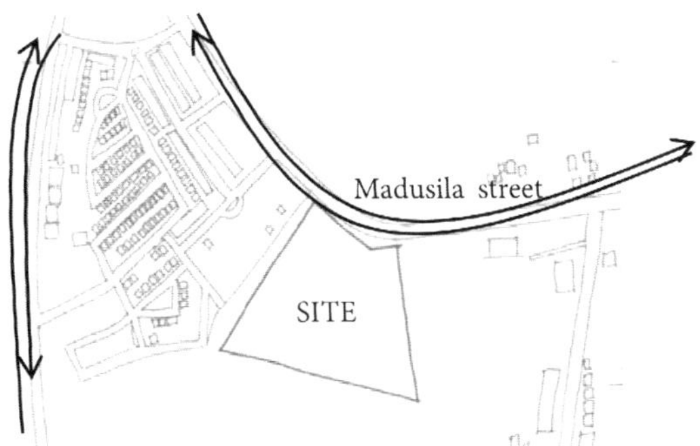

Figure 8. Site Circulation

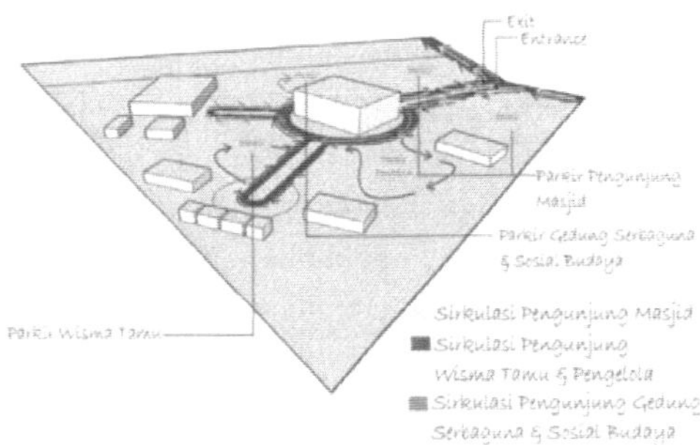

Figure 9. Circulation Response

Circulation inside the site is divided into three, namely the circulation of mosque visitors, the visitor's circulation of multipurpose buildings and socio-cultural, circulation of managers and guest houses. This circulation arrangement is based on zoning. View in the north of the site in the form of roads and ponds, the west view of the site in the form of rivers and settlements, the view on the south of the site is an empty land in the form of weeds, the view in the east of the site is empty land in the form of mangroves and swamps.

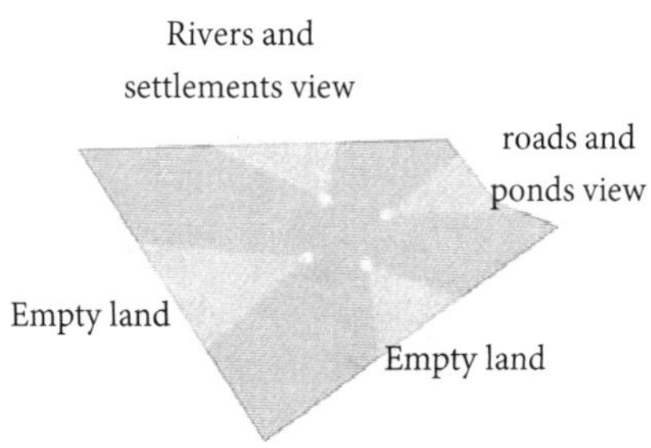

Figure 10. View Existing

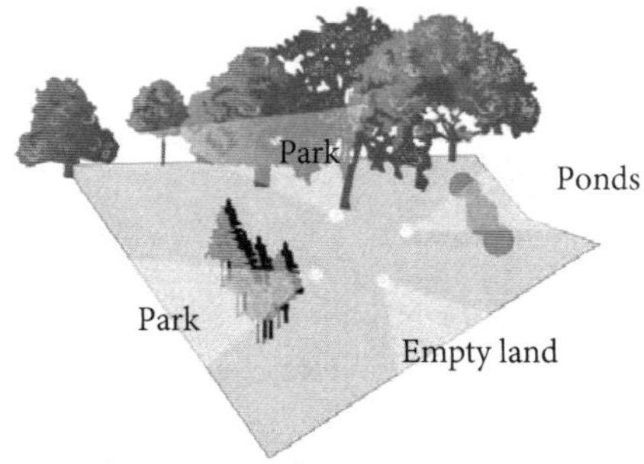

Figure 11. View Response

View from inside to outside of the site will be directed to the artificial view in the form of ponds and parks, while the view from the outside into the site will be focused on public buildings (mosques).

4. Conclusion

Based on theoretical studies, analysis and discussion, the location that is suitable for Islamic Center Regional Planning with the Ecological Architecture Approach is the District of Poasia. Poasia Subdistrict was chosen because it is a City Strategic Area and City Development Center, in addition Poasia Subdistrict also fulfills the criteria which become a consideration of location determination, among others, the allocation of its location in accordance with government regulations and directives, has many empty site alternatives, and has a city utility network. The application of the Ecological Architecture approach to Buildings in the Islamic Center Area is emphasized on micro and macro site processing which has minimal impact on the environment. The output obtained from the site processing is a form of building that is responsive to climate and environment, namely a rectangle as a building shape and a sloping roof as a building cover. This form can later maximize natural lighting and ventilation that affect the thermal comfort of building users.

References

[1] Kota Kendari Dalam Angka 2016
[2] Badan Pusat Statistik Kota Kendari Tahun 2016
[3] Gazalba Sidi 1983, *Mesjid Pusat Ibadah dan Kebudayaan Islam*. Penerbit Pustaka Antara
[4] Kristanto, Ir. Philip 2002. *Ekologi Industri*, Edisi I, Penerbit ANDI, Yogyakarta
[5] Krushe, Per et sl 1982, *Oekologisches Bauen,* Wiesben, Berlin
[6] Frick H, Suskiyanto B 2007, *Dasar-dasar Arsitektur Ekologis*, Penerbit Kanisius, Yogyakarta
[7] Akmal Imelda, *20 Rumah Ekologis Atelier Riri*. Penerbit PT Gramedia Pustaka Utama, Jakarta

IOP Conf. Series: Materials Science and Engineering **797** (2020) 012028 doi:10.1088/1757-899X/797/1/012028

Application of Under-Frequency Relay (UFR) in Load Shedding Scheme for System Stability (A Case Study on PT. PLN Kendari)

Tachrir[1*], Samuel Jie[1], Bunyamin[1], Abdul Djohar[1]
[1]Department of Electrical Engineering, Faculty of Engineering, Universitas Haluoleo, 93232 Kendari, Sulawesi Tenggara, INDONESIA

*Corresponding author's e-mail: tachrir@gmail.com

Abstract. A reliable electric power system must maintain its operation in stable conditions with the support of various protection system during normal operation or when a fault occurs. To minimize the frequency of blackouts in the system due to interference, transient conditions, or the increased load, this study simulated the load shedding scheme using under-frequency relays (UFR). The analysis was done in ETAP 12.6.0 software using the Transient Stability Analysis feature. In this simulation, discontinuation occurs in the CB (Circuit Breaker) generator which causes the generator to escape from the system in two conditions, namely Peak Load Time (PLT) and Off-Peak Load Time (OPLT). The study simulated scenarios where circuit breakers (CB) tripped due to a fault, which causing a power plant to be disconnected from the grid. The scenarios were simulated in two conditions, during the peak load and off-peak load. The results show that during the peak load, the simulated Load Shedding Scheme 1 to 3 was occurred on the frequency of 49.3918 Hz, 48.72 Hz, and 48.0195 Hz, respectively. The respective disconnected loads on Scheme 1 to 3 were 4.529 MW, 41.84MW, and 59,072 MW while the recovery times were 2.432 s, 0.081 s, and 0.626 s, respectively Meanwhile, in the off-peak load, the UFR tripped at the frequency of 49.3958 Hz, 48.5855 Hz, and 47.75 Hz, the disconnected loads were 3.652 MW, 30.627 MW, and 43.93 MW, and the recovery time were 2.934 s, 0.588 s, and 0.592 s for Load Shedding Scheme 1 to 3, respectively.

1. Introduction

The increasing use of electricity could create problems for the system when it is unable to provide sufficient electricity to consumers, such as the overload condition.

The overload condition can be caused by the release of several generating units or the addition of a load on the system. When this condition exists, it will cause the frequency to drop from its nominal value which could lead to a total system blackout. Thus, load shedding schemes are developed to maintain the system so frequency and to avoid a total blackout.

The load shedding scheme uses UFR (under frequency relay) when it operates. In this study, we discussed the UFR application in the load shedding scheme developed by the State Electricity Company – West, South, and Southeast Region, Kendari Area. The application of UFR in the load shedding scheme is expected to stabilize the system as quickly as possible. The power capacity of several plants in Kendari is displayed in Table 1.

IOP Conf. Series: Materials Science and Engineering **797** (2020) 012028 doi:10.1088/1757-899X/797/1/012028

Table 1. Power capacity in Kendari system.

Site	Power (MW)	
	Peak Load	Off-Peak Load
Wua-Wua Diesel Power Plant	30.1	19.25
Poasia Diesel Power Plant	28.1	24.90
Nii Tanassa Steam-Electric Power Station	19.7	19.10
BGP Tanassa Diesel Power Plant	9.0	7.00
TOTAL	86.9	70.25

The backbone of the interconnected Kendari electricity system is the 70 kV transmission lines connecting the Tanassa Substation to the Puuwatu Substation. Meanwhile, in the Kendari City, the electricity distribution uses a 20 kV distribution lines, which connect the Wua-wua Diesel Power Plant (DPP), Poasia DPP, and Puuwatu Substation.

2. Literature review

[1] conducted a research on the simulation of load shedding using frequency relays on the Cnooc Ses Ltd. electric power system. This study discusses a load shedding scheme caused by the overload condition by minimizing the amount of load released from the system while optimizing the recovery speed using UFR. The simulations were done in ETAP 7.0 software, with the results show that the UFR could properly detect the frequency drop due to the overload condition caused by the insufficient supply when a generator was disconnected from the system. The UFR also had a fast recovery time. Meanwhile, [2] simulated load shedding schemes on the 150/20 kV Tragi Sibolga distribution network system. In this simulation, two analyzes were carried out, namely load shedding variations and calculation of the ENS (energy not supplied) value using ETAP 7.0 software. The results show that the frequency recovery time took 4-10 seconds after a fault occurs, which depended on the 150/20 kV Tragi Sibolga feeders overload.

An electric power system should have the power generated by the plants at least equal to the load they cover including the power losses that might occur on the system. However, for the security and reliability of the system, the power plants should also have a power reserve. When a generator is disconnected from the system, it could cause insufficient supply for the electricity demand. This will lead to a frequency drop as the turbine rotation is slower because of the higher load in a generator. As the frequency decreases, it will cause a total system blackout. [3] explains the solutions to overcome the total system blackout are as follows.

- Optimizing generator capacity
  The generator can be optimized by increasing the governor speed; thus, generating more power. In this solution, the reserve power of each generator is supplied to the system.
- Islanding operation
  To avoid a total system blackout when the frequency occurs drastically and the load shedding cannot overcome this condition, the generating system and the load that is still capable of being supplied can be divided in into small groups. This is commonly known as islanding operation. The operation aims to save several sections of the electric power system that can still operate normally.
- Load shedding
  When the overload condition occurs in the electric power system that has optimized all of its generating power capacity, some loads should be disconnected to normalize the frequency. This process is commonly known as a load shedding. The load shedding has several stages according to the level of decreasing frequency. The criteria of a load shedding are as follows.

a. Load shedding is done gradually. If the first load shedding stage cannot recover the frequency, then it proceeds to the next stage.
b. The amount of load released should be as minimal as possible based on the required amount to improve the system frequency.
c. The load shedding starts from the load with the lowest priority in the electric power system. Therefore, all load should be classified initially based on specific criteria.
d. The load shedding must be done appropriately. Therefore, the relay delay time must be determined to detect whether the system frequency decreases due to the overload or other factors, such as the sudden addition of a large load into the system.

The generated power can be defined as the moment of force (torque) experienced by a generator that rotates at a certain angle per a period of time.

$$P = \tau \frac{d\theta}{dt} \quad (1)$$

with:
P = the generated power (Watt)
τ = moment of force/mechanical generator coupling (Nm)
θ = rotation angle (rad)
t = time (s)

The amount of change in the rotation angle of the generator that rotates per a period of time is the value of the generator's angular velocity.

$$P = \tau\,\omega = \tau.2\pi f$$
$$P \approx \tau \quad (2)$$

As can be seen in Equation (2), the change in active power generated by the generator changes the working torque. This change then subsequently affect the system frequency. Therefore, to maintain the magnitude of the system frequency, it is necessary to adjust the magnitude of the mechanical drive of the generator. Mechanical coupling driving the generator is related to the amount of fuel used to drive a prime mover or a turbine. To obtain the desired frequency that is constant in the allowable range, the amount of fuel used in the governor should be regulated as follows.

$$T_G - T_B = J\frac{d\omega}{dt} = j2\pi\frac{df}{dt} \quad (3)$$

with:
T_G = the mechanical driving coupling of the generator (Nm)
T_B = load torque (Nm)
J = moment of inertia of the generator mechanical driver (kg.m^2)
ω = rotation angle velocity of the generator (rad/s)
t = time (s)

The frequency change rate is based on changes in the amount of generated power and the power needed by the load, which can be described by the swing equation of a generator as expressed in Equation (4).

$$\frac{GH}{\pi f_0} X \frac{d^2\delta}{dt^2} = P_A \quad (4)$$

with:
G = generator rating (MVA)
H = generator inertia constant (MJ/MVA)
δ = generator torque angle
f_0 = generator nominal frequency (Hz)
P_A = accelerated power (MW)

IOP Conf. Series: Materials Science and Engineering **797** (2020) 012028 doi:10.1088/1757-899X/797/1/012028

The generator rotation velocity can be formulated using Equations (5) and (6).

$$\omega = \omega_0 + \frac{d\delta}{dt} = 2\pi f \tag{5}$$

$$\frac{d\omega}{dt} = \frac{d^2\delta}{dt^2} = 2\pi\frac{df}{dt} \tag{6}$$

with:
ω_0 = generator velocity at the nominal frequency (rpm)

Then, it can be concluded that:

$$\frac{df}{dt} = \frac{P_A f_0}{2GH} \tag{7}$$

with:
$P_A = P_M - P_E$
P_A = accelerated power (MW)
P_M = generator mechanical power (MW)
P_E = power demanded by the load (MW)

Meanwhile, the energy generated on the synchronous velocity per volt-ampere of the generator rating is called the inertia constant. The generator's kinetic energy can be expressed by:

$$GH = \frac{1}{2} I\omega^2 \tag{8}$$

with:
G = generator rating (MVA)
H = inertia constant (MJ/MVA)
I = Moment of inertia (kg.m^2)
ω = rotation angle velocity (rad/s)

3. Method

This study used ETAP 12.6.0 software to simulate the load shedding scheme due to a decrease in frequency with the help of transient stability analysis feature. The data used in this study were obtained from the State Electricity Company, Kendari Area. The data included the single line diagram of Kendari system, parameters of machines and generators, transformers, cables (transmission lines and feeders), peak load and off-peak load, UFR settings, and load shedding priority.

4. Results and discussion

Before the load shedding simulation was run, first we prepared the single line diagram of Kendari system. Then, the simulation was run to observe the power flow as displayed in Figure 1.

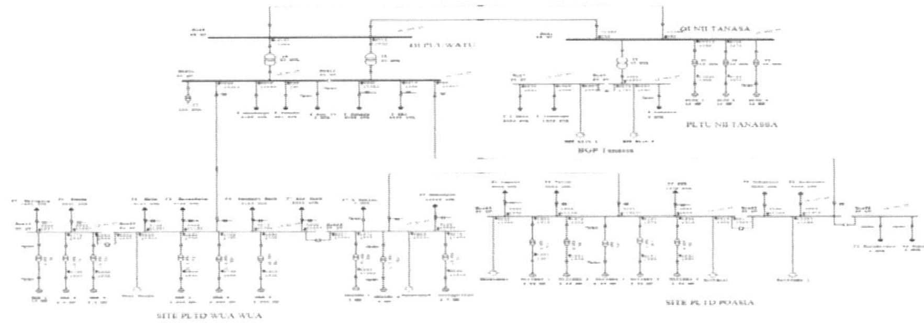

Figure 1. Power flow of the Kendari system.

The simulation results show that the voltage on each feeder is as follows.

1. The voltage on the Wua-wua feeder was 19.51 kV.
2. The voltage on the Poasia feeder was 19.781 kV.
3. The voltages on the Puuwatu feeder, for the 66kV and 20 kV system, were 65.453 kV and 19.87 kV, respectively.
4. The voltages on the Tanassa feeder, for the 66kV and 20 kV system, were 65.595 kV and 19.783 kV, respectively.

Table 2 shows the frequency, the amount of load disconnected from the grid, and the recovery time. Meanwhile, Figures 2 and 3 illustrate the simulation results in each scenario during the peak load and off-peak load, respectively.

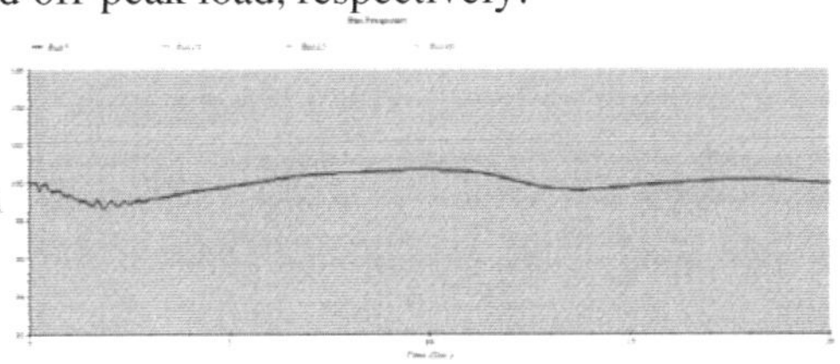

(a)

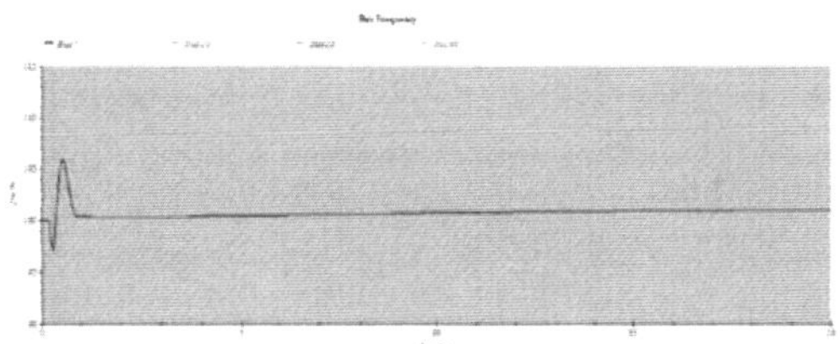

(b)

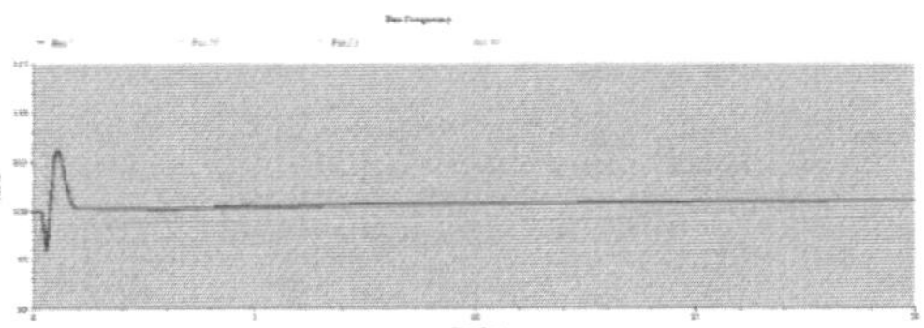

(c)

Figure 2. The frequency change in: (a) Scenario 1 (peak load), (b) Scenario 2 (peak load), Scenario 3 (peak load).

Table 2. Simulation results.

	Peak load			Off-peak load		
	f (Hz)	LS (MW)	Recovery time(s)	f (Hz)	LS (MW)	Recovery time(s)
Scenario 1	49.3918	4.529	2.432	49.3958	3.652	2.934
Scenario 2	48.72	41.84	0.801	48.5855	30.627	0.588
Scenario 3	48.0195	59.072	0.626	47.7525	43.93	0.592

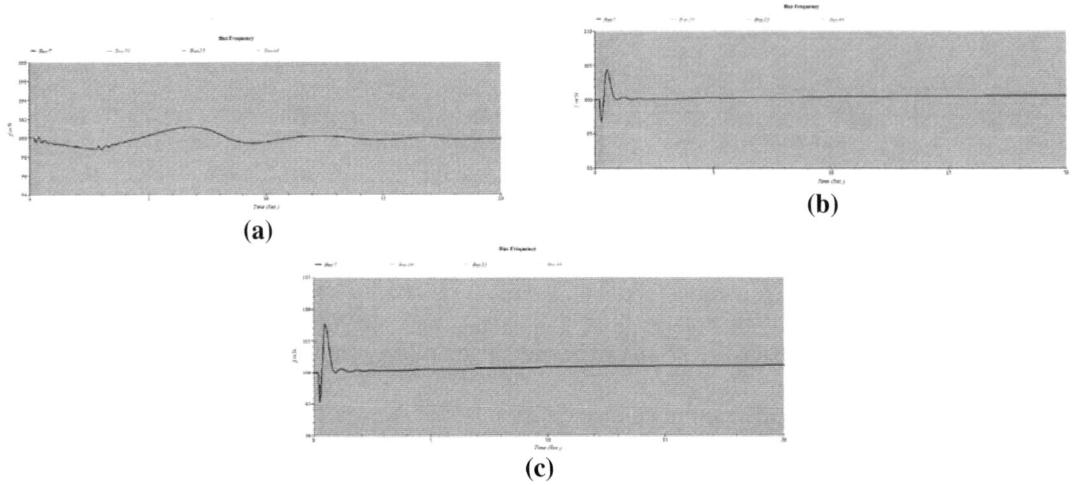

(a) (b) (c)

Figure 3. The frequency change in: (a) Scenario 1 (off-peak load), (b) Scenario 2 (off-peak load), Scenario 3 (off-peak load).

When a fault occurred on a block of Sewatama#1 generators, which disconnected it from the grid at t = 0.2 s. This caused a power loss of 6.1 MW. The disconnected generators caused the frequency to decrease. When the frequency reached 49.3958 Hz at t = 2.786 s, the UFR worked in Stage 1 of the load shedding scheme to disconnect the load of 3.652 MW. The system returned to normal at 7.274 s after the fault occurred.

5. Conclusion

The results show that the UFR worked properly to detect a frequency decrease due to the overload condition caused by disconnected generators. However, the use of swing generator equations to get the optimal relay working frequency, the amount of disconnected load, and the selection of relay delay time should be determined appropriately. The UFR also had a good performance in the load shedding scheme to normalize the frequency on the overload condition during peak and off-peak load. The recovery time during the peak load and off-peak load ranged from 0.626 to 2.432 seconds and 0.592 to 2.934 seconds, respectively.

6. Acknowledgment

We would like to express our gratitude to the management of State Electricity Company – West, South, and Southeast Region, Kendari Area, including all staffs and operators who provided all data and information for this study.

References

[1] Nugraheni A 2011 Simulasi pelepasan beban dengan menggunakan rele frekuensi pada sistem tenaga listrik Cnooc Ses Ltd. *Departeman Elektro Universitas Indonesia*

[2] Tambunan R P T 2014 Simulasi pelepasan beban (load shedding) pada sistem jaringan distribusi Tragi Sibolga 150/20 kV (studi kasus pada penyulang Tragi Sibolga, Sumatra Utara) *Jurusan Teknik Elektro Universitas Diponegoro*

[3] Mahmud S 2014 Simulasi pelepasan beban dengan menggunakan relay frequency pada sistem tenaga listrik Cnooc Ses Ltd. North Business Unit menggunakan software ETAP 7.5 *Jurusan Teknik Elektro, Universitas Muhammadiyah Semarang*

IOP Conf. Series: Materials Science and Engineering **797** (2020) 012029 doi:10.1088/1757-899X/797/1/012029

Space utilization optimization of new campus of Halu Oleo University

La Ode Amrul Hasan[1*], Yudhi Dwi Hartono[2], Arif Saleh[1], Alim Bahri[1]

[1] Architecture Department, Universitas Haluoleo, 93232 Kendari, Sulawesi Tenggara, INDONESIA

[2] Architecture Department, Muhammadiyah University of Kendari, Southeast Sulawesi 93127, Indonesia

*Corresponding author's e-mail: ld.amrul@gmail.com

Abstract. The New Campus of Halu Oleo University (UHO) is one of the state universities included in the term of the largest university in Eastern Indonesia with an area of more than 252 Ha. Located in the central government development area, UHO's New Campus has become an attractive campus, shown by the increasing number of prospective students from year to year. This increasing number causes the university to provide educating facilities as the main need. The existing buildings and future development projects have not considered the concept of good spatial planning. This threatens the availability of land, especially the availability of open space and also affects the arrangement of related elements such as vehicle circulation, building utilities, drainage systems and mechanical-electrical. The purpose of this research is to provide a recommendation about the design of the UHO campus spatial layout which optimizes 60% open space availability proportionally with a more directed structuring concept. This research employed a qualitative method and data was collected through questionnaires and observations. The results indicate that the current space utilization is not well arranged, therefore affects other elements. Through the concept of cluster eco, campus area will be more organized.

1. Introduction

The existing buildings of the New Campus of Halu Oleo University (UHO) and its future development have not considered a good spatial planning. These buildings were constructed following the availability of land, therefore grow and develop without clear direction. It can affect the availability of open space and related elements such as vehicle circulation, building utilities, drainage system, and mechanical and electrical.

The existence of building for education facility at university is crucial to support the learning activity as the number of students is increasing year by year. Without considering good spatial planning, the development can affect the availability of green open space. Green open spaces are important for not only being an ecological balancer but also for the function of shared recreational activity spaces including social activity.

The importance of optimizing the building layout for the New Campus of Halu Oleo University (UHO) is to maintain the effectiveness of the land based on the comparison between the area of open space and built land, where the minimum requirement is 30% of open space available. If the current and future development is uncontrolled and does not have a clear spatial concept. Drainage, cabling systems, and street lighting are affected because of their presence following the pattern of road lanes within the region. As a result, the function of the area as higher education facilities will not be optimal due to the problems arising as the effect of poor conceptual spatial planning. Therefore, this research aims to produce a recommendation of spatial layout for the New Campus of Halu Oleo University by optimizing land use by considering the open space of at least 30% and a maximum of 60%.

2. Literature Study

2.1. Literature Study

"Optimizing the Layout of Buildings for New Campus Space Utilization of Halu Oleo University (UHO)" can be interpreted as an effort to compile, organize, and integrate elements of the building with the aim of maximizing space effectiveness and efficiency. Considering that the new campus space is in the form of an education area with the existence of a building with space in it and the related regional elements, normatively the arrangement of the area is guided by the Regulation released by the Ministry of Public Works [1]. This regulation becomes the General Guidelines for Building and Environmental Planning which main foundation consist of:

a. The Structure of Land Allocation

It is an area design component that plays an important role in the allocation of use and land tenure/land use that has been determined in a particular planning area.

b. *Intensity of land use*

It is the level of allocation and distribution of the maximum floor area of the building to the land/site of its designation.

c. Building arrangements

It is a product of building construction and its environment as a form of space utilization, covering various aspects including the formation of the physical image/character of the environment, the size and configuration of block elements, plots of land, buildings. The other aspect is the height and elevation of building floors which take place in public spaces.

d. Circulation systems and connecting lines

It consists of road and movement networks, circulation of public vehicles, circulation of private vehicles, circulation of local informal vehicles and bicycles, pedestrian circulation (including disability and elderly) transit systems and facilities, parking systems, environmental service planning, and connecting network systems.

e. Open Space and Green System

It is a regional design component that is not only formed as additional elements or residual elements after the architectural design process is completed, but also created as an integral part of a wider environment.

f. Environmental quality management

It refers to efforts to engineer regional elements in such a way as to create an area or sub area with an informative environment, distinctive character, and having a specific orientation.

g. Environmental infrastructure and utility systems

It relates to completeness of the physical basis of an environment whose procurement allows an environment to operate and function properly.

As a supporting theory, [2] stated that a theory of environmental and building elements as elements of regional design consists of land use, building form and massing, circulation and open space, pedestrian ways, activity support, signage, and preservation.

3. Method

This research was conducted in the New Campus area of Halu Oleo University with an area of 277 ha. A qualitative research approach was employed, and the data was collected through observations and questionnaires. The observed objects included the layout of buildings, vehicle and human circulation, utility infrastructure, mechanical, electrical and drainage system. The indicator to see the priority of the condition of the area is based on seven spatial parameters in accordance with Ministerial Regulation [1]. The questionnaires were distributed randomly to 40 students and 10 employees whose activities are within the campus area.

The collected data were then tabulated to find out the tendency of user perceptions of the priority of building problems that exist within the campus environment. The information was then categorized based on its priority elements in accordance with 7 parameters [1] as the indicator. The next step is

IOP Conf. Series: Materials Science and Engineering **797** (2020) 012029 doi:10.1088/1757-899X/797/1/012029

analyzing the area based on the existing problems and then provide responses and design directives with the aim of optimizing space in the form of short-term to long-term design concepts.

4 Results

4.1. Analysis and Finding

4.1.1. Land Allocation Structure

Based on the existing condition, the UHO new campus land allotment is divided into macro and micro. The macro condition shows zoning function is based on each zoning. The western side is a Public Activity Zone (ZAP), an area dominated by public activities such as exercise, and sports games like football, futsal, basketball and tennis.

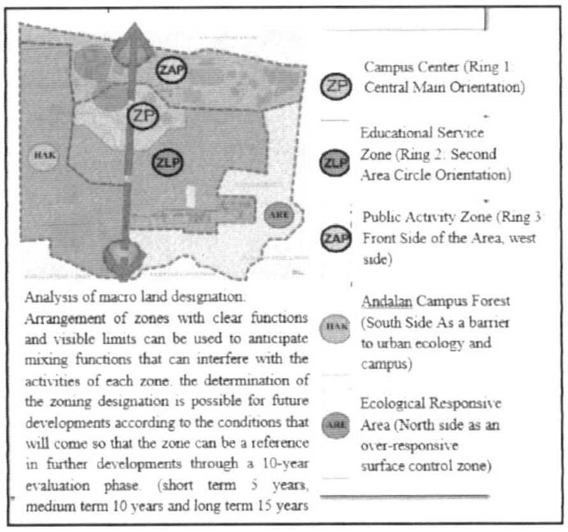

Figure 1. Macro Land Allocation Zoning
Source: Author's analysis (2015)

The Central Zone (ZP) is a campus core zone in the form of rectorate buildings and buildings that have similar functions as private status (have high privacy). Educational Service Zone (ZLP) is a zone of the education area containing educational faculties that have semiprivate status. The South is the *Andalan Hutan Kampus* (HAK) which is an ecological barrier, a forestry research area and the development of private gardens. The Ecological Responsive Zone (ZRE) is a control zone for surface water movement, allowing for retention lakes to accommodate excess water with semi-private status.

Based on the visible pattern, the region has a tendency to be centrally oriented so that it can potentially organize the form to be ring in ring system (a circle in a circle). This circle will later function as a road as well as barrier of the area that can later separate the public, semi-public and private areas including regulating circulation public transport.

The area has the main road character forming an imaginary axis line from the west, which is the main gate, then the core area of the campus to the road in the east. This has the potential strength of the regional character that can be used as a guideline for structuring the main building.

4.1.2. Intensity of Land Use

Based on the results of observations and data obtained, the following information is found:

Table 1. Open Space and Building Comparison Ratio
Source: Author's analysis (2015)

Open space (OC)	Size (M2)	Built Area (BC)	Size (M2)
Experimental Garden	30000	Office building	33346
Arboretum	50000	lecture hall	26666
City Forest	460000	Seminar room	5920
Foot ball field	17600	lecturer rooms	7067.66
Outdoor Futsal field	5000	Laboratory	37791
Tennis field	782.4	Auditorium	1600
Basketball field	728	Student dorm	4398
Green open space	1579851	Sports hall	260.87
Rock climbing area	25	Sport center	1755
		Mosque	6587
		Prayer room	1450.2
		Workshop room	500
		Power house	275
Semi Total	2143986.4		627117
Total	**2,771,103.40**		
OC/BC ratio:	2143986.4	/627117	77.4%:22,6%
The minimum size of open space of 30 % area has been met			

The table shows that the allocation of land use for buildings is 627,117 m^2 or 22.6%, while the open space area is still very large, with 2,143,986 m^2 or 77.4% of the total land owned by UHO. Therefore, the minimum requirements for the current condition are more than 30% so that they have been fulfilled.

4.1.3. Building arrangements

Based on the data obtained, the number of students in 2015 was 36,701 while the building area of education facilities was 70,377 m^2 so that the building area per student ratio was 1.92 m^2 (rounded to 2 m^2). With such results, the UHO's New Campus conditions are not in accordance with UNESCO standard which should be 8 m^2/student. Each building is found to have a different appearance in the sense that the unity of the building character has not been seen properly. The buildings of the faculties tend to grow and develop forming a stand-alone area.

4.1.4. Circulation systems and connecting lines

The circulation conditions of vehicles in the area at this time are still found to be mixing the function of the road lanes where public vehicles are still free to enter and access the lane in the campus area including the rector's core zone.

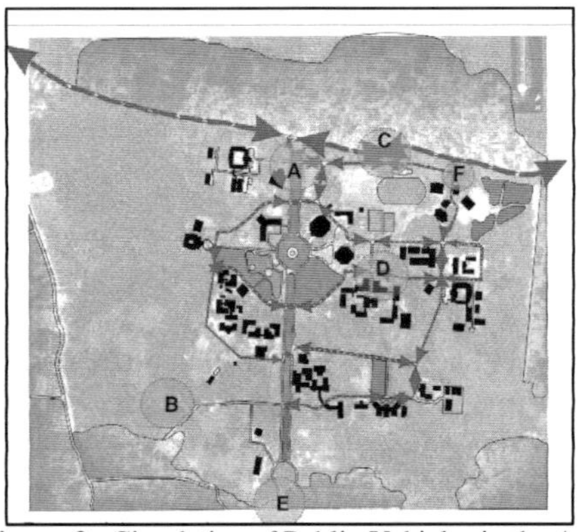

Figure 2. Circulation of Public Vehicles in the Area
Source: Author's analysis (2015)

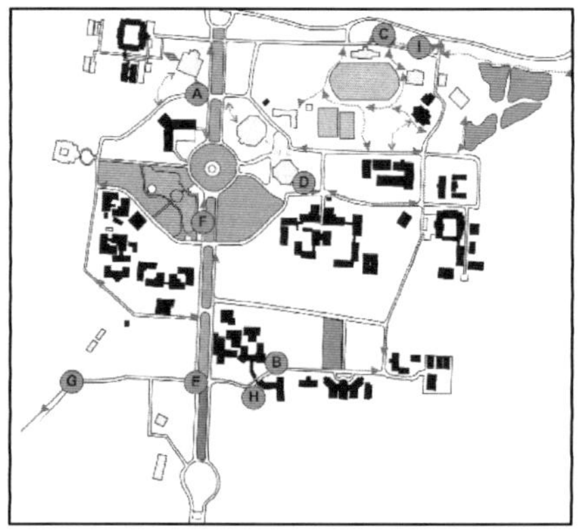

Figure 3. Pedestrian Circulation in Areas that Form Unofficial Lines
Source: Author's Analysis (2015)

The majority of street users are students who use motorized vehicles and the rest are pedestrians. Pedestrian facilities at some points have not been optimal, most are damaged due to rarely being used. It is found that unspecified pedestrian pathways are mostly formed from user reactions to the need for pedestrian paths in the place. Most of the users who are running want facilities that are equipped with shade and security to protect them from vehicles and time efficiency meaning that the closer destination is, the closer pathway they tend to create.

a. Open Space System and Green Governance

The existing open space system nowadays taking up 77.4% is dominated by green open space area which is not well-ordered. In this sense, the green area is dominated by forests growing wild and swamp. The existence of public open space is dominated by sports areas. The number of parks is significantly limited which is not well-organized for recreation and discussion points which include unorganized open public place in faculty area.

b. Environmental Quality Governance

Halu Oleo University's area does not have the concept of spatial direction. The strong character of university as an education place cannot be seen clearly through the main gate. The public transportation can easily access any gates to enter university. This makes university's privatization and characters as an education area decrease.

The university's forest becoming a botanical garden as a part of future planning is one form of investment to make the area well-ordered and spatial directed. The forest will not only be worth investment but also worth ecological area.

c. Infrastructure system and environmental utility

Infrastructure system in this case, including the management of waste is not clearly visible. Although the every building has its own trash bin, the people do not separate the waste based on organic and inorganic categories. Toxic and hazardous waste are also not separated properly.

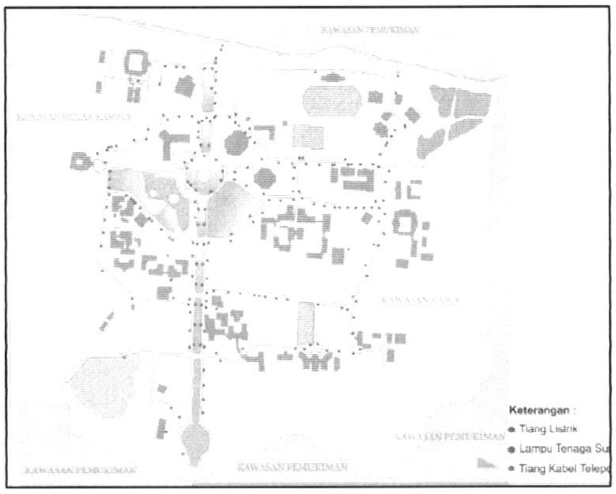

Figure 4. Existing Electricity Stilts and Telephone
Source: Author's analysis (2015)

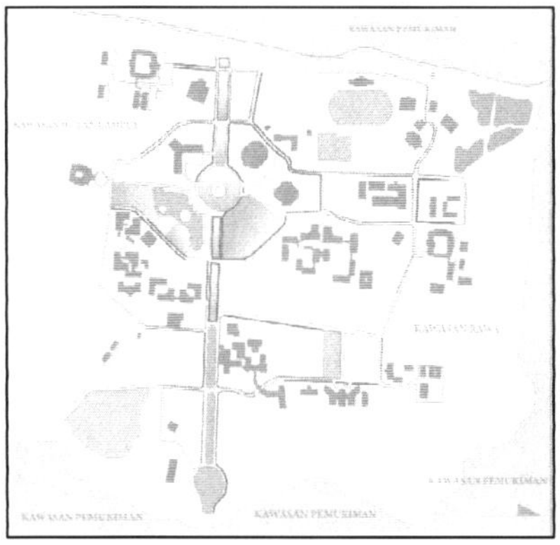

Figure 5. Existing Drainage System
Source: Author's analysis (2015)

In the area, treet lightings are no longer active although its existence is in the prone area. Most lightings use solar power which will become an obstacle when the nights.

The existence of a conventional cabling system using stilts in some points may harm pedestrians. It is because the existence of cross-street electrical wiring is directly beneath a shade tree branches which can be more likely broken which may disconnect the electricity wires and harm the pedestrians.

The drainage system at some points is damaged by the deposition of soil covering the channel so that the flow of water that should be through the pathway overflows into the street and the surrounding area especially when the rain comes.

The swamp area in the northern part of the university which collects the surface water has not been optimally processed so that the tendency of water moving over the surface just move naturally toward the swamp area and only a small portion is across the line in the drainage area.

5. Discussion

The optimal use of an open area of Halu Oleo University can happen through the gradual building arrangement with an adaptive and fixed program and timeline. Instead of making radical changes, preparing individuals to change the concept together is crucial. Therefore, not only the buildings are well-constructed, but also the users' characters can be way better. To reach the optimal use of minimum 30% of open space area, or 60% for the most of it, the optimal and consistent use of the area needs to do. It aims to make the use of open space area felt not only in the environment level, but also the city level. Therefore, the existence of Halu Oleo University as a green campus can stabilize the ecological condition of the city and as the city lung to reduce city pollution.

To create internationally standardized proper education facilities, there should be a commitment to provide $8m^2$ of the education building standard per student. By adhering to the ideal green campus which provides 60% of open space and the provision of educational facilities characterized by structuring the region with the fixed concept included if the condition still requires buildings construction. However, if the construction can take up the 60% of the open space limit, the following buildings need to be constructed vertically in its final stages.

To utilize the space potential and support the space utilities, a green eco university can be applied as the basic guideline to develop the university area. It needs three governance strategies. First, Green Blue eco Belt (GBB) meaning that Halu Oleo University can be a green campus which is environmentally friendly. It means that the university is designed with the woods (green open space in the southern part) as the ecological barrier, while blue area represented by the swamp area as responsive ecology to create the balance between surface water and rivers to be used as recreation and water research. Second, micro-cluster campus system means to cluster the faculties based on the knowledge similarities in one complex cluster by naming clusters based on the figures who contributed to Halu Oleo University. Therefore, it is expected to create many collaboration programs or research in the future, such as conducting a relevant international conference. Third, a thematic landscape is the arrangement of outdoor space or landscaping the area which has unity of design and concept through thematic design creative. It is manifested through the use of façade appearance of the identical building. In other words, the appearance of the building will be united by the use of the same material that the terms of use of material in the form of a natural material which is durable and easily found such as marble or natural stone, second skin or double facade as the building character in order to have the same patterns, especially in some critical areas. Besides, thematic design can be achieved by using the existing elements in the area such as park benches, garden lights, etc which have the similar concept and form, included the tropical roof which suits the weather in Southeast Sulawesi.

The governance of street circulation needs to consider the existing circular pattern, so the concept of the ring in ring; creating circular as an alternative to the arrangement of the function zone as a commitment and regional control in order to create the core area of the campus that is free from public transportation can be applied. So that the university can have three zones with core area called as center zone (ring 1) in the form of regional rectorate, ring 2 is in the form of clusters of education that is free from public transportation, and ring 3 is a service area and public activity center where the public transportation is allowed to operate.

6. Conclusions

From the discussion above it can be concluded that

a. Halu Oleo University uses a center-oriented pattern with a rectorate as the center and surrounded by faculties as education facilities and the outer layer is the green open (university's woods) and public open space (football field and parks).
b. The governance of Halu Oleo University is applied with a green eco concept. It applies with the area optimization through the concept of GBB or Green Blue Belt and the concept of regional zones by function and proximity of each faculty putting into particular clusters.

References

[1] Ministry of Public Works and Housing Number 06/PRT/M/2007, On 16 March 2007, about General Guidelines for Building Management and Environment
[2] Shirvani, Hamid, 1985, The Urban Design *Process*, Van Nostrand Reinhold, New York

Evaluation of Bottom Ash Composition on Modified Hollow Brick Design with Sago Husk as Filler

Kurniati Ornam[1*], Masykur Kimsan[2], La Ode Ngkoimani[3], Siti Belinda Amri[1], Santi[1], Syafrianto Amsyar[1]

[1] Halu Oleo University, Architecture Department, 93232, Kendari, Indonesia
[2] Halu Oleo University, Civil Engineering Department, 93232 Kendari, Indonesia
[3] Halu Oleo University, Geology Engineering Department, 93232 Kendari, Indonesia

*Corresponding author's e-mail: kurniati.ornam@yahoo.co.id

Abstract. Improvement of hollow brick models demands profound analysis in terms of their behavior on mechanical properties as some shape variations provide different characteristics on stress concentration. Additional waste filler i.e. sago husk and utilization of bottom ash are expected to deliver better results based on strength, weight and economic factor. This paper aims to analyze experimentally and analytically those 7 shapes variations of modified hollow bricks design with additional admixtures which consists 1,1% of sago husk and 4 composition of bottom ash i.e. 0% ; 10% ; 20% ; and 25%. Some works were conducted in laboratory, and followed by mathematical stress analysis of hollow shape characteristics. The results show that 10 – 20% bottom ash composition share maximum value of compressive strength, with model 1 is the greatest (17.8 Mpa). Mathematical analysis provides similar results as well, where the stress concentration of model 1 was not amplified with subsequent hollow position. In conclusion, all model still appropriate to be utilized for housing, especially in element selection for lightweight bricks that meet the requirements for strength.

1. Introduction

Improvement of power plant capacity in South East Sulawesi, Indonesia, has involved some natural resources e.g. coal for coal fired power plant in which the coal fly ash and coal bottom ash are generated. This power plant is located in Ni Tanasa, Konawe. Despite immense amount of their production every year worldwide, reusing product of coal ash, mainly in civil engineering applications e.g., is hardly to achieve 30% of its total production [1].

Some studies have been conducted to evaluate utilization of fly ash and bottom ash as partial replacement or substitutive material in modified conventional bricks, in addition to other waste material used as filler and inner burning at firing process. Although fly ash and bottom ash reduced workability of fresh concrete due to their higher water absorption, the compressive strength of concrete indicated better results [2]. It is also obtained that none of the study demonstrated negative impact of bottom ash utilization in concrete. 30% fly ash was argued as best ratio as cement replacement [2] while the code [3] limits for 20% fly ash in concrete. As replacement both fine and coarse aggregates, fly ash and bottom ash did not share negative effects, where 80% ratio was illustrated as the best composition.

In terms of clay bricks, non-plastic component for body formulation is required. Bricks tend to be high in drying and firing shrinkage, hence cracks occur during those processes. Bottom ash has drawn better performance to meet the requirement, as the substitution of sand, where shrinkage might be reduced and the body is successfully stabilized during the process, by creating higher percentage of median pore [4]. Furthermore, through dehydroxilation, degree of moisture decreased due to the presence of bottom ash as effective flux, with its combination with sandy clay that might result in pre-heat and cooling cracking provided self-act. Therefore, presumption of utilization of bricks with bottom ash will secure the advantages of improvement of bricks performance in their physical and mechanical properties.

Inclusion of some local wastes e.g. reed and sago husk in brick production has promising objective in adapting to environment and sustainability issues, and gained positive results in matching the criteria in codes for wall bricks following some experimental and numerical studies [5 – 12]. This might generate economic impact in housing sectors, not only in private sectors, but also in public sectors and support economic growth without ignoring environmental impact i.e. emission and energy used in building compared to building with concrete block. Furthermore, firing process of clay bricks generates a range of gas emissions into the atmosphere, where higher heating rates will significantly reduce emissions, provide insignificant effect on the water absorption and initial rate of absorption, and however, decrease the compressive and tensile strength [13]. Combining those recycled admixtures with bottom ash will also reduce CO emission significantly, with careful allocation of their proportion to get compromise between mechanical and thermal effect that leads on cracking on bricks surface [14].

As strong – light building becomes more preferable in light of major earthquake recently occurring in Indonesia, total mass of house should gain interest to reduce lateral force impact due to strong earthquake. This paper is to scrutinize variation of bottom ash composition with modified 1.1% of sago husk - hollow clay bricks. Previous study [15] demonstrated that the compressive strength of bottom ash – sago husk clay bricks can hit 24 Mpa maximum. Provided proper gap with minimum requirement of compressive strength (10.6 Mpa), this study is supposed to evaluate the best composition of bottom ash.

2. Sample preparation, hollow brick design and methodology

The procedures of this experimental study were as follows:

- Preparation of clay, sago husk and bottom ash. The clay is taken from Punggolaka, Kendari, categorized as silty clay with sand following sieve analysis and atterberg limit test [16]. Sago husk was obtained from Konawe and bottom ash was from Nii Tanasa power plant.
- Sago husk was dried by direct sunlight.
- Clay, sago husk and bottom ash were weighed in accordance to their mass in composition planned i.e. 1,1% sago husk with 4 (four) variation of bottom ash (0%, 10%, 20% and 25%).
- Setting up 6 (six) mold medium as models of hollow bricks, as can be seen on figure 1. All models are 22 cm in length, 11 cm in wide, and 5 cm in thickness. The hollow section used in the designs are circle, square and triangle which the area reductions are 9%, 19%, 15%, 20%, 27% and 30% respectively.
- Brick's making was started at laboratory, at room temperature (25-32 °C) and at humidity 76-81%, firstly to mix up clay with fine sago husk, bottom ash and certain amount of water hence they could be formed and unattachable.
- Next step was molding the bricks in certain size hollow model for specific composition in presenting actual scale at the work shed in the drying process, molded bricks were exposed to fan with certain amount speed, direct sunlight and oven. Additionally, to secure uniform heat

IOP Conf. Series: Materials Science and Engineering **797** (2020) 012030 doi:10.1088/1757-899X/797/1/012030

transfer on bricks surfaces, bricks were covered with alluminium foil and burning process was performed with stove and zinc plate where the firing temperature was measured and maintained at approximately 550°C on bricks, with medium heating rate.

Once the procedures were meticulously performed, some physical and mechanical tests were precisely conducted to obtain accurate result of those properties.

2.1. Density and water absorption

Density measurement was conscientiously performed every steps of molding, drying, prior to and after firing process. Since density and water absorption are alike related to weight and dimension measurement, the values could be established in the same time. Bulk density is defined as sample mass in one unit of its volume and water while water absorption is percentage of dry weight to its saturated weight. Water absorption test was carried out based on [17] for every age of curing specimens. Normally, density of clay brick is from 1,60 gr/cm^3 to 2,00 gr/cm^3, and it can be lessen by some engineered purpose i.e. additional inner burning admixtures.

2.2. Colour, texture and shape

Brick colour will vary based on its raw material and firing process. Normal red/brownish orange colour for clay bricks is affected as the bottom ash is mixed with clay hence the colour might be dark brown. Bricks shape should meet proper length, width and thickness as mentioned in code. Bricks surface is maintained plan and rough, square, producing loud sound when it is tapped, no crack, and unbreakable [18].

2.3. Initial rate of absorption and salt content

Initial rate of absorption is defined as the number of grams of water absorbed in one minute over 30 square inches of brick bed area [19-22]. Brick must not be very dry to secure bond capacity due to high amount of water absorption. Salt content in brick is prevented to avoid reduction of its preservation as its crystallization covers more than 50% of bricks surface.

2.4. Compressive strength

Compressive strength, as mechanical properties, was tested in according to ASTM C67-11 only on bricks stretcher face arae, where the minimum requirement is 10,40 Mpa as stated in code [17]. Based on previous research [5-11], hygroscopic material e.g. sago husk, reed, might accelerate the time and increase the compressive strength.

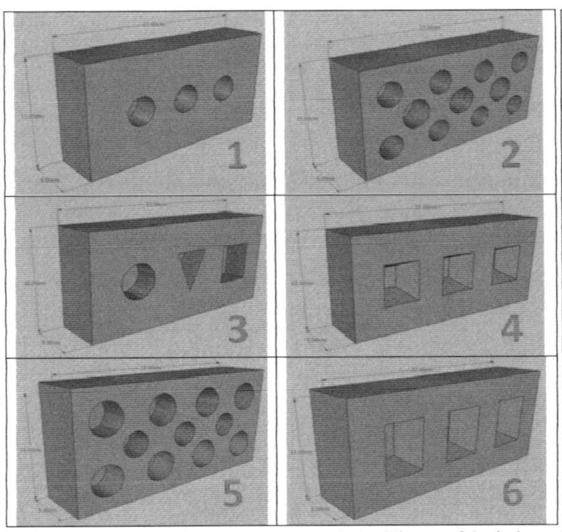

Figure 1. Design models and labelling of bricks.

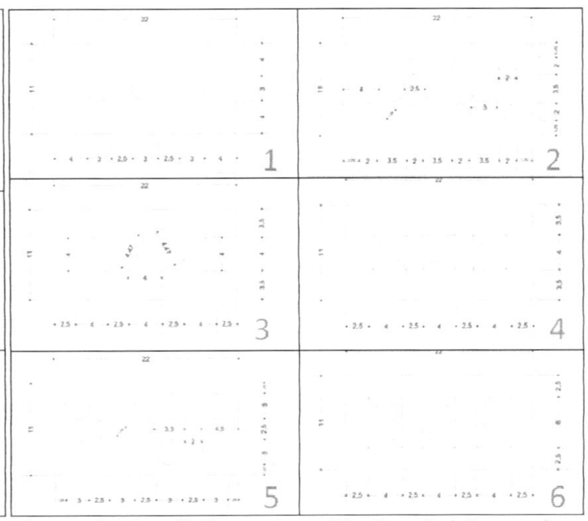

Figure 2. Detail dimension of hollow bricks section.

Figure 3. Modified bricks after molding.

3. Results and discussion

3.1. Physical and mechanical properties

Based on visual inspection, some cracks occurred in pre-heating due to inconsistent water distribution in bottom ash position. Moreover, number of hollow section is also proportional to number and width of cracks since stress concentration of the hollow section that can be expanded for sequence hollow. Drying process of modified bricks took 5-6 hours, which is faster than conventional bricks that took 5 to 7 days. Therefore, production rate is increased by utilization of bottom ash. Area reduction of hollow to non hollow section follows by reduction in weight and compressive strength.

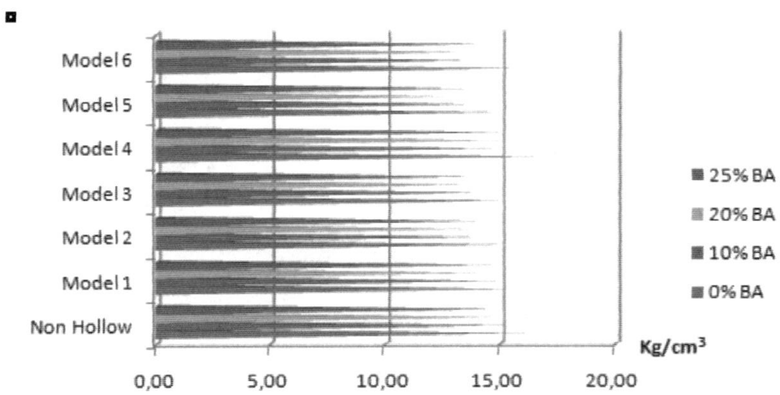

Figure 4. Bulk density of modified bottom ash brick with 1,1% sago husk

The density measurement demonstrated values between 1.29 up to 1.66 g / cm^3 (figure 4). It can be seen that higher bottom ash composition yield lower bulk density due to its water absorption capacity of bottom ash in certain time with 8 – 15% difference between 0% bottom ash and 25% bottom ash. Based on the graph and difference percentage, it is indicated that the bulk density will remain steady after 25% bottom ash after significant (logarithmic) decrease from 0 and 10% bottom ash.

All models, either hollow or non hollow section share similar color i.e. darkest brown. Consistent values of linear shrinkage in all axis (length, width and thickness) were demonstrated by all models with approximately 90%, as stated in the standard to guarantee isotropic properties of bricks. Furthermore, the initital rate of absorption (IRA) of all models is 0.18% in general and it is still considered suitable with ASTM C67-14.

Figure 5 illustrates that 1,1% sago husk bricks with all variation of bottom ash meet requirement for structural bricks i.e. more than 10.4 Mpa. For hollow section, model 1 deliver best value for compressive strength. Model 3 and 4 share lowest value, respectively. Nevertheless, either structural or non structural purpose, their performance is deemed suitable for building element. 20% bottom ash yield excellent result than other composition, where it increases from 0% and decreases after 20%. 10% bottom ash also share fine result in terms of compressive strength.

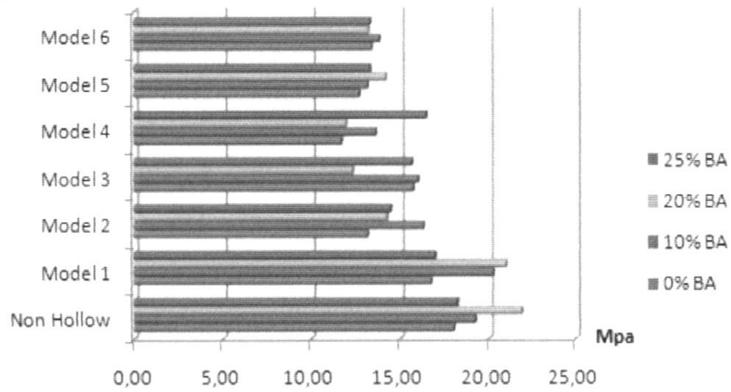

Figure 5. Compressive strength of modified fly ash brick.

4. Conclusion

Utilization of bottom ash with local waste i.e. sago husk could be established in housing sectors, since all properties are well satisfied, both structural and non structural components. 10% and 20% composition of bottom ash in this admixtures might yield better requirement for light and strong building due to their result on bulk density and compressive strength. However, workability in molding process should be vigilantly executed as cracks are comfortable to occur provided water distribution is not uniform in bricks body.

References

[1] M L D Jayaranjan, E D V Hullebusch and A P Annachhatre 2014 *J. Rev. Environ. Sci.* **4** 9336

[2] L Taofiq, A L Aamyem, E Essediqi, M Monkade and A Zradba 2018 J. *J. Mater. Environ. Sci.* **9** 4

[3] ACI 318-14 2014 *Building Code Requirements for Structural Concrete and Cementary*.

[4] G V Tayler, and W Daydone May 9-12 2011 *Proc. Int. Conf. on World of Coal Ash* (Denver, Co, USA)

[5] K Ornam, M Kimsan and C Edy 2015 *Proc. Int. Conf. on Energy Systems, Environment, Entrepreneurship and Innovation (Dubai)* vol. 1 (Uni Emirat Arab) p 373-37

[6] K Ornam, K Masykur, L O Ngkoimani and E Cahyono May 8-10 2015 *Proc. Int. Conf. on The Celebes International Conference on Diversity of Walacea's Line vol 1 (*Kendari*)* p 362-367

[7] K Ornam and M Kimsan August 11-12 2015. *Proc. Int. Conf. on the Second of Makassar International Converence of Civil Engineering* The Effect of Reed as Filler upon Brick Performance of Building Material (Makassar)

[8] K Ornam, M Kimsan and L O Ngkoimani October 22-23 2015 *Proc. Int. Conf. on* The *Second International Conference on Engineering of Tarumanagara* Performance Analysis in Home Industry Scale Production of Modified Traditional Brick as Green Building Material with Reed as Filler (Jakarta)

[9] K Ornam, Noraduola, and Santi March 9-10 2010 Proc. Int. Conf on *The First of Makassar International Converence of Civil Engineering* The Aplicability of the Usage of Hollow Brick Filled Husk for Low Cost Housing Project (Makassar)

[10] K Ornam, M Kimsan, L O Ngkoimani and E. Cahyono 2016. *J of Mathematical and Computational Methods*. **1** 87-91
[11 S P Raut, R V Ralegaonkar and S A Mandavgane 2011 *J Construction and Building Materials*., **25** 4037-4042
[12] V Banhidi and L A Gomze 2008. *Improvement of Insulation Properties of Conventional Brick Products* Materials Science Forum 589 pp 1-6
[13] Abdul Kadir, Aeslina, Mohajerani and Abbas 2015 *J. Applied Clay Science*. 1**04** pp 269-276
[14] Cecile, Bories, Aouba, Laila, Vedrenne, Emeline, Vilarem and Gerard 2015 *J. Construction and Building Materials*. **92** pp 158-163
[15] K Ornam, M Kimsan, L O Ngkoimani and Santi 2017. *Study on Physical and Mechanical Properties with Its Environmental Impact in Konawe - Indonesia upon Utilization of Sago Husk as Filler in Modified Structural Fly Ash – Bricks*. (Proceedia Computer Science) p 420–426
[16] ASTM D 2487 2006 *Standard Practice for Clasification of Soils for Engineering Purposes:* American Society for Testing and Material
[17] ASTM C140 2018 *Standard Test Methods for Sampling and Testing Concrete Masonry Units and Related Units:* American Society for Testing and Material
[18] ASTM C67 2014 *Standard Test Methods for Sampling and Testing Brick and Structural Clay Tile:* American Society for Testing and Material
[19] ASTM C216 2015. *Standard Specification for Facing Brick Philadelphia:* American Society for Testing and Material
[20] ASTM C62 2012 *Standard Specification for Building Brick (Solid Masonry Units Made from Clay or Shale)*: American Society for Testing and Material
[21] ASTM D 2487 2006 *Standard Practice for Clasification of Soils for Engineering Purposes:* American Society for Testing and Material
[22] SNI 15-2094 2000 *Bata Merah Pejal untuk Pasangan Dinding*. Balitbang Kimpraswil (Bandung)

IOP Conf. Series: Materials Science and Engineering **797** (2020) 012031 doi:10.1088/1757-899X/797/1/012031

Dynamics of Community Based Coastal Resource Management in Wabula Village, Buton Regency

La Nalefo

Department of Agriculture Economic, Faculty of Agriculture Universitas Haluoleo, 93232 Kendari, Sulawesi Tenggara, INDONESIA

E-mail: lanalefo@rocketmail.com

Abstract. Coastal Resources Management was carried out by the community have been displaced by government's role which tends prefer to capital investors. The consequence of controlling local communities was becoming weak, and competing got benefit from capital entrepreneurs. The aims of this study were (1)the dynamics of the implementation of coastal resource management in the community, (2)the application of the basic principles of traditional-based coastal resource management and programs, (3)key factors that influence the existence of traditional resource-based resource management systems and programs. The results of the study found that the dynamics of the implementation of community based coastal resource management was constructed by a socio-cultural order that continued to develop in accordance with the demands of the times. The creation of a management system based on its dynamic program was based on the level of needs and sustainability of the ecosystem determined by the Ministry of Fisheries included in the COREMAP program. The application of the basic principles of community-based coastal resource management is very dynamic and manifested on a local conceptual basis. The key factors these influencing the existence of a community based coastal resource management system rested in the weakness of the authority exercised by the *parabela*.

1. Introduction

Coastal is a unique area in the context of landscape, the region is the meeting place of land and sea [1, 2, 3]. The first comprehensive tool kit for coastal planners and those aiming to achieve effective coastal management worldwide. *Coastal Planning and Management* provides a link between planning and management tools and thus includes all stages in the process, from development through evaluation to implementation [4].

Drawing on examples of successful coastal planning and management from around the world, the authors provide clear and practical guidelines for the people who make daily decisions about the world's coastlines. *Coastal Planning and Management* is an invaluable resource for professionals in environmental and planning consultancies, international organizations and governmental departments, as well as for academics and researchers in the local and international fields of geography, marine and environmental science, marine and coastal engineering and marine policy and planning.

Coastal areas are an important area in terms of planning and management points of view. Viewed from its potential, it is divided into four categories, namely (1) renewable resources, such as fish, shrimp and so on, coastal and marine aquaculture activities, (2) non-renewable resources, such as petroleum and gas, (3) marine energy, such as tides and waves, (4) environmental services, such as tourism and marine transportation.

Nevertheless the reality of some serious issues has also become strategic issues in the management of coastal and marine resources such as the condition of coastal resources and sea that are common property with access that is quasi open access, more lead to understanding the nature of resources that are public domain and no owner. This condition produces a phenomenon called the dissipatet resource rent, ie loss of borrowing of resources that should be obtained from optimal managemen [5]. With quasi open access then the actions of one party that harms the other can not be corrected by the market (market failure). This creates economic inefficiency because all parties will try to exploit the resources as much as possible, otherwise the other party will get benefit. Reinforced with technology, capitalists are able to excessively exploit resources resulting in the law of the jungle (who is strong, he wins) and the natural production power becomes disrupted.

This study focuses on the basic principles of ideal CBRM as required by [6], such as (1) mobilization and community participation, (2) social capital and partnerships, (3) resources and justice, (4) dissemination of information and communication, (5) research and development, (6) empowerment and delegation of authority, (7) legitimacy and community trust, (8) monitoring, feedback and accountability; (9) collaboration management and adaptive leadership; (10) participation in making decisions, (11) the possibility of the environment optimizing the initial conditions, and (12) cooperation and conflict resolution.

2. Research Method

2.1 Data Source

The data obtained from this study were sourced from primary and secondary data [7, 8]. Primary data was obtained from the results of interviews and direct observation of researchers in the field with informants, namely: Parabela, Imam, Moji, Village Heads, Religious Figures, Community leaders, Fishermen, Collectors, Cultural observer, Lecturers, Non-Governmental Organizations, Government, Facilitators who had working on Coremap, and others. Secondary data comes from relevant agencies, namely the Department of Marine and Fisheries, Buton District.

Qualitative data collection is done through the determination of informants based on initial information about community members involved in resource utilization and management, including fishermen.

2.2 Procedure for Data Retrieval

In this study, a number of certain effective data were collected and integrated in the analysis process, and were presented in such a way as to support the main themes that were the focus of the research, so that it is a separate construction as a product of interaction between field informants and researchers [9] and in depth interview and focus group [10].

2.3 Data Analysis

The main analysis used in this research was analytical qualitative data analysis through interpretation and comprehension. In order to obtain accurate data, field notes were refined, coded data and problems. Coding of data based on the results of criticism made, the appropriate data separated by certain code. The data obtained were analyzed component by three stages:

The first stage was the process of rough data reduction of field notes. In the process, selected data were relevant to the research focus. Data reduction was done by creating data summaries, scattering themes, and creating a basic data presentation template.

The second stage was the presentation of data, namely the preparation of a collection of information into a statement for subsequent withdrawal of conclusions. The data were presented in the form of narrative texts, initially scattered and separated on various sources of information, then classified according to theme and analysis needs.

Third stage, drawing conclusions based on reduction and presentation of data. The gradual conclusion drew from the general conclusion at the time of data reduction then becomes more specific at the actual conclusion step. This series of processes indicated that the qualitative data analysis in this study combines the data reduction, data presentation, concluding and recurring conclusions.

3. DISCUSSION

3.1 Dynamics Implementation of Community-Based Coastal Resource Management

Community Based Coastal Resource Management (PSPBM) is a resource management system in place where local communities are actively involved in the management of natural resources contained therein, so that the coastal communities to manage their resources together and produce a typical social interaction.

From some of the decisions the result of violations of the rules on the DPL one of these cases has been discussed about the process of completion, which report the violation by taking the octopus in the DPL. Deliberation troubleshooting on Sunday March 2, 2008 in *Galampa* Wabula village. La Duwi Rapporteur is also as Motivator Village (MD) program COREMAP II Village Wabula who caught a villager Wabula the initials SR who took octopus in DPL area Wabula - Wasampela. In a statement reprimanding LD SR for entry in the area of SR precisely DPL but still take an octopus and argued that its existence outside some mast step on which he thought was the limit DPL. However, because the witness so that LD remains that what SR includes elements of intent.

Problem solved with the ceremony kaleo-leo, in accordance with the rules "in the event of an internal conflict between residents in one village and not equipped with witnesses or evidence the settlement kaleo-leo and the results must be accepted by both the guilty party shall pay a fine in accordance with the offense ". Kaleo-leo ceremony won by LD where SR cannot survive in water for 3 minutes while the LD able to survive longer than SR.

Therefore, *kaleo-leo* process won by LD then the SR is obliged to pay a fine of one million, as disclosed in village regulations.

The main course of conflict was the local fishermen between fishermen from the outside, is as said by [11], that the dominance of the doctrine of open access was still strong so that the fishermen consider that marine and coastal waters as belonging to everyone. It marked setting by local people who had been hereditary not got the recognition that strong, it is often violated. Because the doctrine of open access community stems from the tradition of the North Atlantic countries.

In addition, in the form of system to accommodate new entrants, Wabula society provided opportunities from the outside to catch fish in the waters around the coast if requested permission from the parties agreed that Traditional means. Fishermen were allowed to obey the rules that have been agreed upon by the entire community through village regulations or other regulations. However, fishermen are accommodated for arrests region Wabula waters more disciplined than other regions. It caused by firmness/Parabela supported by the police.

3.2 Implementation of Basic Principles of Community-Based Coastal Resource Management

The dynamics of community based coastal resource management and context management were not independent of the motion of society as a continuous (dynamic) that cause changes in the governance of coastal resources. Therefore, the history of the coastal resource management system is a construction that can explain the dynamics of the ongoing process itself. This is similar with the stated that approach today must be understood by the underlying dynamics, and it was not obtained without assessment or at least reveal about past history.

Thus, the dynamic analysis approach management of coastal resources in the study area that were traditional management, nature management and creation of management programs that combine both (traditional and program) was analyzed based coastal resource management systems that occur in the region of cases based on the sequence of events (period). Community based coastal resource management in the region fluctuated case.

Practice management of marine resources in Takimpo and Wabula, in the past, managed limited only fish or by species / species specific, while the types of resources that others even ecosystem too freely could be exploited, because only of the fish only marine resources at that time used to meet the food needs of the community, while other commodities had not had economic value. Fishing ban was reinforced by the agreement and customary regulations which covers an area aligned with the customary authorities. Kaombo effort when it was ecologically very well due to the utilization of fish would be controlled, even the results of the tender used for the public interest in the improvement of roads, mosques and other improvements. The fishermen catch the fish.

Figure 1. Illustration Mechanism Kaombo

Kaombo implementation process began with an event at the consultative meeting Baruga / galampa led by Parabela attended by sarano Traditional, community leaders and the general public. In the meeting there were discussed three main topics: (1) the boundaries of the area to be closed; (2) be exploited marine resources; and (3) sanctions for violations. After it was made public by the Imam and devices, especially on Friday before the Friday prayers. Furthermore, the installation of a line in the form of a flag of sticks and palm leaved on a lie within the region were closed. Installation kaombo limit sign carried by Waci / wati (member councils; sara charge of general rules), and Moji (members of the councils in charge of the rules of religion). Diombo region length according to your needs and *sarano* Traditional society, if a party took a lot of fish took a long time anyway. Kaombo old generally ranges from six to twelve months.

Another thing that was interesting in Wabula is dynamic resource management systems of systems kaombo the supervision of commodity based, namely the prohibition of the exploitation of species of sea cucumbers, trochus or abalone and japing-japing during kaombo (temporary) then entered and fused collaborating with resource management system society based that DPL (Marine Protected Area). Kaombo system in this research report would then be a simple intellectual property rights administered by the researcher.

4. CONCLUSION

Implementation of community based coastal resource management was constructed by the socio-cultural order which continues to develop in every time. While the creation of the program-based management system, the dynamics based on the level of needs and sustainability of ecosystems established by the Ministry of Fisheries are included in the Core map program. The application of the basic principles of community-based coastal resource management was very dynamic and manifested on a local conceptual basis. A key factor influencing the existence of a community based coastal resource management system lies in the weakness of the authority given to the community, and vice versa, in the program was community support in managing coastal resources in its territory. The adaptive/ accommodative coastal resource management scenario refers to the capacity of the community supported by the access and control it had in managing the coastal resources in its territory. The combination of Community Program Management and the Program mutually contributes to the sustainability of coastal resources with increasing coastal area, namely *Ombo* and Marine Protected Areas.

IOP Conf. Series: Materials Science and Engineering **797** (2020) 012031 doi:10.1088/1757-899X/797/1/012031

REFERENCES

[1] Key, R, and Jaqueline Alder, 1999. *Coastal Planning and Management.* E & FN Spon, an imprint of Routlegde, London.

[2] Elmer M. Ferrer and Cristi Marie C. Nozawa 2016. *Community-Based Coastal Resources Management in The Philippines: Key Concepts, Methods and Lessons Learned.* University of the Philippines at Diliman.

[3] M. Gleason, 2018. *Ocean & Coastal Management.* The Nature Conservancy California Coastal and Marine Program, Monterey, California, USA.

[4] California Environmental Association (CEA) 2018. *Trensds in Marine Resources and Fisheries Management in Indonesia.*

[5] Isaac Boateng, 2018. *Policy Process in Coastal and Marine Resource Management in Ghana.* Department of Geography, University of Portsmouth Buckingham Building.

[6] Gruber, J.S, 2010. *Key Principles of Community-Based Natural Resource Management a Synthesis and Enterpretations of Identified Effective Aproaches for Managing the Commors.* Enviromental Management (2010) 45 p. 52-68.

[7] Dagmar. H., Xavier Majo Roca., Ferenc Marvanykövi., Joszef Razc (2008). (Editor). *Development of a Data Collection Protocol for Specialist Harm Reduction Agencies Foundation Regenboog AMOC.*

[8] Stephanie Zaza., Linda K. Wright-De Agu., Peter A. Briss,, Benedict I. Truman, David P. Hopkins., Michael H. Hennessy., Daniel M. Sosin., Laurie Anderson, Vilma G., Carande-Kulis., Steven M. Teutsch., Marguerite Pappaioanou., 2000. *Data Collection Instrument and Procedure for Systematic Reviews in the Guide to Community Preventive Services.* American Journal of Preventive Medicine, 18 (1):44-47.

[9] Alfgeir Logi Kristjansson, Jon Sigfusson, Inga Dora Sigfusdottir, John P. Allegrante, 2017. *Data Collection Procedures for School-Based Surveys Among Adolescents: The Youth in Europe Study.* Journal of School Health Vol. 83, No. 9, pp 662-667.

[10] Adams, Anne and Cox, Anna L. (2008). *Questionnaires, in-depth interviews and focus groups.* In: Cairns, Paul and Cox, Anna L. eds., *Research Methods for Human Computer Interaction.* Cambridge, UK: Cambridge University Press, pp. 17–34.

[11] Matulada, 1997. *Kebudayaan Bugis-Makassar.* Penerbit, Jambatan. Jakarta.

IOP Conf. Series: Materials Science and Engineering **797** (2020) 012032 doi:10.1088/1757-899X/797/1/012032

Groundwater Change Detection by Gravity Measurement on Northern Coast of Java: A Case Study in Semarang City of Central Java of Indonesia

Laode M Sabri[1*], Bambang Sudarsono[1], Yadi Aryadi[2]

[1]Departement of Geodetic Engineering, Universitas Diponegoro, INDONESIA
[2]Pusat Geodesi dan Geodinamika, Badang Informasi Geospasial, INDONESIA

[*]Corresponding author's e-mail: lmsabri@ft.undip.ac.id

Abstract. Northern part of Semarang city is affected by tidal flood due to land subsidence. Land subsidence is triggered by soil compaction and massive groundwater extraction. One of famous methods in analysing groundwater is gravity method. This research aim to present groundwater change in Semarang city based on gravity measurement and to analyze problems related to equipment and groundwater behaviour. Gravimeter Scintrex CG-5 was applied for gravity measurement. Geographic coordinates were measured using GNSS receiver namely TopCon HiperII. This research occupied SRTM90 Plus and regional geology map at 1:100.000 scales. Groundwater change was detected by differentiating simple Bouguer anomaly of 2014 and 2016. Data in gravity unit were then converted by coefficient of 0.012 mgal per meter. Simple Bouguer anomaly in 2014 ranged from -3.423 mgal to 19.874 mgal, while 2016 data ranged from -3.427 mgal to 19.835 mgal. Discrepancy of gravity anomaly ranged from -0.326 mgal to 0.209 mgal. The data confirmed that groundwater change ranged from -27.160 m to 17.408 m. Groundwater depletion occurred commonly on alluvial plain in northern area, while recharging could be found in some locations in southern part. This research concluded that gravity measurement had potential to be applied for groundwater detection in coastal areas.

1. Introduction

Java Island is located on the meeting plate of the Euroasia Plate with the Australian Plate. Java Island has a narrow and elongated shape that can be divided into three zones, namely the southern zone, the middle zone, and the northern zone [1]. The Southern Zone consists of folded hills and mountains stretching along the southern coast of Java. The Central Zone consists of rows of volcanoes and fluvial volcanic plain. The Northern Zone is generally dominated by the alluvial plain with a sloping topography. Major cities in the northern coast of Java region generally grow on the alluvium plain.

The alluvium plains in the region were formed from the erosion and sedimentation process by the river and coastal streams. The compositions still allow consolidation or compaction of upper soil layers [1]. On going soil compaction and groundwater extraction lead to land subsidence in northern part of the city [2]. The alluvium plain in the northern part of Semarang reaches around 40% of the area, while in the southern part it is dominated by volcanic rocks and several fault structures [3]. The thickness of the alluvium layer can reach 50 m or more [4].

Semarang City is a proper case study area to study natural and socio-economic phenomena in the northern coast of Java. The Terboyo Industrial Zone and Wijaya Kusuma Industrial Zone which are located in northern part of the city have a important role in socio-economic development, both regionally and nationally. Contrary, the industrial activities have a strong correlation to the occurrence of land subsidence and decrease in groundwater quantity and quality [5]. Due to low capacity of regional water supply company, industries occupied deep wells for groundwater extraction. At present, there are more than 1000 registered wells operating in the city [6]. On the alluvial plain in the northern part of Semarang City, massive groundwater extraction leads to sea water intrusion.

Groundwater investigation are usually carried out by geo-electric or seismic methods. In the case of large groundwater reservoir, the gravity method may be applied [7]. The complicated problem in applying gravity method is the phenomena of land subsidence and groundwater fluctuations that affect the gravitational value recorded by gravimeter [8]. Based on the free air approach, land subsidence will give a positive impact on the value of gravity, while a decrease in the groundwater level has a negative impact on the value of gravity [9]. Another problem in groundwater research in the city of Semarang is the occurrence of seawater intrusion and tidal flood infiltration filling groundwater reservoir in the alluvial region. Another problem that must be considered in groundwater research is seasonal variations that affect groundwater level fluctuations. This study aims to detect changes in groundwater in the city of Semarang based on topographic, geomorphological, and seasonal variations as a case study for investigating groundwater change of northern coast of Java.

2. Method of research

2.1. Data collection

Groundwater changes were detected by gravity measurements on common points in two periods. The first period was September 2014, while the second period was March 2016. Both periods applied Scintrex CG-5 Gravimeter which has ± 5 μgal accuracy. Applying International System of Units, 1 μgal or 1 microgal as gravity unit is equal to 10^{-8} $m.s^{-2}$ while 1 mgal or 1 milligal is equal to 10^{-5} $m.s^{-2}$. Gravity measurement was carried out on 50 common points in Semarang City. Overlaying observation stations to topography map can be seen in figure 1 while figure 2 shows geological formation below the stations. The geographic positions of the gravity points were measured by TopCon HiperII GNSS receiver applying rapid static method to obtain centimeter-level accuracy [9]. This study also occupied topographic data from SRTM90 Plus and actual rock density data [10].

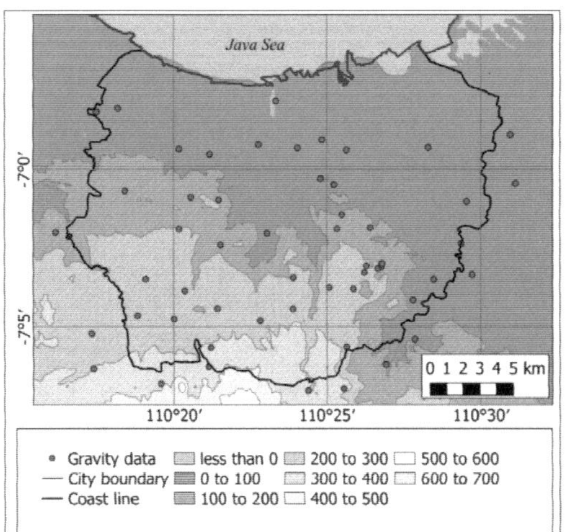

Figure 1. Map of gravity data and topography in Semarang city.

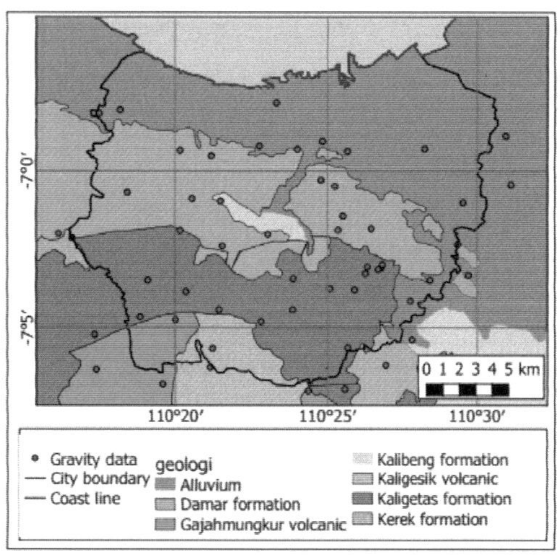

Figure 2. Map of gravity data and geology in Semarang city.

Gravity measurement points were distributed regularly from 0 to 400 meters above sea level as shown by figure 1. Repetitive gravity observations were also arranged to various sediment and volcanic formations as shown by figure 2. This research applied rock density data which was derived from previous interpretastion of geological map and gravity data (war). Density of alluvium, volcanic rock, Damar formation, Kaligetas formation, Kalibeng formation, Kerek formation, and basement were 1.85, 2.00, 2.20, 2.30, 2.45, 2.60, and 2.85 $g.cm^{-3}$ respectively [10].

2.2. *Gravity reduction*

Gravity measurements in 2014 and 2016 refered to the absolute gravity point in Department of Geophysical of Universitas Diponegoro. It is located at 7.048383836 S, 110.4424687 E and 212.707 m above the WGS84 reference ellipsoid. The absolute gravity value of the point is 978078.22865 mgal. Earth tide effect was corrected automatically inside the gravimeter using Longman formula [11]

This study implemented Bouguer anomaly to predict changes in groundwater which based on reduction of terrestrial gravity data by normal gravity in ellipsoid, free air, and Bouguer plates [12]. This study did not use terrain reduction differentiation because this study assumed that the effects of terrain changes during the study period were relatively small and could be ignored. Gravity anomaly (Δg) is calculated by equation (1) as follow

$$\Delta g = g - \gamma \tag{1}$$

where g is actual gravity on surface of the earth, γ is normal gravity on ellipsoid at measurement point. Normal gravity is computed using formula of Somigliana [13] as shown by equation (2) as follow

$$\gamma = \frac{\left(a.\gamma_e.\cos^2\varphi + b.\gamma_p.\sin^2\varphi\right)}{\left(a^2.\cos^2\varphi + b^2.\sin^2\varphi\right)^{\frac{1}{2}}} \tag{2}$$

where γ_a is normal gravity at equator, γ_b is normal gravity at pole, a is semi-major axis, and b is semi-minor axis. The constant of WGS84 parameter can be seen in table 1.

Table 1. Constant of WGS 84 reference ellipsoid.

Symbol	**Value**	**Description**
a	6378137 m	Semi-major axis
b	6356752.3142 m	Semi-minor axis
f	1/298.257223563	Geometrical flattening
m	0.00344978650684	Physical flattening
GM	$3986004.418 \,.\, 10^8\ m^3.s^{-2}$	Geocentric gravitational constant of the earth
M	$5.9733328 \,.\, 10^{24}$ kg	Mass of the earth
γ_a	$9.7803253359\ m.s^{-2}$	Normal gravity at equator
γ_b	$9.8321849378\ m.s^{-2}$	Normal gravity at pole

Effect of elevation on gravity data must be reduced using free air reduction [13]. Free air reduction (FA) is derived from gravity gradient against elevation change as shown by equation (3) below

$$FA = \frac{\partial\gamma}{\partial h} = \frac{2.\gamma_a}{a}\left(1 + f + m - 2f.\sin^2\varphi\right)h - \frac{3\gamma_e}{a^2}.h^2 \tag{3}$$

where f is geometrical flattening, m is physical flattening, φ is geographic latitude of measurement point, h is elevation above reference plan. For this research, free air effect is computed using elevation above WGS 84 elipsoid reference. Geometrical and physical constant of WGS 84 can be found in table 1.

Effect of mass below observation point is reduced using Bouguer reduction [13]. Formula for computing Bouguer reduction (B) can be seen in equation (4) below

$$B = 2.\pi.G.\rho.h \tag{4}$$

where G is gravitational constant of the earth, ρ is rock density, h is elevation or thickness of bouguer plate. Formula of simple Bouguer anomaly is shown in equation (5) below [13].

$$\Delta g_B = g - \gamma + FA - B \quad (5)$$

2.3. Conversion of gravity anomaly to groundwater

Gravity measurements at the same point in different periods may give different value of gravity. In this study, changes in the value of gravity were identified due to changes in the depth of groundwater level. The effect of changes in groundwater level on gravity was calculated by modifying equation (4) into equation (6) as follow

$$\Delta B = 2.\pi.G.\rho.\alpha.\Delta h \quad (6)$$

Using the WGS 84 reference ellipsoid, the change effect is calculated by the following equation (7)

$$\Delta B = 0.04192769.\rho.\alpha.\Delta h \quad mgal \quad (7)$$

With the assumption of rock porosity of 30% and standard density of 2.67 g.cm, the gravity response function is obtained for the following groundwater derivation in the form of equation (8)

$$\frac{\Delta B}{\Delta h} = 0.0125783069 \, {}^{mgal}\!/_{m} \quad (8)$$

3. Result and discussion

Gravity measurements in 2014 were actually carried out at 92 points in Semarang city and surroundings covering more than 400 km^2. Gravity measurements in 2016 consisted of 174 points covering 10000 km^2. For preventing bias and inaccurate results, this study only applied 50 points that both were measured in 2014 and 2016. After earth tide correction and drift correction and instrument height correction, gravity data of 2014 were ranging from 978050.505 mgal to 978119,826 mgal, while data of 2016 were ranging from 978050,544 mgal to 978119,815 mgal. The difference in gravity values from field measurements was generally smaller than 1 mgal and could not be identified visually on the map, as shown by figure 3 and figure 4. The gravity values decreased gradually from coastal area at north part of Semarang city to Ungaran Volcano at southern part of the city.

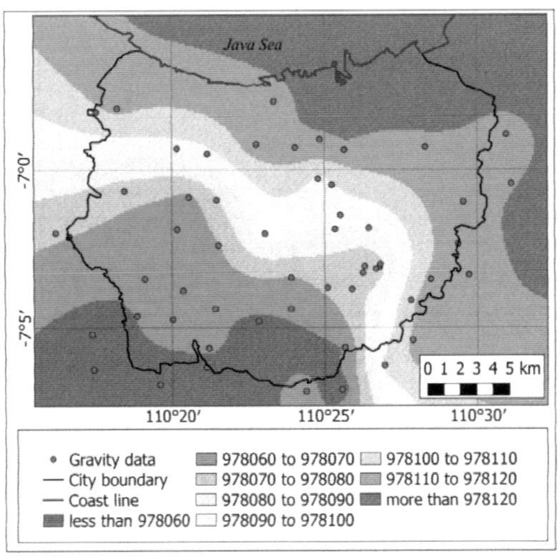

Figure 3. Map of gravity data of 2014.

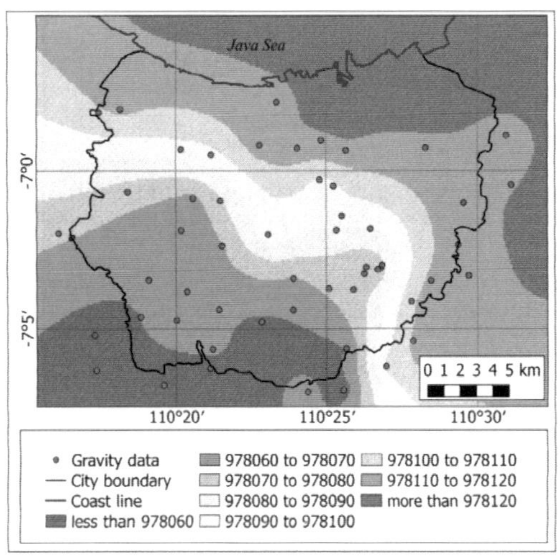

Figure 4. Map of gravity data of 2016.

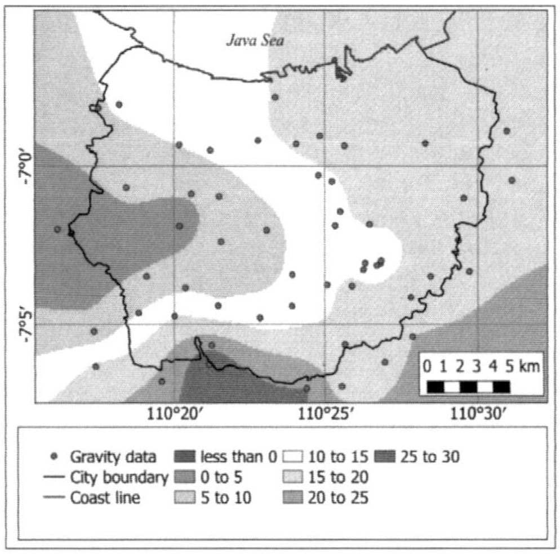

Figure 5. Map of Bouguer gravity anomaly of 2014 in Semarang city.

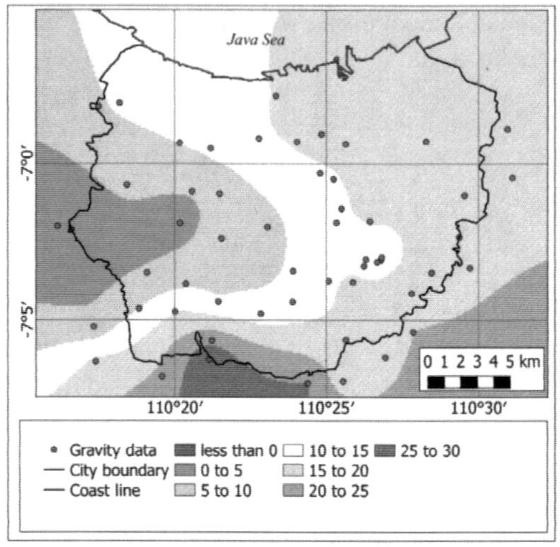

Figure 6. Map of Bouguer gravity anomaly of 2016 in Semarang city.

Trend of gravity data in figure 3 and figure 4 showed that data on earth surface have strong correlation to topography. Effect of topography were eliminated by simple Bouguer reduction to obtain gravity values on ellipsoid reference of WGS84. Gravity anomaly of 2014 were ranging from 0.023 mgal to 29.792 mgal, while gravity anomaly of 2016 was ranging from 0.017 mgal to 29.519 mgal. The differences of gravity anomaly values of 2014 and 2016 were generally smaller than 1 mgal and could not be identified visually on the map, as shown by figure 5 and figure 6.

Change in gravity anomaly were calculated by subtracting data of 2016 to 2014. Gravity anomaly change in period of 2014 and 2016 were ranging from -0.317 mgal to 0.211 mgal as shown by figure 7. Converting gravity anomaly change to groundwater change applied rock and sediment density derived from geological map, and 30% porosity assumption. Groundwater levels were declining in south part of Semarang city ranging from 10 m to 20 m as shown by figure 8. Groundwater levels were increasing at alluvial plain of Semarang city.

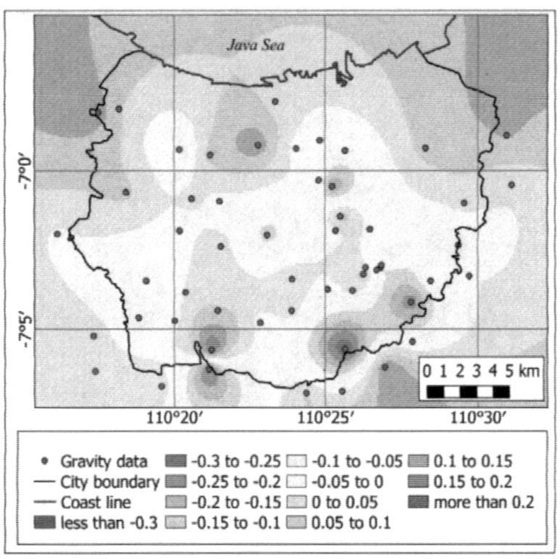

Figure 7. Map of bouguer anomaly change from 2014 to 2016 in Semarang city

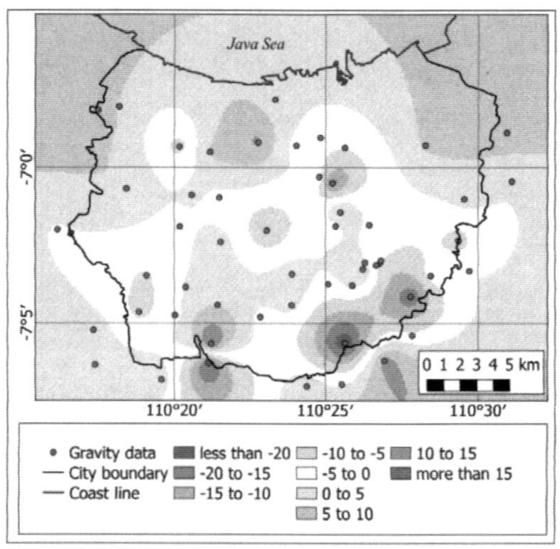

Figure 8. Map of groundwater change from 2014 to 2016 in Semarang city

4. Conclusion

Gravity measurements in 2014 and 2016 were able to detect changes in gravity in the Semarang city. In this study, gravity changes in rocks and sediments with porosity of 30% were caused by changes in groundwater level that change the geological density. The decrease in groundwater generally occurred in the southern part of Semarang City, while the increase in groundwater occurred in the northern part of Semarang City. It was suspected that uplifting of groundwater level was caused by seasonal variations due to gravity measurements in September and March. For the future research, it is recommended to combine tidal flood data to groundwater analysis.

References

[1] Poedjoprajitno S, Wahyudiono J and Cita A 2008 Reaktivitas Sesar Kaligarang, Semarang. *Jurnal Geologi Indonesia* **3** pp 129-138
[2] Abidin H Z, Andreas H, Gumilar I, Sidiq T P and Gamal M 2015 Environmental Impact of Land Subsidence in Urban Areas of Indonesia. *Proceeding of FIG 2015* Sofia
[3] Abidin H Z, Andreas H, Gumilar I, Sidiq T P, Gamal M and Murdohardono D 2010 Studying Land Subsidence in Semarang (Indonesia) using Geodetic Methods *Proceecing of FIG 2010* Sydney
[4] Thanden R E, Sumadirdja H, Richards P W, and Sutisna K 1996. Peta Geologi Lembar Magelang dan Semarang, Jawa, Skala 1:100.000 Bandung
[5] Putranto T T, Hidajat W K and Alexander K 2018 Controlling seawater intrusion beneath Semarang coastal urban city using geophysical survey and hydrochemistry data *International Journal of Geomate* **15** pp 173-179
[6] Putranto T T 2011 Hydrogeology of Semarang Demak groundwater basin: an overview and its challenges in preliminary groundwater flow modelling *Proceeding of JCM Makassar 2011.* Makassar
[7] Supriyadi S 2000 Membangun filter berdasarkan model amblesan dan dinamika muka air tanah untuk memisahkan anomali gayaberat mikro antar waktu *Jurnal Berkala Fisika* **13** pp. 1-10
[8] Sarkowi M 2011 Perhitungan defisit air tanah daerah Semarang berdasarkan inversi anomali 4D microgravity *Jurnal Sains MIPA*
[9] Leick A 2010 GPS satellite surveying (New York: John Wiley & Sons)
[10] Wardhana D D, Harjono H and Sudaryanto 2014 Struktur bawah permukaan Kota Semarang berdasarkan data gayaberat *Jurnal Riset Geologi dan Pertambangan* pp. 53-64
[11] Torge W 1989 Gravimetry (Berlin: Walter de Gruyter)
[12] Longman I M 1959 Formula for computing the tidal acceleration due to the moon and the sun *J. Geophysic Res* **64** pp. 2351-2355
[13] Hoffman-Wellenhof B and Moritz H 2005 Physical Geodesy (Berlin: Springer)

Characteristics of Solid State Sintered Silica Ceramic Derived from Rice Husk Ash

I N Sudiana[1], L Aba[1], L Lestari[1], H Aripin[2], K G Suastika[3], M Z Firihu[4], S Mitsudo[5]
[1]Laboratory of Material Physics, Department of Physics, Faculty of Mathematic and Natural Science, Universitas Haluoleo, 93232 Kendari, Sulawesi Tenggara, INDONESIA
[2]Center for Material Processing and Renewable Energy, Faculty of Learning Teacher and Education Science, Siliwangi University, Jl. Siliwangi 24 Tasikmalaya 46115, West Java, Indonesia.
[3] Department of Education Physics, Universitas Palangka Raya, Indonesia, 73112
[4]Laboratory of Theoretical and Computational Physics, Department of Physics, Faculty of Mathematic and Natural Science, Universitas Haluoleo, 93232 Kendari, Sulawesi Tenggara, INDONESIA
[5]Research Center for Development of Far-Infrared Region, University of Fukui, 3-9-1 Bunkyo, Fukui-shi 910-8507, Japan

*Corresponding author's e-mail: sudiana75@yahoo.com

Abstract. Silica xerogel produced from rice husk ash (RHA) taken from South East Sulawesi Indonesia has been successfully sintered by using a microwave oven 2.45 GHz as well as a millimeter waves (MMW) heating system with a 28 GHz gyrotron. The samples were also heated by using an electric furnace where served as a comparison. Physical characterization of the samples before and after sintering were then investigated by using an Archimedes densification measurement method device, a X-ray diffraction (XRD) and a Scanning Electron Microscopy (SEM), respectively. Effect of microwave energy on the characteristics of silica ceramic after processing were analyzed and compared to conventionally sintered results. The effect of frequency to properties of the silica was also revealed and reported previously in separated paper. In this paper, depedency of microwave power and heating rate to grain growth and density of the ceramic are reported. The results suggested that the microwave power affect to grain growth rate of silica xerogel as well as to their density. In other hand, grain growth is also strongly depend on heating rate during fabrication.

1. Introduction

Silica xerogel is one of the useful materials in the world. It is because its properties can be modified to desired properties by special treatment such as heating or doping by another material. Its structure can be shrinks during heating because the aqueous phase in the pores is removed by evaporation, capillary contraction, condensation – polymerization reactions, structural relaxation, or by viscous flow [1,2]. If the material processing by using heat treatment until below melting point, the process known as solid state sintering [2,3]. Moreover, the silica xerogel ceramic have several industrial applicable characteristics such as high transparency and good chemical durability [3–5]. Its properties can be modified by change its microstructure where can be done by change their chemical composition and physical parameters as well as adjust thermal process parameters during heating [6]. For application to waveguide, it is expected the silica glass ceramic have a high density, good mechanical properties with desired optical properties.

Previous results of processing silica xerogel shows that silica glass-ceramic with maximum density of 2.2 g/cm^3 can be obtained at processing temperature 980°C with low heating rate. It is far below its melting point where about 2000°C [2]. However, processing by low heating rates in conventional heating method needs long time which promotes grain growth result in decreasing its optical tranparancy. In contrast to the conventional heating such as by using electric furnace the heating by using a microwave can be performed in a high heating rate. It is because in microwave heating, the electromagnetic energy is dissipated simultaneously in the whole irradiated volume not only in surface as in conventional heating. Additionally, the microwave radiation offers higher densities of the heating power, selective heating, self heating, etc. Microwave 2.45 GHz up to 1 KW power output avaliable everywhere. But for high frequencies such as millimeter and sub-millimeter waves are rarely avaliable. For a high microwave frequencies, gyrotrons are considered as the most appropriate technology for radiation source up to now.

Apllications of the such electromagnetic energy show several unique physical phenomena compared to that conventional method. A significant densification, mechanical properties, and improvement of the reaction rate of several materials have been reported [7-9].The most materials processed by microwave up to now is alumina.There are reported that shorter processing times and lower sintering temperatures can be acheived compared to the conventional processing. Moreover, in this case the samples were denser and thus stronger than the samples conventionally sintered at the same temperature for the same time duration. For silica xerogel, the time evolution of the density, porosity, and the crystallization of samples observed in the microwave process is similar to the one observed in the conventional process but during the microwave treatment, this evolution occurs at temperatures approximately 200°C lower [9,10]. the materials such as boron carbide and silicon nitride has been also reported recently [11-12]. The effect of microwave frequency are also reported found in these materials [13-15]. The frequency depedency of dielectric loss of materials are indicated as the main factor in this phenomena.

In this paper, we present experimental results of characerization of silica xerogel extracted from local rice husk ash. We study and discuss the influence of the sintering rate and effect of microwave power on the characteristic of the samples after processing.

2. Experimental Method

2.1. Sample preparation

Raw material rice husk was obtained from the rice field taken from South East Sulawesi, Indonesia. Rice husk ashes found by burning rice husk. Rice husk ash es were then washed using 1 M hydrochloric acid (HCl) at 100°C and mixed with 60 ml of 2 M NaOH. After filtered by Whatman No. 41, the silicate was inputted to an ion-exchange resin. The detailed extraction following procedures as described elsewhere [17, 18]. The powder of silica xerogel found from this experiment was characterized by X-ray fluorescence find out its characteristic. The powder was formed to cylinrical shape by applied polyvinyl alcohol as a binder before pressed.

2.2. Sintering experiment

The samples after pressed to cylindrical form were sintered by using a microwave oven 1 KW, 2.45 GHz in Halu Oleo University and a 28 GHz gyrotron heating system in FIR Center University of Fukui, Japan. The sample holder was a hollow cylinder made of thermo-isolating material. Some results of 28 GHz have been reported previously [19]. The power of microwave oven was varied and all samples heated up to 1200 °C. The heating rates were calculated based on time consumption for each microwave power. The cooling was performed by a natural convection after the microwave off and leaving the samples inside. The temperature profile was measured using an infrared thermometer. Figure 1 shows the 28 GHz gyrotron sintering system.

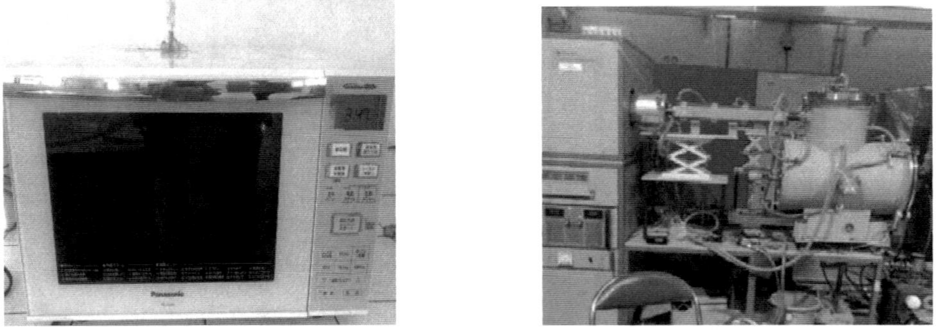

(a) (b)

Figure 1. (a) 2.45 GHz Microwave and (b) Gyrotron 28 GHz heating system

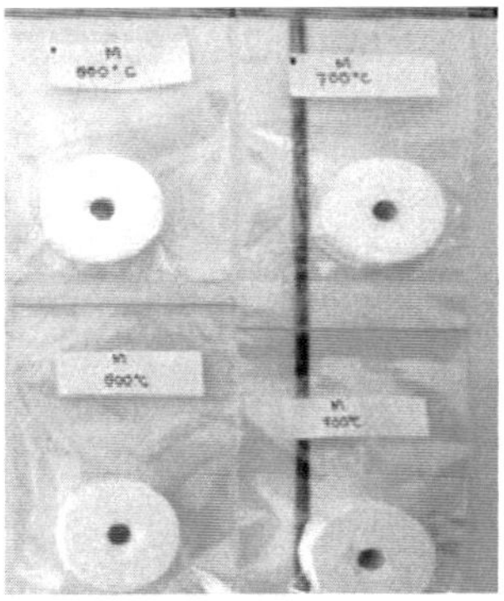

Figure 2. Silica samples after sintering

2.3. Characterization

All silica samples after sintering were then characterized by using Archimedes's method to calculate their density, open and closed pores. The procedure is followed a standard test method of the American Society for Testing and Material Specification (ASTM) C373-88 [20]. To determine their grain growth, the SEM photos of fracture surface of samples were performed by using a scanning electron microscope (SEM) at different magnification and at accelerating voltage of 10-15 kV in Halu Oleo University. SEM photos were processing by using image processing to calculate average grain size, and pores. Another experiments were performed to investigated effect of microwave energy to samples such as XRD, FTIR, and Vicker Hardness testing. They will reported in separated papers.

3. Results and Discussion

Figure 3 shows the grain size of silica with various heating rates in microwave (2.45 GHz). All sampels wer sintered at 1200 °C. The conventional sintering result served as a comparison. The quantitative analysis of microstructure from SEM photos for determining grain size for each samples shows that grain growth of silica strongly depend on heating rate. Furthermore, Fig. 3 also shows that grain growth can be depressed by using fast heating. This is may because of the grain growth depend more strongly to time than to temperature The conventional sintering samples was also calculated at same temperature i.e. 1200°C.

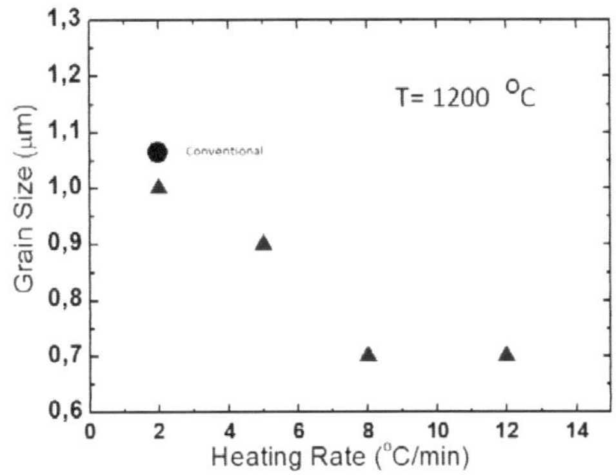

Figure 3 Grain size of silica xerogel as a function of the heating rate at 1200 °C

It indicates that increasing the heating rate will affect to mass transport which contribute to grain growth during sintering. From theory of sintering, there are several atomic diffusion paths may accurs during sintering. Some paths contribute to grain growth and the others are not [21]. The change in diffusion path from surface to grain boundaries and vice versa will change grain growth rate. The atomic diffusion D follow the equation,

$$D = D_0 \exp\left(-\frac{Q}{RT}\right) \qquad [1]$$

where D_0 is pre-exponential of D which represents number of atomic jump per time, Q is activation energy, R is constant, and T is temperature. From this equation we can see that heating rate s one of the main parameter of atomic diffusion.

The atomic diffusion rate and path are also depend on microwave power applied for heating source during sintering. It can be seen in Fig 4. However, changing the microwave power also changes the heating rate byself. It indicates that the changes in grain growth rates affected by both parameters i.e. microwave power and heating rate.

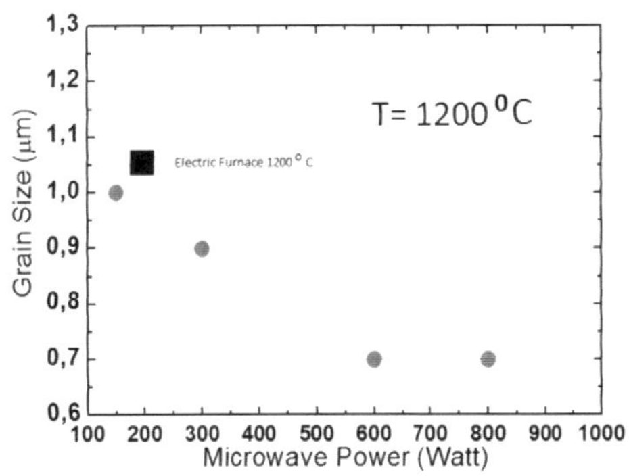

Figure 4 Grain size of silica xerogel as a function of the microwave power at 1200 °C

Figure 5 shows the density depedency to heating rate for samples sintered at 1200 °C. The Archimedes method applied for each sample to determine the density resulted that not only grain growth but also densification also depend on the heating rate. For densification (ρ) of material which processing by a thermally activated process such as in solid state sintering of ceramics, the densification is follows the equation [21],

$$d\rho/dt = C \exp(-Q/RT) \qquad [2]$$

where C is function of grain size and density and Q is apparent activation energy.

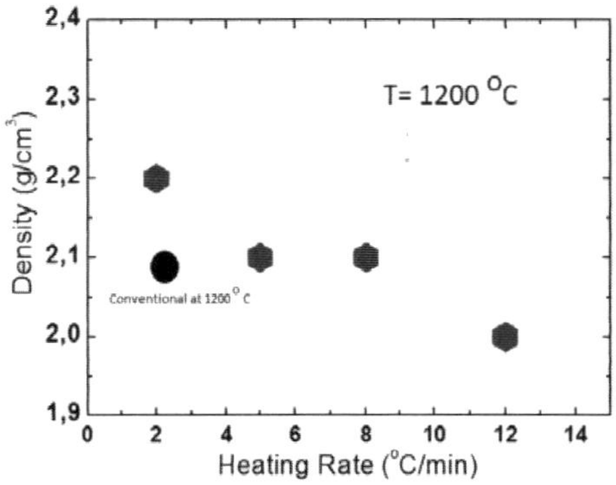

Figure 5. The densification of silica as a function of the heating rate at 1200 °C

However densification rate is not only depend on atomic diffusion only. There are several mechaniss during sintering may affect to this physical property. The appearance of pores is ascribed to the empty sites of the evaporated water and to some residues of alcohol binder as well as to combusted residual organics when the sintering process is one factor to change density. In the temperature range up to 800°C, the increase in density is because of the condensation reactions occurs on the surface of the silanols groups (Si-O-H) [1]. It cause the silica xerogel structure tends to shrink due to the loss of the hydroxide (OH) groups and densification rate increase. In the temperature range more than 1200°C, cannot be achieved by using microwave oven used in this experiment. However for 28 GHz sintering results have been reported in previous paper [19]. Moreover a crystallization of silica glass that affects the grain growth where was also observed in 28 GHz of silica xerogel ceramic will reported in separated paper. The increase in the densification explained by atomic diffusion where also can applied In the case of effect of microwave power to densification as shown in Fig. 6.

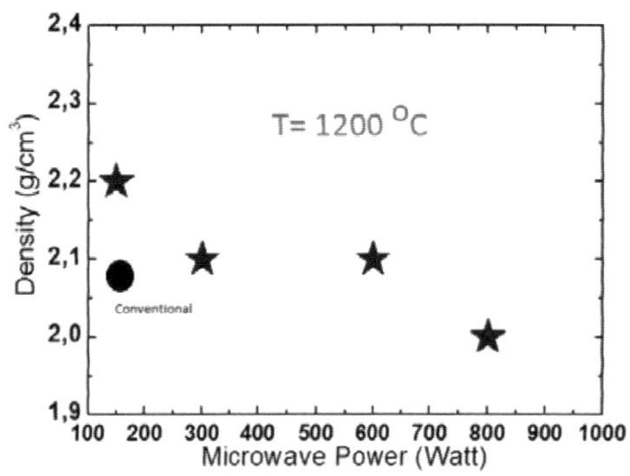

Figure 6. The densification of silica as a function of the microwave power at 1200 °C

For isothermal sintering methods, where the desired temperature keep during heating, we strongly can applied the Equation 2 to estimted the activation energy, Q. The Q describes atomic diffusion rate itself. we estimate the time required to reach a target density at a specific temperature. The rate of densification is recorded at each individual temperature. The experiment commonly is using a dilatometer. In the absence of a dilatometer, sintering runs are performed at regular intervals of time and the resulting densities are recorded. It will be performed in future experiments.

4. Conclusion

A series of experiments to investigate effects of heating rate and microwave power to silica xerogel were successfully performed. The experiments were performed by using a 2.45 GHz microwave sintering system. The different grain growth rate as well as densification rate were found for variation of heating rate and microwave power. It has been found that the high sintering rate depressed the grain growth of silica. But the sintering rate also decreasing densification. It may be because of change the path of atomic diffusion. As a whole, the experimental results obtained in this study demonstrate that the microwave sintering of silica glass-ceramics using a 2.45 GHz shows a microwave power and heating rate. It can be considered as an appropriate technology for control industrial production of silica ceramics to desired properties for applications.

Acknowledgments
This work has been partially supported by the Directorate of Higher Education, at the Ministry of Research Technology and Higher Education, Republic of Indonesia by KLN Research Grant 2018.

References

[1] H. Aripin, S. Mitsudo, I. N. Sudiana, S. Tani, K.Sako, Y. Fujii, T. Saito, Toshitaka Idehara, 2011, J. Infrared, Millimeter, and Terahertz Waves, 32, 867-876.
[2] A. Venkateswara Rao, P. B. Wagh, D. Haranath, P.P. Risbud, S.D. Kumbhare, 1999, Ceram. International, 25, 505 – 509.
[3] H. El Hamzaoui, L. Courtheoux, V. N. Nguyen, E. Berrier, A. Favre, L. Bigot, M. Bouazaoui, B.Capoen, 2010, Mat. Chem. And Phy., 121, 83–88.
[4] N.A. Melosh, B.J. Scott, C. Steinbeck, R.C. Hayward, P. Davidson, G.D. Stucky, and B.F. Chmelka, 2004, J. Phys. Chem. B, 108, 11909–11914.
[5] H. Xiu, Y. Pan, H. Kou, Y. Zhu, J. Guo, 2010. J. Alloy and Comp., 502, L6 – L9
[6] Y.T Kim, D.S Kim, D.H. Yoon, 2005, Thin Solid Film, 475, 271 – 274
[7] I. N. Sudiana, Ryo Ito, S. Inagaki, K. Kuwayama, K. Sako, S. Mitsudo, 2013, J. Infrared, Millimeter, and Terahertz Waves, 34, 627-638.
[8] S. Mitsudo, K. Sako, S. Tani, I.N. Sudiana, High Power Pulsed Submillimeter Wave Sintering of Zirconia Ceramics, The 36th Int. Conference on Infrared, Millimeter and Terahertz Waves (IRMMW-THz 2011), Hyatt Regency Houston, Houston, Texas, USA, October 2-7, 2011.
[9] C. Folgar, D. Folz, C. Suchicital, D. Clark, 2007, J. Non-Crystalline Solid, 353, 1483 – 1490
[10] H. Aripin, S. Mitsudo, E.S. Prima, I.N. Sudiana, H. Kikuchi, Y. Fujii, T. Saito, T. Idehara, S. Sano, S. Sabchevski, 2015, Ceramics International, 41, 6488–6497
[11] S. Mitsudo, H. Hoshizuki, K. Matsuura, T. Saji, T. Idehara, M. Y Glyavin, A. G. Eremeev, T. Honda, Y.Iwai, A. Kitano, J. Ishibashi, H. Nishi, 2010, Proc. of 15-th Int. Conf. on High Power Particle Beams, Saint-Petersburg, Rusia, 528 – 531
[12] V. P Paranosenkov, Y. V Bykov, V. V Kholoptsev, A. A Chikina, I. L Shkarupa, A. V Merkulopa, 1997, Refractories and Industrial Ceramic, 38, 13–15
[13] I.N. Sudiana, S. Mitsudo, T. Nishiwaki, P. E. Susilowati, L. Lestari, M. Z. Firihu, H. Aripin, 2015, Contemporary Engineering Sciences, 8[34], 1607-1615
[14] Y. V Bykov, O. I Get'man, V. V Panichkina, I. V Plotnikov, V. V Skorokhod, V. V Kholoptsev, 2001., Powder Metallurgy and Metal Ceramic, 40, 112 – 120
[15] H. Liu, C. Hu, X. Zhu, H. Hao, J. Luo, J. Zhou, S. Ouyang, 2004, Mat. Chem. and Phys., 88, 290
[16] H. Aripin, S. Mitsudo, E. S. Prima, I. N. Sudiana, S. Tani, K. Sako, Y.Fujii, T. Saito, T. Idehara,S. Sano, B. Sunendar, S. Sabchevski, 2012, J. Infrared, Millimeter, and Terahertz Waves, 33 1149-1162.
[17] H. Aripin, S. Mitsudo, I.N. Sudiana, T. Saito, S. Sabchevski, 2015, Transaction of The Indian Ceramics Society, 74[1], 11-15.
[18] S. Affandi, H. Setyawan, S. Winardi, A. Purwanto, R. Balgis, 2009, Advance Powder Technology, 20, 468 – 472
[19] I Nyoman Sudiana, Muhammad Zamrun Firihu, Seitaro Mitsudo, La Aba, Haji Aripin, Characterization Wave Guide Material Based on Silica Extracted from Rice Husk Ash, Key Engineering Materials, Vol. 266, (2017) pp. 148-152
[20] ASTM, Directory of testing laboratories, commercial-institutional/compiled by the American Society for Testing Material, 1975, ASTM: Philadelphia, C373 – C388.
[21] S..L..J..Kang, Sintering, Densification, Grain Growth and Microstructure, Elsevier, 2005

IOP Conf. Series: Materials Science and Engineering **797** (2020) 012034 doi:10.1088/1757-899X/797/1/012034

Seicmic Load on the Wharf

Abdul Kadir
Departement of Civil Engineering, Universitas Haluoleo, 93232 Kendari, Sulawesi Tenggara, INDONESIA

E-mail: kadir12340@yahoo.com

Abstract. Earthquake is a natural event that shocks can cause casualties and damage public facilities or infrastructure. The wharf is an infrastructure built with large investments. The purpose of this paper is how to design a open wharf and closed wharf that can minimize the risk or prevent collapse due to earthquake. The design uses a approach or force based with reference to the Earthquake Code of the world.

1. Introduction

Earthquakes are natural events that occur not only in earthquake-prone zones but also occur in zones with less seismic activity. In addition to causing fatalities, earthquakes can cause infrastructure damage from minor damage to severe damage. The Wharf is a maritime infrastructure that has the potential to be affected by the earthquake. Damage to wharf in addition to causing loss of large investments can also have an impact on the macro economic sector, especially in the field of import trade and exports. Seismic disaster mitigation has been an interesting field of study in the past decade.

One method of earthquake mitigation is the design concept of earthquake resistant structures. This method has long been developed and implemented in various Codes and Standards in many countries. This method is based on force and is the simplest among other methods. This paper will outline the earthquake resistant design concept on the wharf structure.

2. Type Wharf

Wharves are infrastructure maritimecan be classified into open type and closed type.Open type wharves are constructed in such a way that sea-water can run below the platform which generally rests on piles or columns embedded in an embankment. Closed type wharves are the common construction type where the depth to firm soil is shallow as well as the depth of water which results in a limited wave action against the wharf. These conditions can be found in rivers, and shallow depth shorelines. Closed type wharves are generally made of avertical wall which divides the sea from the shore, and backfilled with soil material and toppedwith a concrete deck or platform. The choice of type of wharves is influenced by geographical conditions, geology, soil conditions, economic and functional.

3. The Wharf design approach

The wharf structure design can be done with two approaches with two approaches namely; forced based design and displacement based desain. Forced based design or called to seismic coefficient method or equivalent lateral. Stiffnessis a design approach where the structure or components are designed to be able to withstand loads up to the allowable limit. This method is widely adopted on many codes in several countries. Ecording [1] this method may be used for preliminary design. Code [2] provide application conditions force based that is:

a. Force-based design is permitted for "low" design classifications.
b. Force-based design is permitted for all design classifications when $S_{DS}< 0.33$

c. Force-based design is permitted for all design classifications where it can be demonstrated that the capacity of all primary structural members exceeds the elastic earthquake demand force when using a value of R =1.

Displacement based design: A design approach whereby an anticipated displacement of the structure is determined for design earthquake motions and the various elements are designed to perform to certain standards under the imposed displacements. This method is known by several approaches namely; 1) Nonlinear Static Pushover, 2) Elastic Stiffness Method, 3) Substitute Structure Method, 4) Modal Response Spectra Analysis, 5) Nonlinear Time-History Analysis.

4. Seismic Mass of Wharf

Seismic mass is the mass of other structures and masses that occur in an earthquake. Ecording [2] and [3] the seismic mass shall include: 1) The mass of the wharf deck, 2) 1/3 of the mass of piles with the tributary length measured from the bottom of deck soffit to 5D below the surface of pile embedment, 3) The mass of any permanently attached equipment/ fixtures, 4) Mass due to the minimum of 10% of the design uniform live load or 100 pounds per square foot, 5) The hydrodynamic mass, if the pile diameter is greater than 2 feet, 6) The part of crane mass not less than $m_{crane,deck}$or 0.05 m_{crane}, where$m_{crane,deck}$; Part of the crane mass positioned within 10 feet above wharf deck or m_{crane} ; Mass of crane.

Some researchers [4] and [5] suggest to calculate contributions from the mass of additional water for piles. Additional mass contributions are estimated at 24%.

5. Forces when an earthquake occurs

5.1. Seismic Load

Base shear earthquake load according to some Code [6], [7] and [2] is calculated as shown in Table 1. Code [6] and [7] does not account for seimic reduction factor (R) due to the influence of ductility in the wharf structural system

Table 1. The horisontal seismic force

Code	Base shear method	Equivalent seismic	R	I
PIANC (2001) [6] OCDI (2002) [7]	V = αW	$\alpha = \frac{\alpha_{max}}{g} I;\ \alpha_{max<0,2g}$ $\alpha = \frac{1}{3}\left(\frac{\alpha_{max}}{g}\right)^{\frac{1}{3}} I;\ \alpha_{max\geq 0,2g}$		[1]; I = 1 [2] I; 0,8 < I < 1,5. For high seismic, I = 1,5
ASCE (2014) [2]	V = αW	$\alpha = \frac{S_{DS} I}{R}$	Prestresed concrete and steel pipe piles; R=2, Other structures, R=1	OC: I or II; I=1 OC: III; I=1,25 OC; IV; I=1,5

V; The horizontal seismic force; α : seismic influence coefficient; W: structures weight; α_{max}: Regional PGA at bedrock Site Category (site amplification factor)earthquake partition coefficient; I: importance factor; R: seismic response reduction coefficient; S_{DS} : Structural reaction coefficient; g gravity acceleration; OC: Occupancy Category

Structural reaction coefficient (S_{DS}) is calculated using the Spectum Response diagram according to [8]. Periode time is calculated by;

$$T = 2\pi\sqrt{\frac{W_g}{g\sum K_{Hi}}} \tag{1}$$

Where

T; natural periode of pile-deck system, W_g; surcharger and dead weigh of deck; g; acceleration of gravity; K_{Hi}; equivalent spring stiffnes for lateral reaction by i[th] pile

5.2.Active earth pressures

In the pseudo-static approach, the earth pressures are usually estimated using the Mononobe–Okabe equation (Mononobe-Okabe, 1924) the seismic inertia angle, ψ defined by

$$\psi = tan^{-1}\left[\frac{\alpha}{1-\alpha_v}\right] \tag{2}$$

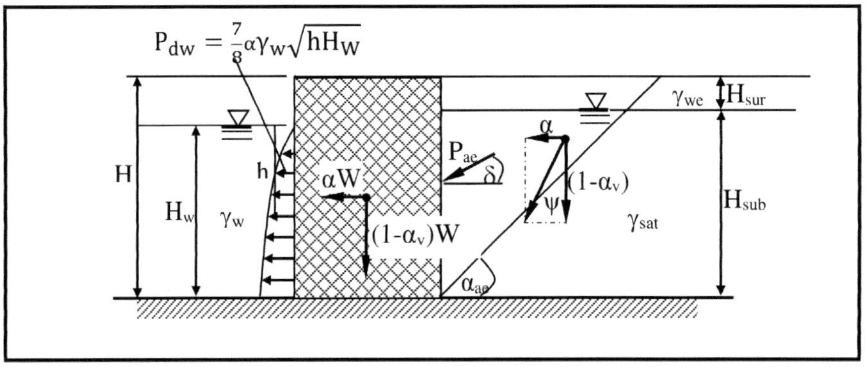

Figure 1. Hydrodynamic pressure and Active earth pressures [6]

For a vertical wall having a friction angle δ, between the backfill and the wall, and retaining horizontal backfill with an angle of internal friction ϕ, the dynamic active earth pressure coefficient, K_{ae} is given by

$$K_{ae} = \frac{cos^2(\phi-\psi)}{cos\psi cos(\psi+\delta)\left[1+\sqrt{\frac{sin(\phi+\delta)sin(\phi-\psi)}{cos(\delta+\psi)}}\right]^2} \tag{3}$$

$$P_{ae} = \frac{1}{2}K_{ae}\gamma_d(1-k_v)H^2 \tag{4}$$

Where γ_d is the unit weight of the dry backfill. In case of a uniformly distributed surcharge *q* sure, γ_d should be substituted with (γ_d + (1/2q_{sur}/H)

The active angle of failure from the horizontal direction for a level surface condition is computed as

$$\alpha_{ae} = \phi + arctan\left[\frac{-tan\phi+\sqrt{tan\phi(tan\phi+cot\phi)(1+tan\delta cot\phi)}}{1+tan\delta(tan\phi+cot\phi)}\right] \tag{5}$$

Considering a fully saturated Coulomb wedge, the modified seismic coefficient as follows

$$\alpha' = \frac{\frac{1}{2}q_{sur}H + \frac{1}{2}\gamma_{wet}H_{sur}^2 + \gamma_{wet}H_{sub}H_{sur} + \frac{1}{2}\gamma_{sat}H_{sub}^2}{\frac{1}{2}q_{sur}H + \frac{1}{2}\gamma_{wet}H_{sur}^2 + \gamma_{wet}H_{sub}H_{sur} + \frac{1}{2}\gamma_{b}H_{sub}^2}\alpha \quad (6)$$

5.3. Hydrodynamic pressure

The resultant hydrodynamic thrust computed by Westergaard (1933), which acts at a height of 0,4H from the base of the wall, is given by

$$P_{dw} = \frac{7}{12}\alpha\gamma_w H_w^2 \quad (7)$$

6. Load kombination and orthogonality effect

The combination of load must be analysis to calculate the worst effect on the wharf. Load combinations are calculated for two conditions, namely a combination of loads for structural strength design or Load and Resistant Factor Design (LRFD) and load combinations for stability design or Allowable Stress Design (ASD).

Table 2. Load factor the horisontal seismic force

CODE		D	L	H	E	B
PIANC,OCADI		1	-	1	1	-
ASCE	LRFD	1+k	0,1	1	1	1.2
		1- k	-	1	1	0.9
	ASD	1+k’	0,1	1	0.7	1
		1- k’	-	1	0.7	0.6

where ; D=Dead loads, including all permanent fixed equipment and structures, and other items expected to be presentfor more than 50% of the time;L = uniform live loads; H = soil pressure loads (e.g., soil pressure on end walls, concrete cutoff walls, steel sheet pile walls on pier orwharf type structures, and/or piles); E = horizontal earthquake loads; k = 50% of Peak ground acceleration (PGA); k’ = 0.7k = 0.35PGA; B = Buoyancy load

7. Design of wharf structures.

The designed structure is an open and closed wharf structure as shown in Figures 4 and 5. The earthquake load structure is calculated based on Code ASCE (2014) and PIANC (2001) and the closed wharf structure is calculated using a combination of Code ASCE (2014) and PIANC (2001). Earthquake Load The closed wharf structure involves service load factors namely 0.7 for earthquake loads and 1.5 for vertical loads for seismic,

Structural data regarding loads, dimensions, pile parameters, property materials, geotechnical parameters and loads are presented in the following Tables and Figure 2 and 3: '

Uniform live load	= 30 kN/m^2; Load seismic 10% x 30 = 3 kN/m^2 or 100 pounds per square foot = 4,88 kN/m^2 (use)		
Detail Crane load (AS4997:2005) [9]	• Container crane on 2,5 m from the waterside • C.t.c. distance of crane tracks: 30,48 m • C.t.c. distance wheelbases: 17,25 m • 8 wheels per leg, c.t.c. 1,2 m, so crane loads acts on 7x1,2=8,40 m		
Design crane load	Per wheel (kN)	Load on Corner (kN)	Load per m (kN/m)
Vertikal	750	6000	750
Horisontal	45	360	43

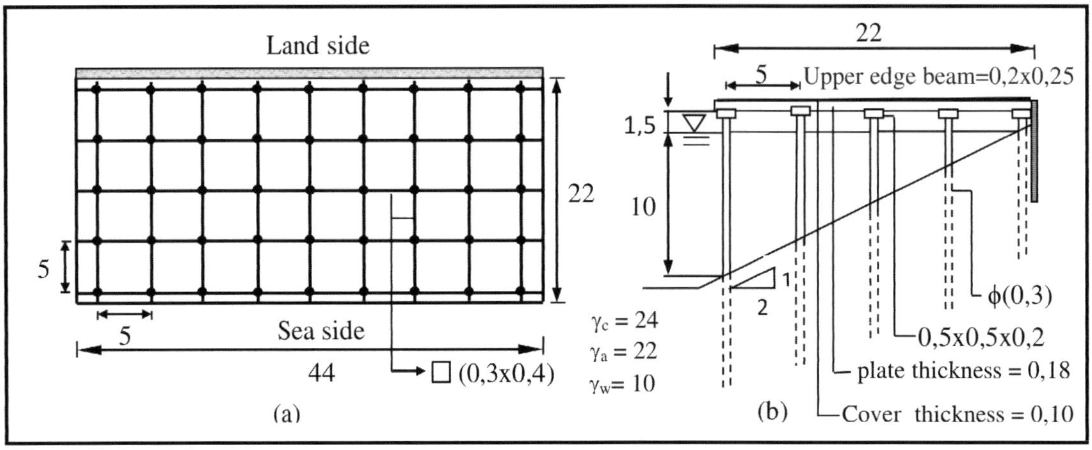

Figure 2. Pile-supported wharf, (a) Plan, (b) Cross section

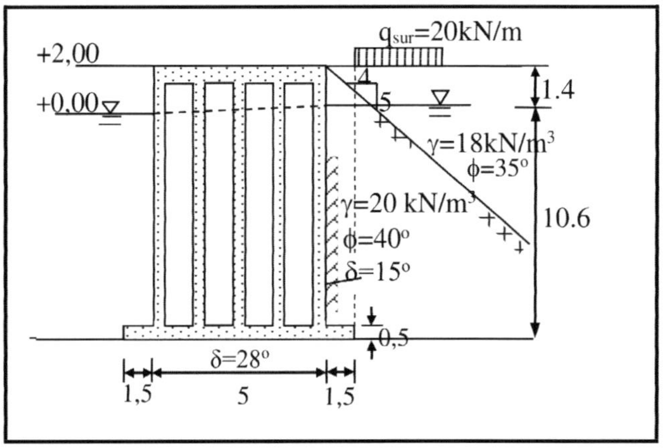

Figure 3. Cross section and dimensions of the wall

7.1.Seismic load design for the open Wharf

(1) Vertical forces

Dead Load; Plate + Cover layer + Girder + Upper edge beam+ Pile Cap + Crane + (Pile + Added mass to pile) + Live load: W = 11363, 53 kN

(2) Horizontal; 1) Border +Fender = 0,5x 20 = 10 kN; 2) Crane = 0,05x360 = 18 kN

Natural periode; equation (2) = 1,36 s; R = 2 I = 1,5
V, in Table 1 **= 3122.76 kN**

7.2.Designed earthquake load for closed Wharf

(1) Active earth pressures and thrust

$\alpha = \frac{S_{DS}}{R} I = 0,6 \frac{0,51}{1} 1,5 = 0,46$

$P_{ae} \cos\delta$ = 1679,53 kN/m at 0.54 x 12 = 5.4 m above bad

$P_{ae} \sin\delta$ = 450,03 kN/m ; 6.5 m from A

(2) Hydrodynamic force P_{dw} = 300, 84 kN/m at 4m above bad
(3) Inertia and other driving force; cover layer, coisson, footing, the backfill inertia force, the static bollard pull (50% reducing its value during earthquake), crane.
(4) Vertical forces; crane, cover layer, caisson dry, caisson wet, footing, backfill material above water table, backfill material under water table.
(5) The hydrostatic pressure; bisade (1,8 and 60) kN, uplift (24 kN)

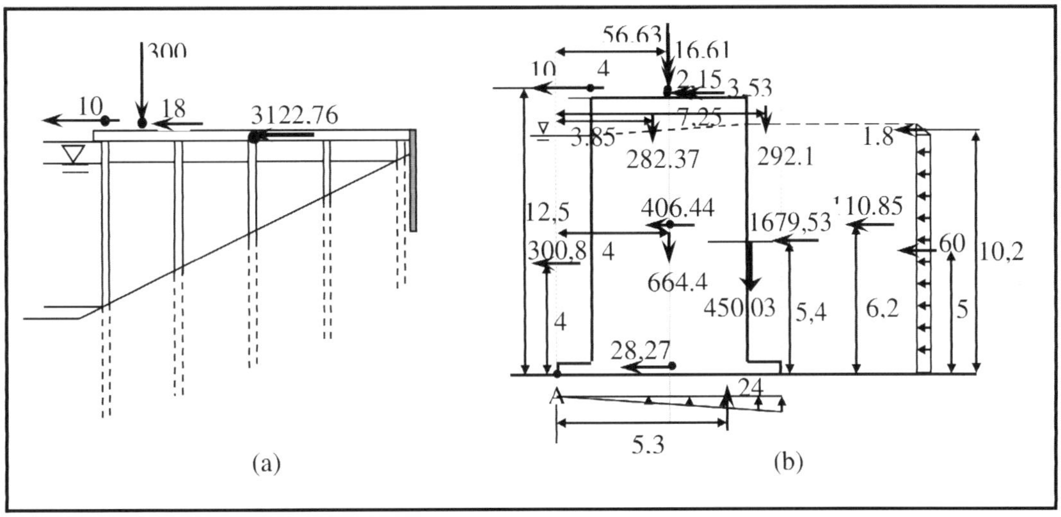

Figure 4. Loads acting on structure wharf during earthquake (a) Open wharf (forces are in kN), (b) Close wharf (forces are in kN/m)

8. Discussion

Base shear earthquake load according to ASCE (2014) and PIANC (2001) for open wharf structures tend to have R = 2 giving adjacent results. Earthquake load for closed wharf structures, ASCE (2014) tends to give results slightly larger than PIANC (2001). For an open wharf structure, it is easier to analysis using software.

Refences

[1] POLB 2015 *Port of Long Beach Wharf Design Criteria WDC* Version 4.0.
[2] ASCE/COPRI 61-14 2014 *Seismic design of piers and wharves* American Society of Civil Engineers
[3] POLA 2007 *Code for Seismic Design, Upgrade and Repair of Container Wharves and Commentary* The Port of Los Angeles, San Pedro, California.
[4] Spijkers JMJ, Metrikine A 2005*Structural dynamics*: college ct4140. TU Delft, Delft
[5] Sarpkaya T S 2010 *Wave Forces on Offshore Structures* Cambridge University Pres. USA.
[6] PIANC 2001 *Seismic Design Guidelines for Port Structures,* International Navigation Association A.A. Balkema Publishers, Tokyo
[7] OCDI 2002 *Technical Standards and Commentaries for Port and Harbor Facilities in Japan,* The Overseas Coastal Area Development Institute of Japan.
[8] ASCE-7-10 2010 *Minimum Design Loads for Buildings and Other Structures* American Society of Civil Engineers
[9] AS 4997-2005 Australian Standard™ *Guidelines for the design of Maritime Structures*Committee CE-030.

IOP Conf. Series: Materials Science and Engineering **797** (2020) 012035 doi:10.1088/1757-899X/797/1/012035

Case Study on Relocation of Port Infrastructure using the Randomized Expert Panel Opinion Marginalizing Procedure

Kiril Tenekedjiev[*#1] **and Natalia Nikolova**[*#]
* Australian Maritime College, University of Tasmania, 1 Maritime Way, 7250 Launceston, Tasmania (Australia)
# Nikola Vaptsarov Naval Academy, 73 V. Drumev Str., 9026 Varna (Bulgaria)

E-mail: Kiril.Tenekedjiev@utas.edu.au
E-mail: Natalia.Nikolova@utas.edu.au

Abstract. We present a case study related to the relocation of parts of port facilities in a Black sea port in Varna (Bulgaria). As the city and the port expanded over the years, the east sector of the port ended up being in the middle of the city center. This causes environmental and spatial planning problems to both the city and the port area. We present several alternatives to relocate part of the industrial activities of the port to another site further to the west sector of the port and utilize the emptied space for recreational, public and social activities. We then apply the Randomized Expert Panel Opinion Marginalizing Procedure (REPOMP) to rank the alternatives. The procedure utilizes a hierarchy of criteria to assess the alternatives from four major aspects – environmental, social, technological and economical. A total of 12 experts provided their estimates on how the alternatives meet all criteria as well as on the relative significance of each criterion. Bootstrap simulation is utilized to not only find the total marginal criterion of each alternative, but also its distribution and best estimate. We also tested whether there is significant difference between the ranking scores of the alternatives.

1. Introduction

Artificial intelligence (AI) has been a major scope of research and development for decades. Its focus is to elaborate techniques and methodologies that can teach an intelligent agent to analyse the environment and choose actions that maximize a certain task or objective [1]. AI has been developed with the specific focus to replicate the human intelligence in tasks that are of higher complexity. AI utilizes the tools of statistics, mathematics, linguistics, computational intelligence and symbolism. It has found application areas in healthcare, military science, finances, economics, etc. In complex tasks of decision making, intelligent agents learn and/or may be able to rely and use external knowledge, which is usually in the form of expert opinion and expert assessments. Hence, despite the development of the AI domain, the reliance on expert knowledge to build complex systems to process knowledge and make decisions is still present.

Making decisions about critical infrastructure (ports, industrial plants, transport networks, etc.) is a complex task, stemming from the fact that alternatives are usually defined and assessed on a large series of (usually contradicting) criteria and uncertainty is rather high [2]. Quantitative analysis through utility theory and subjective statistics is one possible way to handle such complex decisions

[1] To whom any correspondence should be addressed.

(see [3, 4]) and it often has strong reliance on expert judgment. The task of bringing the multi-dimensional subjective estimates, given by the experts, to a total marginal indicator of the quality of each alternative then becomes crucial. The REPOMP method, initially proposed in [5] is a simplified way of tackling this complex task.
Previous works have developed and demonstrated the application of the expert-panel based approach to rank alternatives of various complexity – the REPOMP procedure (Randomized Expert Panel Opinion Marginalizing Procedure) [5]. REPOMP requires that the analysed alternatives are assessed based on a preliminarily defined hierarchy of criteria. Then a panel of experts assesses both the performance of each alternative against the criteria and the significance (weights) of each criterion at its hierarchical level. Due to its generalized nature, the REPOMP is adaptable to problems of varying nature and complexity. Some examples of applications so far include: selection of technology for waste treatment [5; 6], selection of alternatives for modernization and deployment of energy-efficiency measures for public buildings [6; 7], selection of structure and organization of national infrastructure in a European Union's Member State regarding the collection, usage, and dissemination of full-range spatial data (and spatial meta-data) following the INSPIRE directive [8; 6; 9]. Each of those example case studies depended on different number of experts, different hierarchy of criteria and different count of alternatives. The procedure also allows the analyst to use different sets of weight coefficients to the marginal criteria, thus acquire solutions based on different potential structure of experts (i.e. different prioritization of criteria). The mathematical procedures of REPOMP are realized in MATLAB functions, available free-of-charge upon request from the authors.
In this work, we present a case study related to infrastructural plans to modernize and restructure a sea port in the Black Sea city of Varna (Bulgaria). The analysis was conducted in 2012 within the work package 6 of the EU funded project ECOPORT 8: *Environmental management of trans-border corridor ports* (project code SEE/A/218/2.2/X). The information provided in the paper reflects the status of the port and its infrastructure at the time the study was conducted. Some parts of this study were initially presented in earlier works [10].
The way the city of Varna expanded over the years ended up in the port eventually being in the city centre with no ability to expand its industrial activities. This causes environmental, social and economic issues for the city. The restructure of the port has been the scope of political debate for over two decades. Advances in port technologies and logistics are making the problem even more crucial. Hence, the situation can be characterized as follows:

- The Port of Varna is situated within the urban area of the city. It causes traffic and noise pollution for the city area. The port has very limited abilities to expand and develop in future.
- Local and national governmental authorities have discussed the possibilities to relocate the port to a site far from the city. Different expert teams of architects and urban planning specialists worked on the elaboration of strategies to relocate the port and develop recreational and cultural centres in its place.
- At the time of start of this study, there was no official selected procedure to optimize the needs of the city and the port infrastructure;
- The discussion for the relocation of the port evokes serious political and economic confrontation, due to different interests that are affected by this decision;
- While the problem remains unsolved, further activities of local and national governmental institutions regarding the establishment of an environmental management system for the area of the port are impossible and unjustified.

In this study, we describe several alternatives to restructure (relocate) the industrial activities of the port and develop recreational and public centres in their place. A hierarchy of criteria is constructed, and the REPOMP procedure is implemented to select the best option according to a group of experts.

In what follows, section 2 presents the structure and challenges of the Port of Varna and clarifies the objectives of this study. Section 3 presents the implementation and the results from the REPOMP procedure.

2. Characteristics of the Port of Varna

The port of Varna is the largest of the seaports of Bulgaria. It is located on the Varna bay, which is in the western end of the Black Sea, and close to lake Varna and lake Beloslav. The Port of Varna is part of (or services) the pan-European corridor 8, the Rhine-Main-Danube canal, and the TRACECA international transport corridor (Europe-Caucasus-Asia). The port area has significant developmental potentials with a stretch of about 45 km of inland waterfront on the lakes. The location of this zone is easily accessible by rail and road, and it also has proximity to the international airport in Varna. The port has two sectors – Varna East and Varna West, connected by inland canals. Since 2010, the Port of Varna implemented the Integrated Management System, which includes certified management systems to international standards ISO 9001:2008, BS OHSAS 18001:2007, ISO 14001:2004. The Port of Varna is one of the first Black Sea ports to be certified under the requirements of the International Ship and Port Facility Security Code (ISPS Code).

2.1. Port of Varna - East

The Port of Varna East is situated at the inner end of Varna Bay, only 1 km away from the city centre (see Figure 1). It is an important logistics operation link for the transit traffic to and from Central Europe routed via the Danubian Port of Rousse. The main section of this link is the transit railroad link Rousse-Varna. This is also the only passenger terminal along the sea road between Istanbul and Odessa. A special feature of the port is the multi-purpose use of the berths. All quays, except the passenger one, are utilized for handling of various types of cargo: general, containers, sugar, scrap, molasses, etc. The port has 14 berths with a total length of 2378 m and maximum depth of 11.3 m. The open storage area is 49 800 m^2 and the warehouses have the total area of 41 500 m^2. The equipment of the port consists of 28 cranes (up to 32 tonnes) and one gantry crane with lifting capacity of 30.5 tonnes, specialized in handling of containers. Part of the equipment is also the 2 reachstackers, forklift tracks, rail grain discharge station, quay ship loader for grain and unloading facility for molasses. An integral part of the port is the storage base, also known as The Dry Port. It is located 5 km away from the port on a total area of 13 000 m^2 and has 6 warehouses. It is specialized in storage of all types of cereals, soy and general cargoes.

Figure 1. The Port of Varna East

2.2. Port of Varna West

The Port of Varna West is the most modern and perspective facility at the northern Bulgarian coast (see Figure 2). It is located 30 km west of the city of Varna, on the western shore of the Beloslav Lake. It was constructed about 40 years ago and implements modern cargo handling operations. It is adjacent to the chemical factory of Devnya and enables direct handling directly on the ship. Most of the port berths are multifunctional and are used for handling of various general cargo: packed soda, fertilizers and cement, as well as metals, equipment, timber, etc. The port includes 18 mooring berths with total length of 3 223 m and maximum depth of 11 m. The storage area is 191 000 m^2 and the covered storage area is 27 000 m^2. The port is equipped with 27 portal cranes (up to 16 tonnes), 2 gantry cranes and one mobile crane with lifting capacity up to 100 tonnes. The equipment also includes 2 reach stackers, 3 rubber belt conveyors, loading facility for soda and fertilizers and 2 units of quay ship loaders for soda and fertilizers.

The Varna Ferry Complex (situated on the territory of the Port of Varna West) is the only place in the Black Sea region which has a rail ferry terminal with possibility to change the rail car bodies from European to Russian standard and vice-versa. This is a unique advantage to the ferry terminal offering one of the shortest and the cheapest roads for the cargo traffic between Europe and Asia. All berths of the port have a connection with the Bulgarian rail network. This fact provides excellent conditions to develop intermodal transport networks. With modern equipment, excellent transport links, rail ferry service and a crossroad location, the port is a convenient bridge for the cargo flows between Europe, Russia, Ukraine, Middle Asia, the Middle East and the Far East.

2.3. Challenges for the development of the Port of Varna at a national and European perspective

The preliminary analysis demonstrates that the port has no Port Monitoring System and no specific guidelines on main environmental issues regarding water quality, air quality, waste management, odour control, energy saving, rehabilitation of polluted areas, port development, etc. The development and deployment of an effective environmental monitoring system for the entire port infrastructure is a long-term, time-consuming and financially expensive initiative, which also caused political turmoil during discussions about the transport policy of the country. At the time our study initiated, a conclusion of whether to relocate some of the port industrial activities was not reached despite the analyses and consultations at various levels over the years.

Figure 2. The Port of Varna West

The strategy of "Motorways of the Sea" at the European Union (EU) level is another major reason why regional and national authorities in Bulgaria should pay attention to the development of Bulgarian maritime ports. The "Motorways of the Sea" concept, introduced in the "2001 Transport White Paper

– European transport policy for 2010: Time to Decide" [11] aimed to introduce new intermodal maritime-based logistics networks in Europe to improve the organization of transport in Europe. These networks should be sustainable and commercially more efficient than road-only transport. In this way, the EU transport strategy aimed to improve access to markets throughout Europe and provide alternatives to the road network for commercial and non-commercial traffic. The key focus of the strategy was better use of existing maritime transport resources and its adequate linkage to the railroad network and the inland waterways of the community, as part of an integrated transport network. The White Paper also postulated that the Motorways of the Sea should be part of the trans-European network (TEN-T) and funds should be made available for its development. Article 12a of the TEN-T guidelines [12] gives three main objectives for the sea motorways projects as follows:

(1) freight flow concentration on sea-based logistical routes;
(2) increasing cohesion;
(3) reducing road congestion through modal shift.

Four corridors were designated for the setting up of projects of European interest. The one related to Bulgaria is the corridor for south-west Europe (western Mediterranean, connecting Spain, France, Italy and including Malta and linking with the Motorway of the Sea of south-east Europe and including links to the Black Sea) (see Figure 3).

Figure 3. Map of the Motorways of the Sea

The next stage of the European transport strategy was presented in the 2011 Transport White Paper "Roadmap for a single European transport" [13]. It stressed the importance of Motorways of the Sea. This white paper adopted a roadmap of 40 concrete initiatives for the next decade to build a competitive transport system that will increase mobility, remove major barriers in key areas and boost growth and employment. At the same time, its aim is to dramatically reduce Europe's dependence on imported oil and cut carbon emissions in transport by 60% by 2050. The 2013 TEN-T Guidelines [14] redefined the Motorways of the Sea as the maritime dimension of the trans-European transport network and indicated that it should contribute towards the achievement of a European maritime space without barriers by the following:

(a) Maritime links between maritime ports of the comprehensive network or between a port of the comprehensive network and a third-country port where such links are of strategic importance to the Union;

(b) Port facilities, freight terminals, logistics platforms and freight villages located outside the port area but associated with the port operations, information and communication technologies (ICT) such as electronic logistics management systems, and safety and security and administrative and customs procedures in at least one Member State;

(c) Infrastructure for direct land and sea access.

All those European strategies increase the importance of adequate measures to maintain and develop transport infrastructure. However, it is up to industry, Member States and the Community to implement financially and operationally sound projects to use these maritime resources better for new intermodal maritime-based transport systems. All these concerns put Bulgaria under the necessity to comply with the sea motorways strategy and provide the possibilities for effective integration of the maritime ports into the international intermodal chain.

In Table 1, we specify some of the main strengths, weaknesses, opportunities and threads that the development of the port faced at the time our study was initiated. Hence, some of the key factors for the necessity to relocate some of the port industrial activities stem from the vicinity to the city centre, the transport problems, the environmental issues and the lack of free spaces for future port development, as well as the necessity to modernize and increase effectiveness of the port. This is also a topic embedded in the Strategic City plan of Varna, indicating two main issues:

1. Removal of activities and equipment for cargo-handling operations away from the city centre in accordance with the environmental and architectural considerations for urban development. The emptied space in the port should be utilized for recreational, commercial, public and cultural sites to improve the urban outlook of the area.
2. Relocation of the container terminal from the port of Varna East to the north coast of the Varna Lake. In this way larger container ships will have access to the port and the container turnover in the Black Sea would increase.

This problem has been under discussion at a local, regional and national level, and multiple projects have been developed to solve it. The main criteria in those projects were:

- Effective reconstruction of the "old" port into accessible zone with business, commercial and entertainment areas
- Connection with the old city centre – the Roman Baths, the Greek neighbourhood, the Sea Garden, etc.
- Preserving the existing valuable buildings and equipment
- Ensuring convenient communication and transport services
- Innovatory method in the functional and architectural appearance of the area

Our study analyses four alternative plans for the development of the Port of Varna East. Each of those alternatives imposes different scale of change to the current terrain of this section of the port, has impact on different berths of the port and assumes different activities to be established (or relocated) in the emptied areas. We have excluded from our study those alternative plans, which only analysed the first 5 berths of the port as these options were deemed rather ineffective since the future development of those berths is strongly dependent on the development of the whole port area. We utilize the REPOMP procedure to compare the four alternatives, based on a clearly defined hierarchy of criteria and with the involvement of 12 experts who provided their assessments.

Table 1. Strengths, weaknesses, opportunities and threads to the development of the Port of Varna

STRENGTHS	WEAKNESSES
• Key location of the port in the pan-European transport networks • Increase in the container traffic, leading to decrease of the unhealthy impact on environment in the port area • Implemented Integrated Management System, including certified management system to the international standards ISO 9001:2008, BS OHSAS 18001:2007 and ISO 14001:2004 • Certification by the International Ship and Port Facility Security Code (ISPS Code)	• No system for environment pollution risk assessment • Insufficient activities in waste recycling • Existent pollutants in the territory and aquatory of the port • No self-monitoring system for pollutant components • No transparency in the environmental protection initiatives • Outdated infrastructure with limited abilities to be modernized
• Presence of EU strategies for the development of intermodal transport corridors • Development of national environmental legislation and programs harmonized with EU guidelines • Establishment administrative procedures to implement and enforce environmental legislation • Establishment of monitoring system for the port area • Attract EU funding for environmental monitoring and protection activities	• Significantly complicated national environmental legislation • High costs for environmental law enforcement • Poor compliance at regional and national level with EU regulations in the transport and environmental domains • Legislative and administrative obstacles in attracting EU infrastructural • Political and governmental dynamics not in favour of long-term infrastructural decisions and investment
OPPORTUNITIES	THREADS

3. Analysis of alternatives using the REPOMP procedure

In this section we shall present details about the four alternatives to be analysed, and we will also demonstrate the implementation of the stages of REPOMP.

3.1. Description of alternative plans for the development of the Port of Varna East

We analyse four alternatives on how the future Port of Varna East may look and operate.

Alternative 1: Establish the recreational centre at berths 1 to 10 emphasizing on business, shopping and studying areas

Alternative 1 is presented through design images in Figures 4 and 5. It is an alternative oriented to business and entertainment. The project presents a modern exterior with a spectacular view to the city centre and to the sea. One of the landmarks on this option is expected to be a five-star hotel with yacht club. The yacht club, adjacent to the hotel, is a suitable addition that outlines the maritime elements of the recreational centre. The alternative also assumes that the location will host a new City Hall and a Universal Hall. Those facilities shall serve citizens through series of civil, cultural and entertainment activities. A city library is to be established there. Important public events (e.g. conferences, voting campaigns, educational initiatives, etc.) are to be held at the conference facilities of those two buildings. Also envisaged by this alternative is the development of rental office spaces for local businesses, as well as shopping malls.

The transport connections to the other parts of the city will be provided through land (bus, car), rail (with the rail station close to the area) and waterway transport (through the passenger terminal at the port). In addition to using the green areas of the complex for walking and leisure, visitors will also be able to visit the Contemporary Art Museum, the Aquarium and Naval Historical Museum, as well as a swimming pool (which are currently in close vicinity to the area). Part of the current port is envisaged to remain for port activities and shall contain fleet service office and a storage yard.

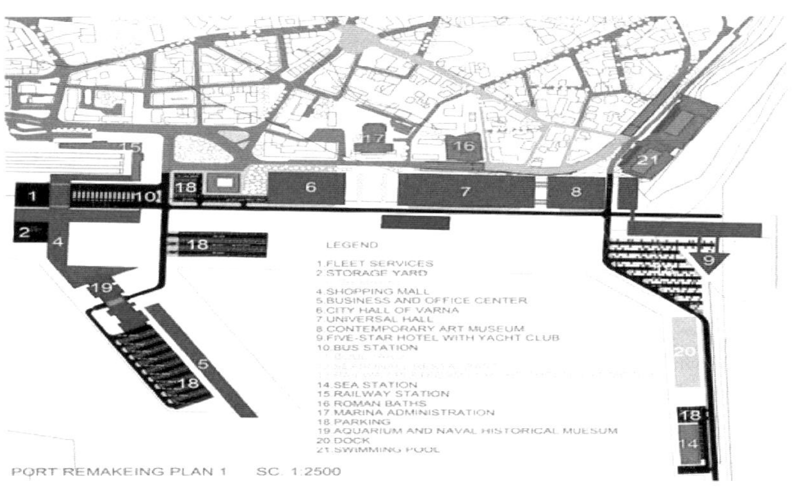

Figure 4. Scheme of Alternative 1

Figure 5. Outlook of the port under Alternative 1

Alternative 2: Establish the recreational center at berths 1 to 10 emphasizing on the new appearance of the City Hall and the city square

In this alternative, a key component is the establishment of a new City Hall for the municipal administration and local government, as well as to the establishment of a new city square. The building where the current local government is located is situated in the very centre of the city. While being iconic in its historical background, it is in an area with heavy traffic, high noise pollution, and limited access. The building has limited renovation capacity due to outdated construction, multiple floors and limited area of each floor. The City Hall in the project is only in four floors, with long corridors and multiple offices at each floor. This provides effectiveness and convenience of work, which in turn facilitates citizens. Additionally, the building shall be in a central position to the new city centre, with facilitated access. Cultural monuments, waterfront and green areas, sidewalks and alleys, streets with limited motor traffic, residential buildings, parking areas, and other new public buildings shall also be situated in the new city square. The overall idea of this alternative is to present Varna as a city, open to the sea and to the world. As it is obviously from the project pictures and schemes (see Figures 6 and 7), the new recreational centre will have modern appearance that improves the urban quality of the city and makes it more attractive to both citizens and visitors.

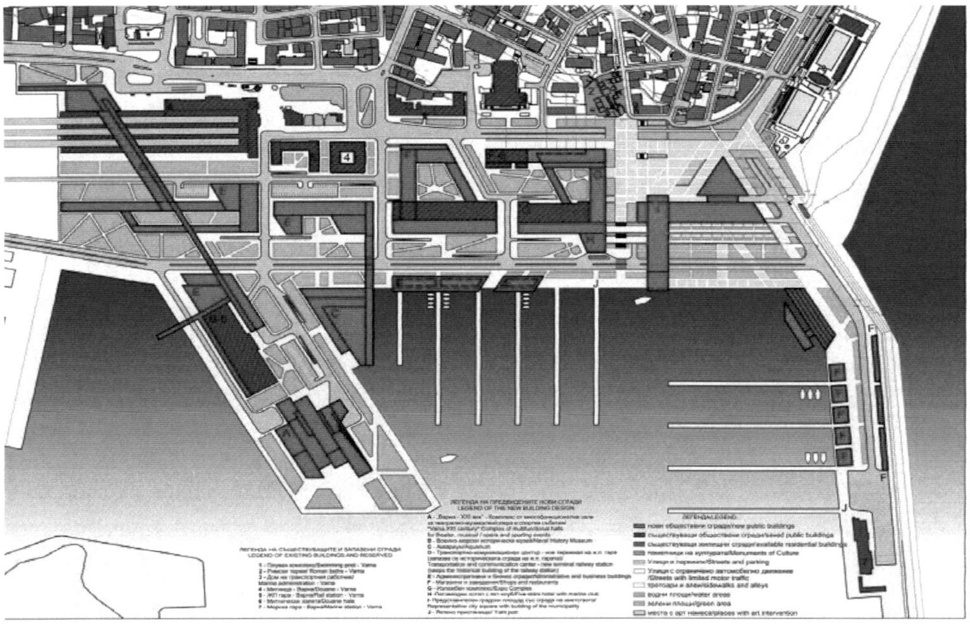

Figure 6. Scheme of Alternative 2

Alternative 3: Move the whole port and establish the recreational centre on the entire territory

In this alternative, the whole port is envisaged to be relocated along the Varna Lake, which is an option way more complicated than the idea for moving parts of the port. This requires larger free areas for the new port location, larger investments in both relocating the port and developing the recreational centre in its place. Hence, this is the alternative that is most complicated in terms of long-term planning and funding. Developing the center on the entire territory of the present port gives much more space for creative urban design, as well as possibilities to establish multiple facilities, cultural and public areas, etc. The ideas in Alternatives 1 and 2 can easily be combined (in terms of new buildings and facilities, new City Hall, new museums, business areas, transport facilities, halls and entertainment areas).

Alternative 4: Keep the port at its present location

This alternative represents the option to keep the Port of Varna East at its current location and do no activities towards the development of a recreational area. The ships will berth in the city area, near the residential buildings, the cargo-handling activities shall also be carried out in this area. The trucks traffic carrying different type of cargo shall continue passing through the city area. The pollution from cargo operations and supporting truck traffic shall continue to have its negative impact on the environmental profile and image of the city. Other solutions shall have to be found for the environmental and transport problems of the city, as well as the issues concerning its future development.

Figure 7. Outlook of the port under Alternative 2

3.2. Implementation of the REPOMP procedure

The REPOMP procedure (Randomized Expert Panel Opinion Marginalizing Procedure) was initially presented in [5] as a procedure to rank alternatives based on the subjective opinion of an expert panel in multi-criteria decision making (MCDM) situations (for further reading on MCDM, please refer to [15; 16; 17]). It allows to use arbitrary hierarchical structure of criteria to compare a certain number of alternatives based on marginal indicators, but also on their probability distribution, analysed using Bootstrap simulation [18]. REPOMP has five implementation stages as described in Figure 8.

The experts who volunteered and involved in this study were from the following organizations: City of Varna, Bulgarian Port Infrastructure Company, Navigation Maritime Bulgar Shipping Company, Varna Free University, and Technical University – Varna. To guarantee anonymity of the experts' opinions, we have elaborated the following procedure (see Figure 9).

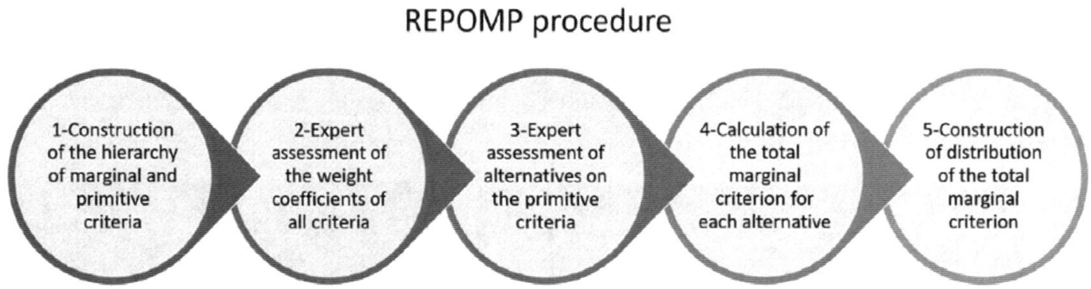

Figure. 8. Stages of the REPOMP procedure

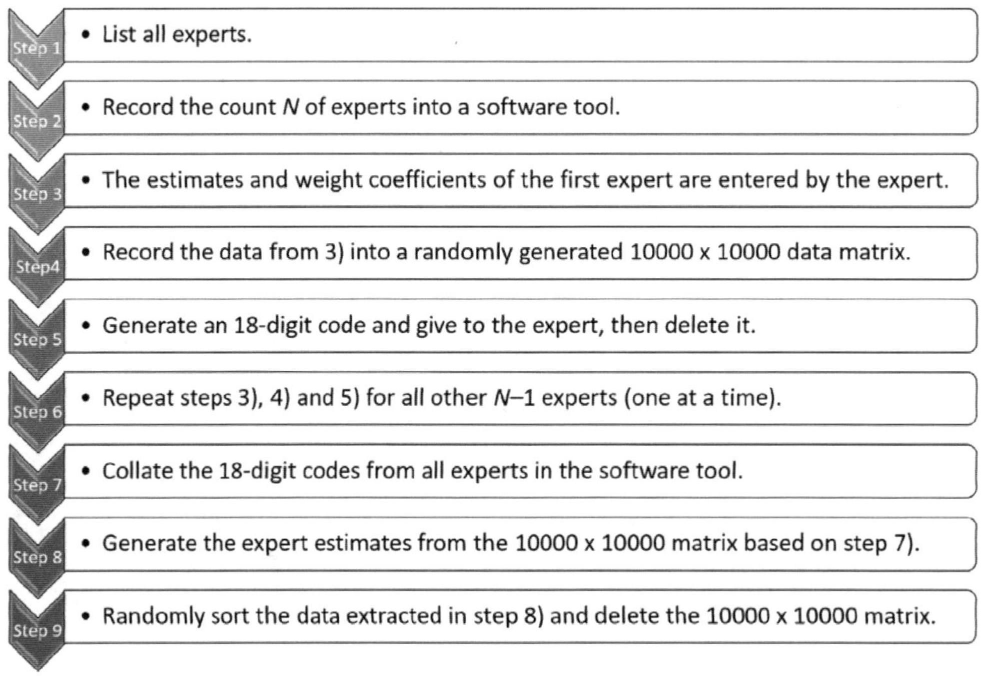

Figure 9. Structure of the algorithm to protect the anonymity of experts

In ***STAGE 1 OF REPOMP*** we need to construct the hierarchy of criteria. The total marginal criterion for assessment K is divided into b=4 base marginal criteria: K_1- Environmental; K_2 – Social; K_3-Technological, and K_4 – Economical. The hierarchy of criteria is presented on Figure 10 with their meaning given in Table 2.

Alternatively, the hierarchy of criteria can be modelled by denoting the index *ind* of any criterion with an index vector containing integer values. The set of sub-criteria for the *ind* criterion is denoted as *S*(*ind*). In our case the index vectors can be 0, 1, 2, 3 or 4 dimensional as shown in the first and third column of Table 3, whereas the correspondent sub-criteria sets are given in the second and third columns of the same table. The dimension of the index vector is also called level of the criteria. The only zero-level criterion is the total marginal criterion and its index is an empty vector, denoted as []. In the same way, for all primitive criteria *S*(*ind*) is an empty set, denoted as { }.

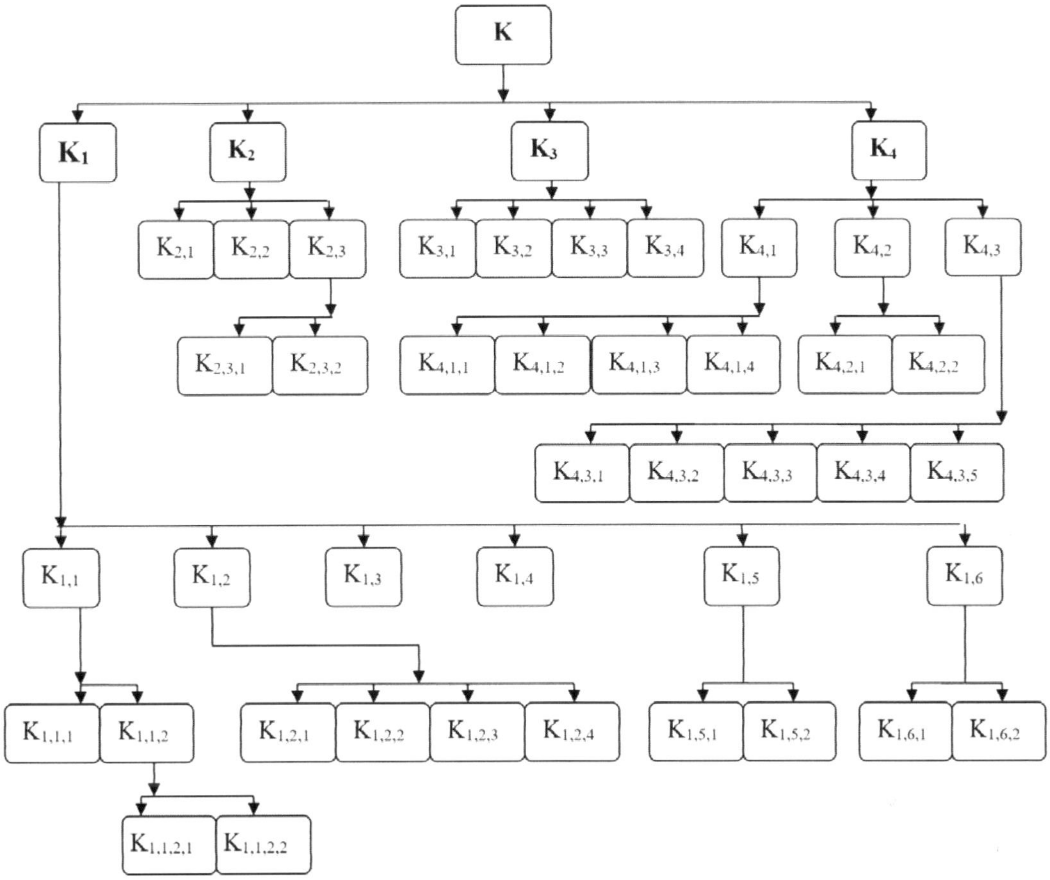

Figure 10. Hierarchy of criteria for the port development problem

In ***STAGE 2 OF REPOMP***, the following scale is utilized to assess the significance of the criteria: 0 – No significance; 1 – Low significance; 2 – High significance; 3 – Very high significance. A group of 12 experts made assessment of the weight coefficients of the primitive criteria (i.e. those which have no sub-criteria, see Table 2) and for the marginal criteria (those which have sub-criteria, see Table 2).

If the weight coefficient of the *ind* criterion is assessed by n_{ind}^{Y} experts, Y_r^{ind} being the mark given by the *r*-th expert, then the opinion of all experts for the weight coefficient of the criterion can be organized in a n_{ind}^{Y} -dimensional vector $\vec{Y}_{ind}=(Y_1^{ind}, Y_2^{ind}, \ldots, Y_{n_{ind}^{Y}}^{ind})$. The opinion of the experts regarding the weight coefficients is given in Table 4. The average of the weight coefficient of criterion *ind* is calculated for each criterion except for the 0-level one:

$$Y_{mean}^{ind} = \frac{1}{n_{ind}^{Y}} \sum_{r=1}^{n_{ind}^{Y}} Y_r^{ind} \text{ for } ind \neq [] \tag{1}$$

Table 2. Meaning of the criteria in the hierarchy from Figure 9. The marginal criteria are bolded.

K_1 – ENVIRONMENTAL	**$K_{1,1}$ – Air pollution in the port area**	$K_{1,1,1}$ – Gas emissions	
		$K_{1,1,2}$ – Dust emissions	$K_{1,1,2,1}$ – Non-organic emissions
			$K_{1,1,2,2}$ – Organic emissions
	$K_{1,2}$ – Water pollution and other impact in the port area	$K_{1,2,1}$ – Pollution from ships	
		$K_{1,2,2}$ – Pollution from cargo handling operations	
		$K_{1,2,3}$ – Impact on the tide levels	
		$K_{1,2,4}$ – Impact on currents near the port	
	$K_{1,3}$ – Urbanization of new areas		
	$K_{1,4}$ – Environmental risk and impact		
	$K_{1,5}$ – Noise pollution	$K_{1,5,1}$ – Levels of noise pollution during the day	
		$K_{1,5,2}$ – Levels of noise pollution during the night	
	$K_{1,6}$ – Increase in waste volumes	$K_{1,6,1}$ – Construction wastes	
		$K_{1,6,2}$ – Household and industrial wastes	
K_2 – SOCIAL	$K_{2,1}$ – Increase of employment		
	$K_{2,2}$ – Employee's health risk		
	$K_{2,3}$ – Ergonomic aspects	$K_{2,3,1}$ – Quality of external design of the new buildings	
		$K_{2,3,2}$ – Urban efficiency of the centre	
K_3 – TECHNOLOGICAL	$K_{3,1}$ – Design time		
	$K_{3,2}$ – Time to arrange administrative issues		
	$K_{3,3}$ – Time to construct the centre		
	$K_{3,4}$ – Meteorological influence and natural hazards		
K_4 – ECONOMICAL	**$K_{4,1}$ – Funding of the project**	$K_{4,1,1}$ – Currency risk	
		$K_{4,1,2}$ – Possibility for funding from international funds	
		$K_{4,1,3}$ – Amount of investment	
		$K_{4,1,4}$ – Time to return of investment	
	$K_{4,2}$ – Cost levels	$K_{4,2,1}$ – Consumption costs for utilities	
		$K_{4,2,2}$ – Depreciation costs	
	$K_{4,3}$ – Income levels	$K_{4,3,1}$ – Income from tourism	
		$K_{4,3,2}$ – Income from passenger maritime transport	
		$K_{4,3,3}$ – Income from concessions in the centre	
		$K_{4,3,4}$ – Increase of investments in the region	
		$K_{4,3,5}$ – Income from land price differences	

The only 0-level criterion is the total marginal criterion, which is the quantity utilized to rank the alternatives. In (1) we assume that each expert is equally important in the assessment process, but if the level of expertise is different, then a weighted average of the experts' estimates can be utilised instead of (1).

Note that Y_{mean}^{ind} should be treated as the relative significance of the *ind* criterion by the *r*-th expert within the marginal criterion that the *ind* criterion belongs to (e.g. the weight coefficient for $K_{4,1,2}$ represents its relative importance for the marginal criterion $K_{4,1}$). So, the first-level criterion is calculated based on the second-level criteria, whose relative coefficients of significance sum to one. Such a setup significantly helps the experts in their estimation of the weight coefficients. The normalized weight coefficient of criterion *ind* is calculated for each criterion except for the 0-level one:

$$\lambda_{ind} = Y_{mean}^{ind} \;/ \sum_{inds \in S(ind)} Y_r^{inds} \quad \text{for } ind \neq [\,] \tag{2}$$

The information in Table 4 is the same for all alternatives. It allows for automatic generation of the criteria hierarchy shown in Table 3.

Table 3. Criteria indices and sub-criteria sets. The marginal criteria are bolded.

ind	*S*(*ind*)	*ind*	*S*(*ind*)
[]	{**[1]**,**[2]**,**[3]**,**[4]**}	[1,1,1]	{}
[1]	{**[1,1]**,**[1,2]**,[1,3],[1,4],**[1,5]**,**[1,6]**}	**[1,1,2]**	{[1,1,2,1],[1,1,2,2]}
[2]	{[2,1],[2,2],**[2,3]**}	[1,2,1]	{}
[3]	{[3,1],[3,2],[3,3],[3,4]}	[1,2,2]	{}
[4]	{**[4,1]**,**[4,2]**,**[4,3]**}	[1,2,3]	{}
[1,1]	{[1,1,1],**[1,1,2]**}	[1,2,4]	{}
[1,2]	{[1,2,1],[1,2,2],[1,2,3],[1,2,4]}	[1,5,1]	{}
[1,3]	{}	[1,5,2]	{}
[1,4]	{}	[1,6,1]	{}
[1,5]	{[1,5,1],[1,5,2]}	[1,6,2]	{}
[1,6]	{[1,6,1],[1,6,2]}	[2,3,1]	{}
[2,1]	{}	[2,3,2]	{}
[2,2]	{}	[4,1,1]	{}
[2,3]	{[2,3,1],[2,3,2]}	[4,1,2]	{}
[3,1]	{}	[4,1,3]	{}
[3,2]	{}	[4,1,4]	{}
[3,3]	{}	[4,1,5]	{}
[3,4]	{}	[4,1,2]	{}
[4,1]	{[4,1,1],[4,1,2],[4,1,3],[4,1,4]}	[1,1,2,1]	{}
[4,2]	{[4,2,1],[4,2,2]}	[1,1,2,2]	{}
[4,3]	{[4,3,1],[4,3,2],[4,3,3],[4,3,4],[4,3,5]}		

In ***STAGE 3 OF REPOMP*** we utilize a ranking scale, given in Table 5, to allow the experts to assess how each alternative complies with the primitive criteria. The alternatives are assessed against the primitive criteria by twelve experts. If an alternative is assessed against a primitive criterion *ind* by n_{ind}^{X} experts, X_r^{ind} being the mark given by the *r*-th expert, then the opinion of all experts for the alternative against the primitive criterion can be organized in a n_{ind}^{X}-dimensional vector $\vec{X}_{ind}=(X_1^{ind}, X_2^{ind}, \ldots, X_{n_{ind}^{X}}^{ind})$. The opinion of the experts for the alternatives against the primitive criteria is given in Table 6.

In ***STAGE 4 OF REPOMP*** we calculate the total marginal criterion for the analysed alternatives. We can present each alternative as a multidimensional vector $\vec{T}$ $(t_1, t_2, \ldots, t_N)$ in the *N*-dimensional space of the experts estimates against the primitive criteria (i.e. $t_i = X_r^{ind}$). So the total marginal criterion X^m of the alternative is an *N*-dimensional real valued function *v*(.) of the coordinates of $\vec{T}$ of the alternative: $X^m=v(t_1, t_2, \ldots, t_N)$. The form of this function depends on the preferences over the criteria. If it is true that an alternative is always better than another if it has better scores, all else being equal, then there is preferential independence over the coordinates of $\vec{T}$. Then the value function has additive form [3]:

$$X^m = v(t_1,t_2,\ldots,t_N) = w_1(t_1) + w_1(t_1) + \ldots + w_N(t_N) \qquad (3)$$

IOP Conf. Series: Materials Science and Engineering **797** (2020) 012035 doi:10.1088/1757-899X/797/1/012035

Table 4. Experts' opinion for the weight coefficients of the criteria

ind	n_{ind}^{Y}	$\vec{Y}_{ind}$	*ind*	n_{ind}^{Y}	$\vec{Y}_{ind}$
1	12	3 3 3 3 3 3 2 2 2 2 1 1	2,3,1	12	3 3 3 3 2 2 2 2 1 1 1 1
1,1	12	3 3 2 2 2 2 1 1 1 1 1 1	2,3,2	12	3 3 3 3 3 3 3 3 2 2 2 2
1,1,1	12	3 2 2 2 1 1 1 1 1 1 1 1	3	12	1 1 1 1 1 0 0 0 0 0 0 0
1,1,2	12	3 3 3 3 3 3 2 2 2 2 1 1	3,1	12	3 2 2 2 1 1 1 1 1 1 0 0
1,1,2,1	12	3 3 3 3 3 3 3 2 2 2 2 1	3,2	12	3 3 3 3 3 2 2 2 2 2 1 1
1,1,2,2	12	3 3 3 2 2 2 2 1 1 1 1 1	3,3	12	2 2 2 1 1 1 1 1 1 1 1 1
1,2	12	3 3 3 3 3 3 3 2 2 2 1 1	3,4	12	3 3 2 1 1 1 1 1 1 1 1 1
1,2,1	12	3 3 2 2 2 2 1 1 1 1 1 1	4	12	3 3 3 3 3 3 3 3 3 3 2 1
1,2,2	12	3 3 3 3 3 3 3 3 3 2 2 2	4,1	12	3 3 2 2 2 2 2 1 1 1 1 1
1,2,3	12	1 1 0 0 0 0 0 0 0 0 0 0	4,1,1	12	1 1 1 0 0 0 0 0 0 0 0 0
1,2,4	12	1 0 0 0 0 0 0 0 0 0 0 0	4,1,2	12	3 3 3 3 3 3 3 3 3 3 3 3
1,3	12	3 2 1 1 1 0 0 0 0 0 0 0	4,1,3	12	3 3 3 3 3 3 3 3 3 3 3 3
1,4	12	3 3 3 3 3 3 3 3 2 2 2 1	4,1,4	12	3 3 2 2 2 1 1 1 1 1 1 1
1,5	12	3 3 2 2 2 2 1 1 1 1 0 0	4,2	12	3 3 3 3 3 3 3 3 3 2 2 2
1,5,1	12	3 2 1 1 1 1 1 1 1 1 1 1	4,2,1	12	3 2 2 2 2 2 2 2 1 1 1 1
1,5,2	12	3 3 3 3 3 2 2 2 2 2 1 1	4,2,2	12	3 3 2 2 2 2 2 1 1 1 1 1
1,6	12	3 2 1 1 1 0 0 0 0 0 0 0	4,3	12	3 3 3 3 3 3 3 3 3 3 2 2
1,6,1	12	3 3 2 2 2 1 1 1 1 1 0 0	4,3,1	12	3 3 3 3 3 2 2 2 2 2 2 2
1,6,2	12	3 2 2 1 1 0 0 0 0 0 0 0	4,3,2	12	3 3 3 2 2 2 1 1 1 1 1 1
2	12	3 3 3 2 2 2 2 2 2 2 1 1	4,3,3	12	3 3 3 3 3 3 3 3 3 2 2 2
2,1	12	3 3 3 3 3 3 3 3 3 2 2 1	4,3,4	12	3 3 3 3 3 3 3 3 3 3 3 2
2,2	12	3 2 2 2 2 2 2 2 2 2 2 1	4,3,5	12	3 3 3 3 3 3 2 2 2 2 2 2
2,3	12	3 3 3 3 3 3 3 3 3 3 2 1			

To find the total marginal criterion, REPOMP utilizes the Churchman-Ackoff model to evaluate a set of alternatives based on multiple criteria [19]. The additive function is presented as

$$X^{m} = v(t_1, t_2, \ldots, t_n) = \mu_1 v_1(t_1) + \mu_2 v_2(t_2) + \ldots + \mu_N v_N(t_N) \quad (4)$$

Table 5. Ranking scale to assess how each alternative meets the primitive criteria

For criteria regarding emissions	0 – emissions below the lower assessment limit 1 – emissions between the lower and the upper assessment limit 2 – emissions on the upper assessment limit 3 – emissions over the upper assessment limit
For all other criteria	0 – excellent 1 – very good 2 – good 3 – satisfactory 4 – non-satisfactory 5 – non-acceptable

Table 6. Expert estimates for the four alternatives against the primitive criteria

ind	Alternative 1	Alternative 2	Alternative 3	Alternative 4
	$\vec{X}_{ind}$	$\vec{X}_{ind}$	$\vec{X}_{ind}$	$\vec{X}_{ind}$
1,1,1	2 2 2 2 1 1 1 1 1 1 1 1	2 1 1 1 1 1 1 1 0 0 0 0	0 0 0 0 0 0 0 0 0 0 0 0	4 4 4 3 3 3 3 3 3 3 3 2
1,1,2,1	2 2 2 1 1 1 1 1 1 1 1 0	2 1 1 1 1 1 1 0 0 0 0 0	0 0 0 0 0 0 0 0 0 0 0 0	4 4 3 3 3 3 3 3 3 3 2 2
1,1,2,2	2 2 2 2 2 1 1 1 1 1 1 1	2 2 1 1 1 1 1 1 1 0 0 0	0 0 0 0 0 0 0 0 0 0 0 0	4 4 4 4 4 3 3 3 3 3 3 2
1,2,1	3 3 3 2 2 2 2 2 2 2 1 1	3 3 2 2 2 2 2 2 2 1 1 1	3 2 2 2 2 2 1 1 1 1 1 1	5 4 4 4 4 4 4 3 3 3 2 2
1,2,2	2 2 2 1 1 1 1 1 1 1 1 1	2 2 1 1 1 1 1 1 1 1 0 0	1 1 1 1 1 1 0 0 0 0 0 0	4 4 4 3 3 3 3 3 3 3 2 2
1,2,3	1 1 0 0 0 0 0 0 0 0 0 0	1 1 0 0 0 0 0 0 0 0 0 0	1 0 0 0 0 0 0 0 0 0 0 0	2 2 0 0 0 0 0 0 0 0 0 0
1,2,4	1 0 0 0 0 0 0 0 0 0 0 0	1 0 0 0 0 0 0 0 0 0 0 0	0 0 0 0 0 0 0 0 0 0 0 0	2 0 0 0 0 0 0 0 0 0 0 0
1,3	3 2 2 2 2 2 2 2 1 1 1 1	3 2 2 2 2 2 2 2 1 1 1 1	5 5 5 5 5 5 4 4 4 4 4 3	0 0 0 0 0 0 0 0 0 0 0 0
1,4	4 3 3 3 2 2 2 2 2 2 2 2	4 3 3 3 2 2 2 2 2 2 2 2	3 2 2 2 2 2 2 1 1 1 1 1	5 5 5 5 5 5 4 4 4 4 3 3
1,5,1	3 3 2 2 2 2 2 1 1 1 1 0	2 2 1 1 1 1 0 0 0 0 0 0	2 2 2 2 1 1 1 1 1 1 1 0	4 4 4 3 3 3 3 3 3 3 3 2
1,5,2	3 3 2 2 2 2 2 1 1 1 1 0	3 3 2 2 2 2 2 1 1 1 1 0	3 3 3 3 3 2 2 2 2 2 1 0	4 4 4 3 3 3 3 3 3 3 3 2
1,6,1	2 2 2 2 2 2 1 1 1 1 1 1	2 2 2 2 2 2 1 1 1 1 1 1	4 4 3 3 3 3 3 3 3 3 2 2	0 0 0 0 0 0 0 0 0 0 0 0
1,6,2	3 3 3 3 2 2 2 2 2 2 2 1	3 3 2 2 2 2 2 2 2 2 1 1	4 4 4 3 3 3 3 3 3 3 2 2	4 4 4 4 3 3 3 3 3 3 3 2
2,1	3 3 3 3 3 3 2 2 2 2 2 2	3 3 3 2 2 2 2 2 2 2 2 1	5 5 5 5 5 5 5 4 4 4 4 3	0 0 0 0 0 0 0 0 0 0 0 0
2,2	3 3 3 3 2 2 2 2 2 2 2 1	3 2 2 2 2 2 2 2 2 2 1 1	3 2 2 2 2 2 2 1 1 1 1 0	4 4 4 3 3 3 3 3 3 2 2 2
2,3,1	4 4 3 3 3 3 3 2 2 2 2 1	4 3 2 1 1 0 0 0 0 0 0 0	4 4 4 4 3 3 3 3 2 2 1 1	5 5 5 5 5 5 5 5 4 4 4 3
2,3,2	4 4 4 3 3 3 3 3 3 3 2 1	2 1 0 0 0 0 0 0 0 0 0 0	4 4 4 4 3 3 3 3 3 2 1 1	5 5 5 5 5 5 5 5 5 5 5 5
3,1	3 3 3 3 2 2 2 2 2 1 1 1	3 3 2 2 1 1 1 1 1 1 1 0	4 4 4 4 4 4 4 4 3 3 2 1	0 0 0 0 0 0 0 0 0 0 0 0
3,2	3 2 2 2 2 2 2 2 2 2 1 0	2 2 2 2 2 1 1 1 1 0 0 0	5 5 5 5 5 5 5 5 5 5 4 3	0 0 0 0 0 0 0 0 0 0 0 0
3,3	3 3 3 3 3 3 3 3 2 2 2 1	3 3 3 3 3 3 3 3 2 2 2 1	5 5 5 5 5 5 5 5 4 4 3 3	0 0 0 0 0 0 0 0 0 0 0 0
3,4	3 3 3 3 3 3 2 2 2 2 2 2	3 3 3 3 3 3 2 2 2 2 2 2	2 2 2 2 2 2 2 1 1 1 1 1	5 5 5 5 5 5 5 5 5 5 4 3
4,1,1	1 1 1 1 1 1 1 0 0 0 0 0	1 1 1 1 1 1 1 0 0 0 0 0	2 2 2 2 2 2 2 1 1 1 0 0	0 0 0 0 0 0 0 0 0 0 0 0
4,1,2	2 2 2 2 2 2 2 1 1 1 1 0	2 1 1 1 1 0 0 0 0 0 0 0	1 0 0 0 0 0 0 0 0 0 0 0	5 5 5 5 5 5 5 5 5 5 5 5
4,1,3	4 3 3 3 3 3 3 3 3 3 2 2	4 3 3 3 3 2 2 2 2 2 1 1	5 5 5 5 5 5 5 5 5 4 4 4	0 0 0 0 0 0 0 0 0 0 0 0
4,1,4	3 3 2 2 2 2 2 2 2 2 2 1	3 2 2 2 2 2 2 1 1 1 0 0	5 5 5 5 5 5 5 5 4 4 3 3	0 0 0 0 0 0 0 0 0 0 0 0
4,2,1	4 4 3 3 3 3 3 3 3 3 1 1	4 4 3 3 3 3 3 3 3 3 1 1	3 2 2 2 2 2 2 2 1 1 1 1	5 5 5 4 4 4 4 4 3 3 3 3
4,2,2	4 3 3 3 3 2 2 2 2 2 1 1	4 3 3 3 3 2 2 2 2 2 1 1	3 3 2 2 2 2 2 1 1 1 1 0	4 3 3 3 3 3 3 3 2 2 2 1
4,3,1	2 2 1 1 1 1 1 1 0 0 0 0	2 2 2 2 2 1 1 1 1 1 1 0	1 1 1 1 1 0 0 0 0 0 0 0	5 5 5 5 5 5 5 5 5 5 5 5
4,3,2	1 1 0 0 0 0 0 0 0 0 0 0	1 1 0 0 0 0 0 0 0 0 0 0	1 1 0 0 0 0 0 0 0 0 0 0	5 5 5 5 5 5 5 5 5 5 5 5
4,3,3	1 1 1 1 1 1 1 0 0 0 0 0	2 2 1 1 1 1 1 1 1 0 0 0	1 0 0 0 0 0 0 0 0 0 0 0	5 5 5 5 5 5 5 5 5 5 5 5
4,3,4	4 3 3 3 2 2 2 2 2 2 1 0	4 3 3 3 2 2 2 2 2 2 1 0	4 3 2 2 2 2 2 2 1 1 1 0	5 5 5 5 5 5 5 5 5 5 5 5
4,3,5	5 5 4 4 4 4 3 3 3 3 3 3	2 1 0 0 0 0 0 0 0 0 0 0	4 4 3 3 3 3 3 3 2 2 1 1	5 5 5 5 5 5 5 5 5 5 5 5

The sum of μ_i in (4) should be equal to 1 for the total marginal criterion X^m to be properly scaled. The formulae (4) is impractical and extremely difficult to use. Instead, the criteria in Table 2 are calculated recursively for each alternative. The average rating of criterion *ind* is calculated for each primitive criterion:

$$X_{mean}^{ind} = \frac{1}{n_{ind}^{X}} \sum_{r=1}^{n_{ind}^{X}} X_r^{ind} \text{ for } S(ind) = \{\} \tag{5}$$

In (5) we assume that each expert is equally important in the assessment process, but if the level of expertise is different, then a weighted average of the experts' estimates can be utilised instead of (4) in the same way as it was proposed for the mean weighted coefficients. The average rating of criterion *ind* is calculated for each marginal criterion:

$$X_{mean}^{ind} = \sum_{inds \in S(ind)} \lambda_{inds} X_{mean}^{inds} \text{ , for } S(ind) \neq \{\} \tag{6}$$

The order of the calculations of the marginal criteria rating is important. Firstly, (5) must be applied for all ratings of the highest level, then for all ratings of the second-highest level, and so on till it is applied for the only criterion of 0-level, which is the total marginal criterion of the alternative:

$$X^m = X_{mean}^{[]} \tag{7}$$

The calculated rating of the criteria for alternative 2 are given in Table 7, where the total marginal criterion X^m is 2.1299. The same table also shows the order of calculation. The resulting values for the four analysed alternatives are given in the second column of Table 10.

Table 7. Criteria ratings for alternative 1. The marginal criteria are bolded.

ind	X^{ind}_{mean}	Calculation order	*ind*	X^{ind}_{mean}	Calculation order
[]	**1.6129**	**46**	**[2,3]**	**0.53571**	**38**
[1]	1.5641	42	[2,3,1]	0.91667	16
[1,1]	**0.76396**	**34**	[2,3,2]	0.25000	17
[1,1,1]	0.75000	1	**[3]**	**1.8108**	**44**
[1,1,2]	**0.77244**	**33**	[3,1]	1.4167	18
[1,1,2,1]	0.66667	2	[3,2]	1.1667	19
[1,1,2,2]	0.91667	3	[3,3]	2.5833	20
[1,2]	**1.2812**	**35**	[3,4]	2.5000	21
[1,2,1]	1.9167	4	**[4]**	**1.7140**	**45**
[1,2,2]	1.0000	5	**[4,1]**	**1.4069**	**39**
[1,2,3]	0.16667	6	[4,1,1]	0.58333	22
[1,2,4]	0.083333	7	[4,1,2]	0.50000	23
[1,3]	1.7500	8	[4,1,3]	2.3333	24
[1,4]	2.4167	9	[4,1,4]	1.5000	25
[1,5]	**1.3095**	**36**	[4,2]	2.5833	40
[1,5,1]	0.66667	10	[4,2,1]	2.8333	26
[1,5,2]	1.6667	11	[4,2,2]	2.3333	27
[1,6]	**1.6738**	**37**	[4,3]	1.0599	41
[1,6,1]	1.5000	12	[4,3,1]	1.3333	28
[1,6,2]	2.0000	13	[4,3,2]	0.16667	29
[2]	**1.4945**	**43**	[4,3,3]	0.91667	30
[2,1]	2.1667	14	[4,3,4]	2.1667	31
[2,2]	1.9167	15	[4,3,5]	0.25000	32

In ***STAGE 5 OF REPOMP*** we apply simulation techniques to construct the distribution of the total marginal criterion. If we were able to include all possible experts in our analysis, we would calculate the true total marginal criterion for each alternative X^{true}. However, we only have a sample of the whole population of experts, hence the calculated X^{m} is only an estimate (the actual one) of X^{true} for each alternative. If we were able to acquire information from another set of experts, then we would have another estimate of X^{true}. If we hypothetically had a great number (*M*) of expert panels, then we would obtain a great number of point-estimates $(X^{m})_i$, for i=1, 2, …, *M*. Then we could also construct the distribution and find numerical characteristics for the total marginal indicator (e.g. mean, standard deviation, confidence intervals, etc.). To generate those *M* number of expert panels' estimates, we can adopt the computer-intensive Bootstrap simulation method [18]. It requires no information about the parameters of the process and needs to use data that consists of "independent and identically distributed data points" [20]. For further details regarding the essence and justification of the Bootstrap method, please refer to [21]. Using Bootstrap, we generate a large number *M* of pseudo-realities, for each alternative. In each pseudo-reality, synthetic learning samples of the same size as the original sample are generated by drawing with replacement from the original data. For example, the 9th synthetic sample of the weight coefficients and the estimates against the primitive criteria for alternative 2 are given respectively in Table 8 and 9.

Then each learning sample is processes as in stage 2 and stage 4 of REPOMP to obtain its synthetic total marginal criteria. As a result, synthetic ranking balls $\left(X^{m}\right)_i^{synt}$, i=1, 2, …, *M* are obtained. For example, $\left(X^{m}\right)_9^{synt}=1.6249$ is calculated as a synthetic ranking ball of Alternative 2 in the 9th pseudo reality using (1), (2), (5), (6), and (7) over the data in Table 8 and Table 9. The main assumption of the Bootstrap method is that the way the actual estimate deviates from the true parameter is the same as the deviation of the synthetic estimates from the actual one. Hence the distribution of the total marginal criterion constructed using the synthetic samples will give us information on what the true parameter X^{true} is based on the actual value X^{m}.

Table 8. Synthetic experts' opinion for the weight coefficients of the criteria in the 9th pseudo-reality for Alternative 2

ind	n^{Y}_{ind}	$\vec{Y}_{ind}$	*ind*	n^{Y}_{ind}	$\vec{Y}_{ind}$
1	12	1 3 1 3 3 2 3 3 2 2 3 3	2,3,1	12	3 1 3 3 1 3 1 1 1 2 2 2
1,1	12	2 2 2 1 2 1 3 3 1 2 1 1	2,3,2	12	3 2 3 2 3 2 3 3 3 3 3 3
1,1,1	12	2 3 3 2 2 2 3 1 1 1 2 1	3	12	0 0 1 0 1 1 0 1 1 1 0 0
1,1,2	12	3 3 3 3 3 3 3 2 2 3 3 3	3,1	12	0 0 2 0 1 3 3 2 1 2 2 1
1,1,2,1	12	2 2 2 2 2 2 2 2 3 3 3 1	3,2	12	3 2 2 3 3 2 1 2 2 2 2 2
1,1,2,2	12	3 1 3 3 2 2 1 1 1 2 3 3	3,3	12	1 1 1 2 1 1 1 1 1 2 1 1
1,2	12	3 3 2 3 1 3 3 3 3 2 2 3	3,4	12	1 1 1 2 1 1 1 1 1 3 1 1
1,2,1	12	3 1 1 1 2 2 1 2 2 1 1 3	4	12	3 3 3 3 3 1 3 3 3 3 3 3
1,2,2	12	3 2 3 3 3 3 2 3 3 3 3 3	4,1	12	2 1 1 1 2 2 3 3 2 1 2 2
1,2,3	12	0 0 0 0 0 0 0 1 1 0 0 1	4,1,1	12	1 0 0 0 1 1 0 1 0 1 0 0
1,2,4	12	0 0 0 0 1 0 1 0 0 0 0 0	4,1,2	12	3 3 3 3 3 3 3 3 3 3 3 3
1,3	12	2 0 1 0 1 2 0 0 0 2 1 1	4,1,3	12	3 3 3 3 3 3 3 3 3 3 3 3
1,4	12	3 3 2 2 2 2 3 3 3 3 2 3	4,1,4	12	1 3 2 2 1 2 1 1 2 1 1 3
1,5	12	1 1 3 2 1 1 1 1 1 1 2 1	4,2	12	3 2 3 3 3 3 3 3 2 2 3 3
1,5,1	12	1 2 1 3 1 1 1 1 1 1 1 1	4,2,1	12	3 3 2 1 1 2 1 1 2 2 1 2
1,5,2	12	2 1 2 2 3 1 2 2 2 3 2 2	4,2,2	12	2 1 1 2 3 3 3 2 2 3 3 2
1,6	12	1 1 0 0 0 1 0 1 0 0 0 0	4,3	12	3 3 3 3 3 3 3 3 3 3 3 3
1,6,1	12	2 3 1 2 1 0 1 3 0 1 1 3	4,3,1	12	3 2 3 3 2 3 2 2 2 3 3 2
1,6,2	12	0 0 3 1 0 0 1 2 3 1 2 0	4,3,2	12	1 1 3 1 1 1 1 2 1 1 2 1
2	12	3 2 2 3 2 3 3 2 3 2 2 2	4,3,3	12	3 2 3 2 2 3 2 3 3 3 3 3
2,1	12	3 1 3 1 1 3 2 2 3 3 3 3	4,3,4	12	3 2 3 3 3 3 3 3 2 3 3 3
2,2	12	2 2 2 2 2 2 2 2 2 2 2 3	4,3,5	12	2 2 2 2 2 3 2 2 2 3 3 2
2,3	12	3 3 2 1 3 3 3 3 3 3 3 3			

Table 9. Synthetic experts' estimates for the alternative 2 against the primitive criteria in the 9th pseudo-reality.

ind	**Alternative 2**	*ind*	**Alternative 2**
	$\vec{X}_{ind}$		$\vec{X}_{ind}$
1,1,1	1 1 1 1 0 0 1 1 0 1 1 1	2,3,2	0 2 0 0 2 2 0 0 0 0 0 0
1,1,2,1	1 0 0 0 1 0 1 0 1 0 0 2	3,1	1 1 1 0 1 3 1 0 2 1 1 0
1,1,2,2	1 1 0 1 2 1 0 2 0 0 0 2	3,2	0 2 2 0 1 0 0 2 0 2 1 1
1,2,1	2 2 2 3 2 2 1 2 3 1 2 3	3,3	3 3 1 2 3 3 3 3 2 3 3 3
1,2,2	0 1 1 1 1 1 1 0 1 1 2 0	3,4	2 3 2 2 3 3 3 2 3 3 3 2
1,2,3	0 0 0 0 0 0 0 0 0 0 0 0	4,1,1	1 0 1 0 0 1 0 0 0 1 0 1
1,2,4	1 0 0 0 0 0 0 0 0 0 0 0	4,1,2	1 1 0 1 0 0 1 1 0 1 0 1
1,3	2 2 3 2 2 1 1 1 2 1 2 2	4,1,3	2 2 4 2 3 3 1 3 2 3 3 1
1,4	2 2 3 3 3 2 3 3 3 4 2 2	4,1,4	2 2 2 3 3 3 2 2 2 2 0 2
1,5,1	2 0 0 2 2 0 2 0 0 0 0 1	4,2,1	4 3 3 3 3 3 3 3 1 3 4 3
1,5,2	3 2 1 2 3 2 2 1 2 1 1 1	4,2,2	3 3 2 1 3 4 3 2 1 1 2 3
1,6,1	2 2 1 1 2 2 2 2 1 1 1 1	4,3,1	1 1 1 2 2 1 2 2 0 0 1 1
1,6,2	2 2 1 1 2 1 2 1 2 2 2 2	4,3,2	1 0 0 0 0 0 0 0 0 0 0 0
2,1	3 3 2 2 2 3 1 2 3 1 2 2	4,3,3	1 1 1 2 0 1 0 0 1 0 1 1
2,2	1 2 2 2 2 3 1 2 2 2 1 2	4,3,4	3 4 3 1 2 2 3 1 3 2 3 2
2,3,1	0 2 0 3 1 1 0 2 1 0 0 1	4,3,5	0 1 0 0 0 0 0 0 0 1 0 2

To alleviate biases in the calculation process, we obtain the best randomized point estimate X_{best}^{m} as the mean of the synthetic ranking balls $\left(X^{m}\right)_{i}^{synt}$, flipped around the actual estimate X^m:

$$X_{best}^{m} = 2X^{m} - \sum_{i=1}^{M}\left(X_{i}^{synt}\right)/M \tag{8}$$

We can easily calculate the standard deviation σ_X of the best randomized point-estimate X_{best}^{m} by:

$$\sigma_X = \sqrt{\sum_{i=1}^{M}\left(X_{i}^{synt} - \sum_{i=1}^{M}\left(X_{i}^{synt}\right)/M\right)^2 \Big/ (M-1)} \tag{9}$$

We generated 1000 pseudo-realities and calculated the synthetic total marginal criterion for each alternative. The results are given in the third and fourth column of Table 10. The results show that Alternative 2 is the best (*Establish the recreational centre at berths 1 to 10, emphasizing on the new appearance of the City Hall and the city square*), followed by Alternatives 1, 3 and 4. The densities of the total marginal criteria X^m for each alternative are also constructed, based on the synthetic rating marks (see Figure 11).

Table 10. Sample estimate, best point estimate and standard deviation of the total marginal criterion for each alternative

Alternative	Total marginal criterion X^m	Best estimate X_{best}^{m}	Standard deviation σ_X
1	2.1299	2.1265	0.0431
2	1.6129	1.6139	0.0445
3	2.1427	2.1397	0.0451
4	3.0392	3.0391	0.0349

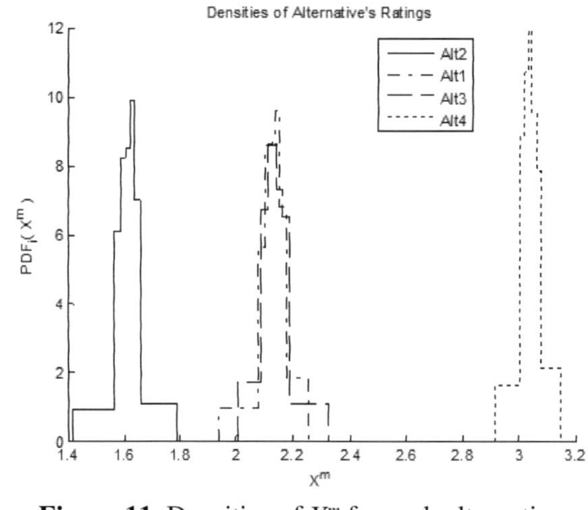

Figure 11. Densities of X^m for each alternative

Previous works [7; 22] extended the REPOMP results to test if the alternatives are significantly different (or else the difference in their ranking scores is negligible). The alternatives are grouped into pseudo-indifference classes, where the alternatives are indifferent to at least one of the other alternatives in the class (unlike classical indifference classes, where all elements in the set are indifferent to each other) [22]. The same work demonstrated that in such cases, the resulting relation of strict preference over pseudo indifference classes may end up being non-transitive. So, it is demonstrated in theoretical discussion and practical cases studies that this relation is strongly dependent on the significance level of comparisons to allocate alternatives into groups. In our study, at a significance level of 0.05, three pseudo-indifference classes are formed, as shown in Figure 12. The p_{value} is given above the connecting lines. Hypothesis testing is applied to each consecutive alternative.

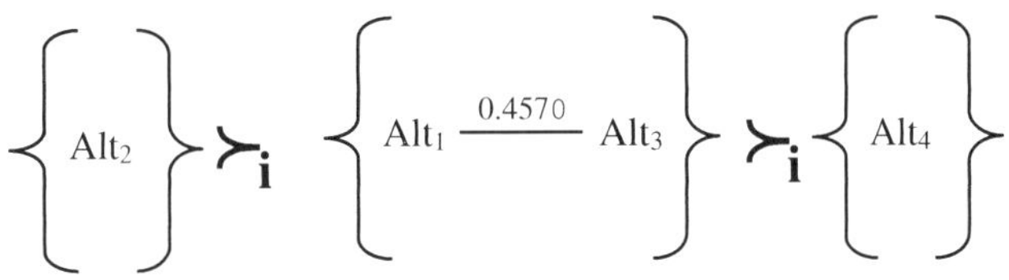

Figure 12. Grouping of alternatives in pseudo-indifference classes at significance level of 0.05

4. Conclusion

The REPOMP procedure replaces the usually-complex multi-criteria utility analysis that a classical decision-making procedure would require with an easy-to-use procedure based on expert estimates. The calculations of the total marginal criteria and their distribution under stages 4 and 5 of REPOMP, as well as the analysis of indifference classes are performed using original MATLAB functions. The functions are available free of change upon request from the authors. At the time of publication of this paper, the Port of Varna East has already undergone part of the renovations and restructures envisaged in the Alternative 2, recommended by our study.

Acknowledgments

This case study was supported by the ECOPORT 8 project, Code SEE/A/218/2.2/X, as well as by the INPORT project (DVU01/0031) of the Bulgarian Science Fund. We would also like to express gratitude to all the experts that participated in the study, as well as to our colleagues Miroslav Nikolov and Nora Ilinkolova for their devoted work on the realization of the case study.

5. References

[1] Luger G. and Stubblefield W. 2004 *Artificial Intelligence: Structures and Strategies for Complex Problem Solving* (5th ed., Benjamin/Cummings)

[2] Hertz D. and Thomas H. 1984 *Practical risk analysis: an approach through case history* (Bath, Avon, John Wiley, Pitman Press Ltd)

[3] French, S. and Insua D.R. 2010 *Statistical Decision Theory: Kendall's Library Of Statistics 9.* (Chichester, UK: John Wiley & Sons Ltd.)

[4] Wright G. and Ayton P. 1994 *Subjective Probability* (Chichester, UK: John Wiley&Sons)

[5] Tenekedjiev K., Kamenova S. and Nikolova N.D. 2004 *J Cleaner Prod.* **12(4)** 353-360

[6] Nikolova N., Ivanova S., Georgieva G. and Tenekedjiev K. 2015 *Bootstrap evaluation of expert panel opinion in case studies solved by REPOMP*, In: Zlateva I., Raykov V., and Nikolov N. (Eds.) *Progressive Engineering Practices in Marine Resources Management* (IGI Global, pp. 48-76)

[7] Parushev P., Nikolova N.D., Kobashikawa C. and Tenekedjiev, K. 2006 REPOMP-Screening of Energy Saving Alternatives in an Education Building *Proc. International IFAC Workshop on Energy Saving Control in Plants and Buildings* (2-5 October 2006, Bansko, Bulgaria) pp. 37-42

[8] European Commission 2007 *Directive 2007/2/EC of the European Parliament and of the Council of 14 March 2007 establishing an Infrastructure for Spatial Information in the*

European Community (INSPIRE) (OJ L 108, 25.4.2007, p. 1–14) https://eur-lex.europa.eu/legal-content/EN/TXT/PDF/?uri=CELEX:32007L0002&from=EN
[9] Ivanova S., Nikolova N.D., Kalinov K., Georgieva G., Mednikarov B. and Tenekedjiev, K. 2013 Selection of Adequate Models for Development of Spatial Data Infrastructure Using REPOMP *Proc. First International Conference of the Black Sea Association of Maritime Institutions; Actual Problems of Marine Shipping* (Admiral Ushakov Maritime State University, Novorossiysk, Russia, 25-27 June 2013) pp. 264-274
[10] Nikolova N.D, Armenski I., Tchikova K. and Tenekedjiev, K. 2012 Expert Assessment of Alternatives to Change the Location of the Port of Varna *Proc. 13th IFAC Symposium on Control in Transportation Systems CTS'2012* (September 12-14, Sofia, Bulgaria) pp. 347-352
[11] European Commission 2001 *White Paper: Roadmap to a Single European Transport Area – Towards a Competitive and Resource Efficient Transport System* (28.3.2011 COM(2011)) https://ec.europa.eu/transport/sites/transport/files/themes/strategies/doc/2011_white_paper/white-paper-illustrated-brochure_en.pdf
[12] Official Journal of the European Union 2004 *Decision No 884/2004/EC of the European Parliament and of the Council of 29 April 2004 amending Decision No 1692/96/EC on Community Guidelines for the Development of the Trans-European Transport Network* (L 167 /1 – L 167 /38) http://ec.europa.eu/ten/transport/legislation/doc/2004_0884_en.pdf
[13] European Commission 2011 *White Paper – Roadmap for a single European transport 2011* (28.3.2011 COM) https://eur-lex.europa.eu/LexUriServ/LexUriServ.do?uri=COM:2011:0923:FIN:EN:PDF
[14] Official Journal of the European Union, *Regulation (EU) No 1315/2013 of the European Parliament and of the Council of 11 December 2013 on Union guidelines for the development of the trans-European transport network and repealing Decision No 661/2010/EU*, https://publications.europa.eu/en/publication-detail/-/publication/58c00e4e-b59c-4b64-ad83-3d1a02b33fe8/language-en
[15] Belton V. and Steward T.J. 2002 *Multiple Criteria Decision Analysis – an Integrated Approach* (Kluwer Academic Publishers, USA)
[16] Baucells M. and Sarin R. 2003 *Man Sci* 49(8) pp. 1105-1118
[17] Hogarth R. and Karelaia N. 2005 *Man Sci.* 51(12) 1860–1872
[18] Efron B. and Tibshirani, R.J. 1993 *An Introduction to the Bootstrap* (Chapman &Hall).
[19] Britton G.A. and McCallion H. 1994 *Sys Pract* **7(5)** 487-521
[20] Press W., Teukolski S., Vetterling W.T. and Flannery B. 2007 *Numerical Recipes: the Art of Scientific Computing* (Third Edition, Cambridge University Press)
[21] Shao J. and Tu D. 1995 *The Jackknife and Bootstrap* (Verlag, New York, NY: Springer)
[22] Nikolova, N.D., Ivanova, S., Georgieva, G., Tenekedjiev, K., Non-Transitive Binary Relation of Preference in the Case of Random Value Functions, Derived from REPOMP *Proc. Fourth International Workshop Proceedings – Studies on Knowledge Discovery, Knowledge Management and Decision Making EUREKA-2013* (26-28 November, Mazatlan, Mexico), pp. 255-261

IOP Conf. Series: Materials Science and Engineering **797** (2020) 012036 doi:10.1088/1757-899X/797/1/012036

Sea level rise impact on mangrove growth and development in Coral Triangle Ecoregion Southeast, Indonesia

Kangkuso Analuddin[1*], Andi Septiana[1], Jamili[2], Idin Sahidin[3], Usman Rianse[4], Sahadev Sharma[5], Kazuo Nadaoka[6], Eka Diaz Anjani[7]

[1]Department of Biotechnology, Faculty of Mathematics and Natural Sciences, Universitas Haluoleo, 93232 Kendari, Sulawesi Tenggara, INDONESIA

[2,7]Study Program of Biology, Faculty of Mathematics and Natural Sciences, Universitas Haluoleo, 93232 Kendari, Sulawesi Tenggara, INDONESIA

[3]Faculty of Pharmacy, Universitas Haluoleo, 93232 Kendari, Sulawesi Tenggara, INDONESIA

[4]Faculty of Agriculture, Universitas Haluoleo, 93232 Kendari, Sulawesi Tenggara, INDONESIA

[5]Institute of Ocean and Earth Sciences, University of Malaya, MALAYSIA

[6]Graduate School of Transdisiplinary Sciences and Engineering, Tokyo Institute of Technology, Tokyo, JAPAN

*Corresponding author's email: zanzarafli@gmail.com

Abstract. The mangroves are among the most at-risk coastal ecosystems, which are especially vulnerable to sea-level rise. This study aimed to know the growth capacity and biochemical defense of mangrove seedlings of Rhizophora mucronata under different sea water logging at experimental condition in Kendari bay, Southeast Sulawesi, Indonesia. Results showed that maximum leaf area, intrinsic rate increase of leaf area and relative growth rate (RGR) of height of the R. mucronata seedlings were significantly higher at treatment of 45 cm from the ground than other treatments, whereas RGR of diameter was the higher at treatment of 30 cm from the ground. Meanwhile, half expansion period leaf area was almost similar for all treatment. However, the carbon and nitrogen ratio in leaves of R. mucronata seedlings was significantly higher at seawater logging of 0 cm from the ground. The flavonoid and vitamin C concentrations in R. mucronata seedlings were higher at higher level inundation indicating protective role of these compounds under seawater logging stress.

1. INTRODUCTION

The mangrove ecosystem is among the most at-risk ecosystems from sea-level rise. Therefore, understanding which mangrove stands are able to survive under sea-level rise, managers can identify and protect refuges that self-seed and act as sources for seeding of future mangrove communities.

However, recent condition of mangrove ecosystems throughout the world are suffered due to global warming and anthtropogenic pressures, and are influence the resistance of mangrove ecosystems, which play very important roles as one of primary producers to suppor the productivity of coastal zone and human life at the coral triangle ecoregion. The mangroves in Southeast Sulawesi consists high diversity including *Rhizophora apiculata*, *R. mucronata* and, *Ceriops tagal* [1]. Other studies in this area highlighted the potential of mangroves bio-prospecting [2, 3], allometric model and biomass [4, 5] and mangroves bio-filtering capacity [6].

Elucidation on resistance of mangrove forest under global climate change is needed for projecting how does this ecosystem will be affected by sea level rise pressure. This study aimed to know the growth capacity and biochemical defense of mangrove seedlings under different sea water logging at experimental condition in Kendari bay, Southeast Sulawesi, Indonesia.

2. LITERATURE REVIEW

The ability of mangroves to adapt growth with sea-level rise depends on accretion rate relative to rate of sea-level change [7]. [8] found that seedlings in the tanks simulating sea-level rise initially grew faster than plants in the other treatments, but grew slowly at sapling stage. [9] found that the mangroves along macro-tidal (>4 m) coastlines, along wet tropical coasts and/or in areas adjacent to significant river input, are the least vulnerable to the impact of sea-level rise.

Mangroves are biochemically unique, producing a wide array of novel natural products to adapt growth and sustain with environmental stress. For example *Excoecaria agallocha* exudes acrid latex that is injurious to the human eye, hence its designation as "the blinding tree". The latex is toxic to a variety of marine organisms [10, 11]. Mangroves are rich in polyphenols and tannins [12, 13].

3. RESEARCH METHODS

3.1. Site description

This study was carried out at the mangrove forest of Kendari Bay located at the Kendari city in Southeast Sulawesi province. The mangrove condition provides an excellent site for establishment of sea level rise effect on mangroves because the growth and development of mangroves showed good condition, though extensive mangrove land conversion are happen.

3.2. Experiment the effect of sea level rise on mangroves

We grew the five mangrove seedling in four different height of pipes, i.e. (1) 0 cm from the ground, (2) 15 cm above ground, (3) 30 cm above the ground, and (4) 45 cm above the ground. During high tide the water levels were 30 cm, 45 cm, 60 cm and 90 cm, respectively. The experiment are monitored every two days. The growth parameters such as stem diameter, and the height of seedlings were measured three times a week. Biochemical and physiological data collection determined at the end experiment and harvesting of plant material. The proline and alcoholic dehydrogenase were analyzed. In addition, biocehmical analyses were done for antocyanin, flavonoid, alkaloid, tannin and vitamin C. Meanwhile, carbon organic and nitrogen content were determined from seedlings in whole treatment.

4. RESULTS AND DISCUSSION

4.1. Mangrove leaf area growth

Leaf area growth of *R. mucronata* seedlings presented in Figure 1. As shown in Figure 1A, the maximum leaf area was significantly higher at seawater logging of 45 cm (24.62 cm^2) from the ground than treatment of 30 cm (12.10 cm^2), 15 cm (17.92 cm^2), and 0 cm (14.21 cm^2) from the ground. Moreover, intrinsic rate increase of leaf area (Figure 1B) was not significantly different among treatments of 45 cm, 30 cm, 15 cm and 0 cm from the ground was estimated as 0.13 $^{d-1}$, 0.25 $^{d-1}$, 0.14^{d-1}and 0.18^{d-1}, respectively. Similar trend showed to half expansion period (Figure 2C) of leaf area of *R. mucronata* seedlings. Therefore, sea level rise seemed to reduce the area of leaf of the mangrove of *R. mucronata* seedlings.

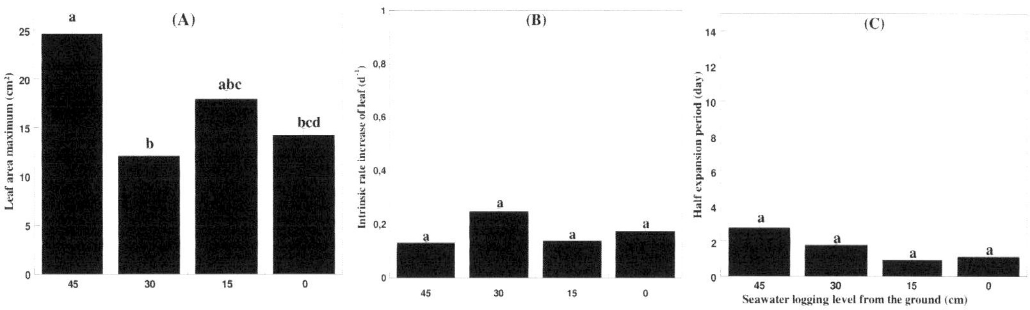

Figure 1. Trends in leaf area growth parameters of *R. mucronata* seedlings based on different seawater logging level. (A) maximum leaf area (*U*); (B) intrinsic rate increase (λ); and (C) half expansion period (t*). Similar letter indicates no significantly different.

4.2. Mangrove relative growth rate

Relative growth rate (RGR) of height and stem diameter of mangroves seedlings is represented by Table 1 below. It shows that the RGRs of height and stem of *R.* lower were the lower at treatment of 0 cm of seawater logging or deeper seawater, but faster at lower level of seawater logging, indicating sea level rise affect the growth of mangrove seedlings. However, mangroves can activate diverse components of their antioxidative system to restrict effect of seawater stresses of salinity and improving the good mangroves growth under tidal influenced varied saline condition. Sea level rise might reduce growth rate of mangrove seedlings at the present study that flooding and salinity might stressed the growth of *R. mucronata* seedlings. As found by [7] that sea level rise might reduce survival and growth performance of mangroves seedlings, expressed as biomass decrement due to variation in flooding and salinity. They also found that relative growth rate of *Bruguiera gymnorrhiza* decreased significantly with waterlogged time, but it was no clear effect of waterlogged to the growth of *Kandelia candel* plants.

Tabel 1. Relative growth rate (RGR) of height and stem diameter of mangrove *Rhizophora mucronata* seedlings under different seawater logging. Similar letter indicates no significantly different.

Mangroves	Seedlings position from the ground	RGR of height (cm)	RGR Diameter (cm)
R. mucronata	45 cm	0.13±0.007*a*	0.034±0.011*a*
	30 cm	0.09±0.009*b*	0.037±0.014*a*
	15 cm	0.08±0.010*bc*	0.036±0.005*a*
	0 cm	0.06±0.007*bcd*	0.014±0.005*b*

4.3. Biochemical adaptation

Table 2 depicts chemical compounds concentration as biochemical adaptation of *R. mucronata* seedlings under different seawater logging level. The flavonoid and vitamin C concentrations were higher at seawater treatment of 0 cm from the ground or higher level inundation indication protective role of these compounds under seawater logging stress. However, concentration of anthocyanin, proline and alcohol dehydrogenase as well as C/N ratio did not show clear trends indicating unclear their function under seawater logging treatment. [7] found that the chlorophyll contents of *K. candel* and *B. gymnorrhiza* increased in response to waterlogging. This is the way for these mangroves seedlings to survive in the environmental stress of sea waterlogging. However, our study shown that the seedlings of *R. mucronata* produce various chemical compounds to adapt growth in the stressing of seawater logging.

Table 2. Chemical compounds production as biochemical adaptation of *Rhizophora mucronata* seedlings under different seawater logging. Similar letter indicates no significantly different.

Mangroves	Compounds	Seawater level from the ground			
		45 cm	30 cm	15 cm	0 cm
Rhizophora mucronata	Flavonoid	0.20 ± 0.03 [a]	0.33 ± 0.03 [b]	0.30 ± 0.07 [a]	0.53 ± 0.11 [b]
	Vitamin C	0.10 ± 0.01 [a]	0.14 ± 0.01 [b]	0.14 ± 0.00 [c]	0.16 ± 0.00 [b]
	Proline	18.34 ± 0.39 [a]	17.95 ± 0.24 [b]	16.68 ± 0.15 [a]	6.48 ± 0.18 [b]
	Alcoholde-hydrogenase	0.61 ± 0.05 [a]	0.74 ± 0.09 [b]	0.99 ± 0.07 [b]	0.74 ± 0.08 [c]
	Rasio C/N	28.28718805 [a]	31.85215243 [a]	31.06056115 [a]	32.2812531 [b]

5. CONCLUSION

The growth of mangrove *Rhizophora mucronata* seedlings affected by sea level rise, although they are able to growth better due to production of various chemical compounds.

ACKNOWLEDGMENT

This research was supported by the Higher Eduation, Ministry of Education and Culture, Republic of Indonesia with grant nos. 0299/E3/2016, 025/E3/2017 and 0045/E3/LL/2018.

REFERENCES

[1] Analuddin, K., Jamili, Septiana A, Raya R, Rahim S. 2013. The spatial trends in the structural characteristics of mangrove forest at the Rawa Aopa Watumohai National Park, Southeast Sulawesi, Indonesia. *International Journal of Plant Sciences*. 4(8): 214-221,

[2] Septiana, A., J. Jamili, W. Harlis, and K. Analuddin. 2016. Mangroves bioprospecting: Antioxidant source and habitat of endemic animal Buballus sp in Rawa Aopa Watumohai National Park, Southeast Sulawesi, Indonesia. *Malaysian Applied Biology* 45(1):23–34.

[3] Analuddin, Andi Septiana, Nasaruddin, Yusuf Sabilu & Sahadev Sharma. 2019. Mangrove Fruit Bioprospecting: Nutritional and Antioxidant Potential as a Food Source for Coastal Communities in the Rawa Aopa Watumohai National Park, Southeast Sulawesi, Indonesia. *International Journal of Fruits Science*. Inpress.

[4] Analuddin, K., Jamili, Septiana, A., Raya, R., Rianse, R., Sahidin, I., Rahim S., Alfirman, Sharma S., Saban R and Nadaoka K. 2018. Trends in allometric models and aboveground biomass of family *Rhizophoraceae* mangroves in the Coral Triangle ecoregion, Southeast Sulawesi, Indonesia. *Journal of Sustainable Forestry*, 37(7) 691–711.

[5] Analuddin, K., Jamili, Septiana, A., Raya, R., Rianse, R., Sahidin, I., Rahim S., Alfirman, Sharma S., Nadaoka K. 2016. Allometric models and aboveground biomass of mangrove *Lumnitzera racemosa* Wild. Forest at Rawa Aopa Watumohai National Park, Southeast Sulawesi, Indonesia. *Forest Science & Technology*, 12(1), 43–50.

[6] Analuddin, K., Jamili, Septiana, A., Raya, R., Rianse, R., Sahidin, I., Rahim S., Alfirman, Sharma S., Saban R and Nadaoka K. 2018.. 2017. Heavy metal bioaccumulation in mangrove ecosystem at the coral triangle ecoregion, Southeast Sulawesi, Indonesia. *Marine Bulletin Pollution*, 125, 472–480.

[7] Ellison, A.M. & Farnsworth, E.J. 1997. Simulated Sea Level Change Alters Anatomy, Physiology, Growth, and Reproduction of Red Mangrove (*Rhizophora mangle* L.). *Oecologia*, 112(4): 435-446.

[8] Semeniuk, V. 1994. Predicting the Effect of Sea-Level Rise on Mangroves in Northwestern Australia. *Journal of Coastal Research*, 10(4) 1050-1076.

[9] McLeod and Salm, 2006. Managing Mangroves for Resilience to Climate Change. IUCN, Gland, Switzerland. 64pp.

[10] Kathiresan, K. and Thangam, T.S. 1987. Biotoxicity of *Excoecaria agallocha* L. latex on marine organisms. *Current Science,* 56 (7) : 314-315.

[11] Kathiresan, K. and Thangam, T.S. 1990. Effect of phenolics on growth of viviparous seedlings of *Rhizophora apiculata*. *Geobios,* 17 : 142-143.

[12] Kathiresan, K. and Ravi, V. 1990. Seasonal changes in tannin content of mangrove leaves. *The Indian Forester,* 116 (5) : 390-392.

[13] Ravi, A.V. and Kathiresan, K. (1990). Seasonal variation in gallotannin from mangroves. *Indian Journal of Marine Sciences,* 19 (3) : 224-225.

Indonesian License Plate Detection Using Morphological Operation

Y P Pasrun[1], M Muchtar[2*], A N Basyarah[3], Noorhasanah[1]

[1] Department of Information System, Faculty of Information Technology, Sembilanbelas November Kolaka University, South East Sulawesi, Indonesia, Kolaka 93517
[*2] Department of Informatics Engineering, Faculty of Engineering, Halu Oleo University, South East Sulawesi, Indonesia, Kendari 93132
[3] Department of Computer Science, Faculty of Mathematics and Natural Science, Halu Oleo University, South East Sulawesi, Indonesia, Kendari 93132

E-mail: [1] yuwanda@usn.ac.id, [*2] muthmainnahm@uho.ac.id, [3]amalia.basyarah@uho.ac.id

Abstract. Technological developments have enabled the use of vehicle number plate recognition in various fields such as parking systems, e-toll payments, handling motor vehicle theft cases, traffic violations and so on. Vehicle number plate characteristics in each country have characteristics that are not the same. Each country has a standard numbering system in its writing method, colour, language and size of the license plate, so that to take the research standards for license plate numbers from other countries is not effective. Special research is needed in accordance with the format of the number plate used by the country, in this case Indonesia. This research applies simple method, morphological operations in detecting vehicle plates in Indonesia. Morphological results will be segmented to get the image of the plate character separately. The test results show that the method applied can provide a maximum accuracy of up to 100% on taking plate images with a distance of 1.5 meters.

1. Introduction

Every country uses specific vehicle license plate recognition system. The vehicle license plate recognition system is one component of the intelligent transportation system [1] and is usually used for the control of traffic, traffic surveillance, monitoring against illegal activities, security control of restricted areas, traffic law enforcements, toll collection and parking management [2]. In Indonesia, this identification system has been used by the police in enforcing traffic law, even though the activities have not been fully carried out in each region due to the limitations of the vehicle's equipment and data that have not been synchronized. Parking system technology in Indonesia is currently growing along with the increasing public awareness of the environment and sustainability. Atomatic parking systems are important part of the environmental support systems. As the main support of parking system in urban areas, it is necessary to develop green infrastructure [3].

License plate recognition is an importan part in automatic parking system in urban areas. The license plate recognition system (LPRS) consists of two main technologies, which are the license plate positioning technology and the plate number recognition technology [4]. Plate number detection is part of license plate positioning technology and is very important to process LPRS effectively. Plate number detection and car image recognition can be a quite challenging task when the image acquisition is taken in different weather conditions, lighting conditions, angles, or various distances. There are several studies that try to find the best solutions in plate number detection, such as using artificial neural networks (ANNs) method in Korean vehicle plate detection [5], using SVM linear classification in china vehicle plate detection and Indonesian vehicle plat detection using combination of sliding window [6], histogram of oriented gradient (HOG) to license plate detection [7]. However, the method is time consuming because the complexity of computation.

This study proposes a simple morphological method for Indonesia license plate detection. Tests using morphological operations are also carried out in various ways to determine the effect of the distance of plate image taking on the results of segmentation. The proposed stage consists of three parts: image preprocessing, plate detection with extraction using morphological operation and character segmentation as stage to determine success.

2. Research Methods

The main process design of the system built is to use morphological operations to detect the plate area by ending the character segmentation process in the detection plate area. The general description of the system is shown in Figure 1.

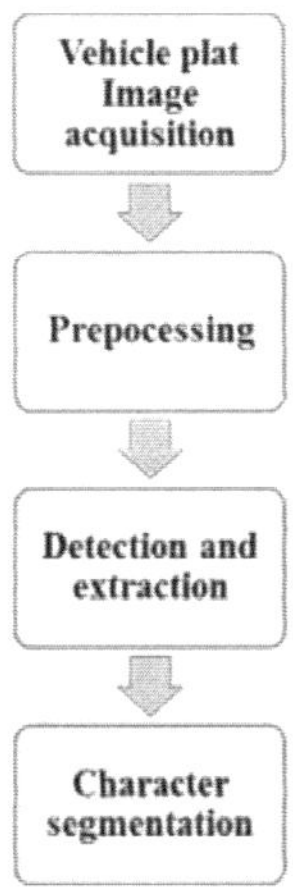

Figure 1. Overview of the System

Image acquisition is done in a parking area with sufficient lighting conditions using a Samsung smartphone camera with 8 MP resolution. To support the research scenario, image acquisition is divided into 3 conditions, namely at different distances (1 m, 1.5 m and 2 m). Each distance consists of 11 plate images that are the same as the total image of all 33 images.

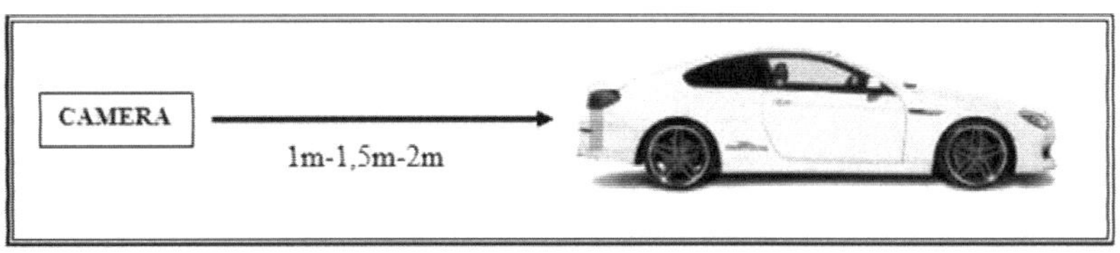

Figure 2. Image acquisition distance

Preprocessing stage is done to determine input according to the needs of the detection process. In this case because only one value is used in the process, three RGB channel values are put together by means of the image size of 640x480 acquisition pixels converted to grayscale images, using the grayscale formula to form the sum of R, G, and B components with the intention of removing hue information and saturation while maintaining luminance [8]. This grayscale formula is shown in Equation (1).

$$0.2989 * R + 0.5870 * G + 0.1140 * B \quad (1)$$

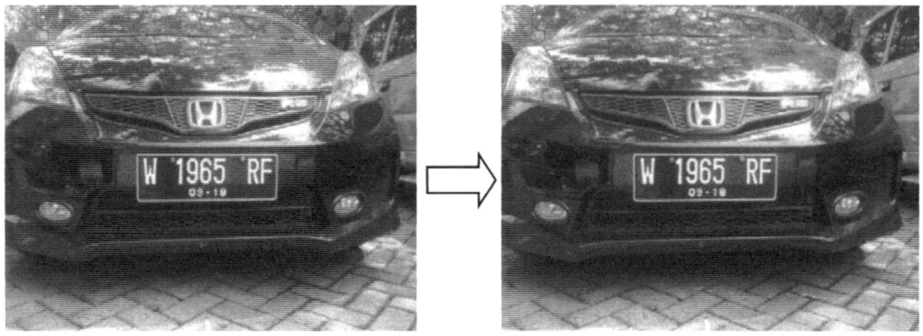

Figure 3. The Results of Converting RGB Images to Grayscale Images

In the process of plate detection and extraction, this grayscale image is then converted to binary images and then analysed using morphological operations that produce the area that has the most potential as a candidate plate. At this stage, the thresholding value is divided by 2. If the detected pixel has a pixel value less than the threshold value, it will be converted as background. If it is greater than the threshold value, then it is converted as foreground.

Figure 4. The Results of Converting RGB Images to Grayscale Image

The purpose of morphological operations is to obtain information about the shape of an image by adjusting the shape and size of a structuring element. In its use, morphology always involves an image with component I (image) and E (structuring element) constituent elements. Morphology has two basic operators, namely Dilation (dilation) and Erosion (erosion) which are commonly used to extract the desired components in an image. Dilation is the addition of pixels to the boundary between objects in a digital image while erosion is the opposite of dilation. This process will make the size of an image smaller. Based on these two operators, two other operators can be derived which are useful for smoothing the subinterval boundaries of the extracted components, namely opening and closing [9].

Opening is a combination of processes where a digital image is subjected to erosion surgery followed by dilation. Opening operations on images have the effect of smoothing the boundaries of objects, separating previously held objects, and removing objects that are smaller than structural sizes. The morphological operation used in this study is image erosion shown in Equation (2) to remove connected components in the thresholding image using the 8x8 pixel strel rectangle. The opening operation is then shown in Equation (4) to remove small objects that are less than 30 pixels in size. Then after the opening, an image dilation operation is performed (Equation (3) which aims to fill holes, thickening objects with the size of structuring elements.

$$(A\,B)(x,y) = \max\{A(x-s, y-t) + B(s,t) \qquad (2)$$

$$(A\,B)(x,y) = \max\{A(x+s, y+t) - B(s,t) \qquad (3)$$

$$A\ o\ B = (A\ B)\ B \tag{4}$$

In all the above operations A (x, y) are binary images and B (x, y) are structuring elements. Structuring elements are structural characteristics and certain features for measuring image shapes and are used to perform other image processing operations. In this approach, a rectangular arrangement element is used to erode the image with a radius of 8 * 8. Furthermore, the analysis of components that are connected to each other to get the plate area, with a ratio of plate length and width is between 0.1 and 0.42. Figure 6 shows the plate that has been detected.

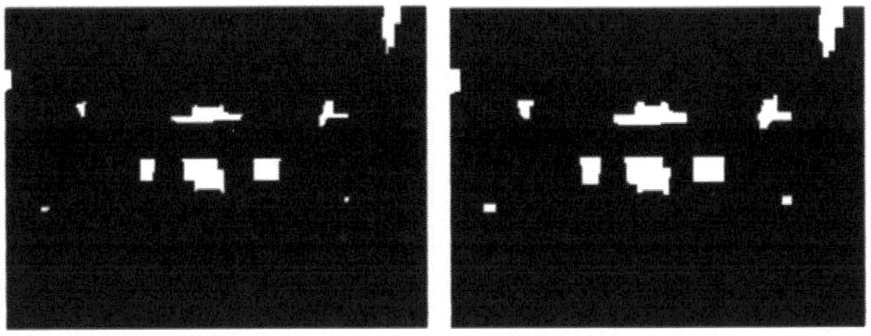

(a) (b)

Figure 5. Process image results, erosion (a) and dilation (b)

After the plate area is found, the plate is converted again to grayscale and binary to segment the character. Labelling is done to mark pixels that are connected to each other in the direction of 8 pixels. After that, the criteria are given that the area that is considered a character is an area with a minimum size of 20 pixels and the ratio of length / width is between 0.1 to 0.5. Furthermore, each character detected according to the criteria will be resized with a size of 30x20 pixels using the interpolation method. Interpolation method uses the average value of a region to represent its region. Figure 7 shows the character of the plate that has undergone segmentation.

Figure 6. Image detection plate

Figure 7. The Results of Character Segmentation in The Plate

3. Results and Discussions

The experiment was conducted to determine the successful rate for implementing the morphological operations on vehicle plate detection and find out the effect of distance in taking pictures on the final results of the study, namely the results of character segmentation in the plate. In this experiment, 33 images taken from the parking area were analysed and the size of all images was 640x480 pixels. All of them were taken by smartphone Samsung 8 MP camera under the same lighting conditions during the day (10 a.m. - 1 p.m.) and varied distance from 1 to 2 meters with the camera focusing on the front part or the rear part of the vehicle in which expected license plate is located.

IOP Conf. Series: Materials Science and Engineering **797** (2020) 012037 doi:10.1088/1757-899X/797/1/012037

Table 1. The Results of Plate Detection Based on Distance

Image	Distance		
	1 m	1.5 m	2 m
1	Successful	Successful	Successful
2	Successful	Successful	Unsuccessful
3	Successful	Successful	Successful
4	Successful	Successful	Successful
5	Successful	Successful	Successful
6	Successful	Successful	Successful
7	Successful	Successful	Successful
8	Successful	Successful	Unsuccessful
9	Successful	Successful	Successful
10	Successful	Successful	Successful
11	Unsuccessful	Successful	Successful
Accuracy	**90.9%**	**100.0%**	**81.8%**

Table 1 shows the experiment result which is done by dividing the acquired distance into 3 category. From the experimental results, it was found that distance affects the accuracy of detection and plate segmentation with an average accuracy of 90.90% and the ideal distance is 1.5 m.

Table 2. Example of Plate Detection with Successful Result in All Distance

Note	Distance		
	1 m	1.5 m	2 m
Image	 1_1.BMP	 1_2.BMP	 1_3.BMP
Plate			
Character			
Status	Successful	Successful	Successful

Table 3. Example of Plate Detection with Successful and Unsuccessful Result

Note	Distance		
	1 m	1.5 m	2 m
Image	 8_1.BMP	 8_2.BMP	 8_3.BMP

Plate			
Character			
Status	Successful	Successful	Unsuccessful

However, some license plates were not successfully detected. Table 3 shows example of plate detection process with succesful and unsuccessful result. In figure 8_3.bmp in Table 3 shows an example of a plate that was detected but was considered a failure because the segmented plate was less than perfect. Unlike the image 8_1 and 8_2 taken at a distance of 1 m and 1.5 m. Even though the image comes from the same vehicle as the image at 8_3.bmp, the plate is detected from both images. This shows that distance affects the accuracy of plate detection. From the Table 1, it can also be seen that the highest accuracy (100%) is at a distance of 1.5 m.

4. Conclusion

We have conducted a study to determine the impact of morphological operation in detecting Indonesian vehicle plate number. Using 3 types of distance conditions which is 1m, 1.5m and 2m to find out the most ideal distance in image acquisition for plate detection, we concluded that morphological operations can be applied to detect and segment characters on Indonesian vehicle plate number. Based on our observation, the ideal distance is 1.5 meters with an accuracy of 100% in a sufficient lighting conditions.

5. Reference

[1] Pandya P and Singh M 2011 Morphology Based Approach to Recognize Number Plates in India *International Journal of Soft Computing and Engineering (IJSCE)* vol. 1 (3) p 107-112

[2] Mai V, Miao D and Wang R 2013 Vietnam License Plate Recognition System Based On Edge Detection and Neural Networks *Journal of Information and Computing Science* vol. 8 (1) p 27-40

[3] Uniaty Q 2018 Park system concept for environmental sustainabilityin urban spatial development *Earth and Environmental Science : Conf. Ser.* **106** 012004

[4] Kim T and Kim E 2017 A Vehicle License Plate Recognition System Using Morphological ROI (Region of Interest) Map Generated from Morphology Operation *Journal of Physics: Conf. Ser.* **806** 012004

[5] Yoon J, Kang B and Kim D 2012 Car License Plate Detection Under Large Variations Using Covariance and HOG escriptors *International Symposium on Visual Computing*, Springer, Berlin, Heidelberg p 636-647

[6] Yuan Y, Zou W, Zhao Y, Wang X, Hu X, and Komodakis N 2017 A robust and efficient approach to license plate detection *IEEE Transactions on Image Processing* vol. 26 (3) p 1101-1114

[7] Astawa I N G A, Caturbawa I G N B, Sajayasa I M, and Atmaja I M A D S 2018 Detection of License Plate using Sliding Window, Histogram of Oriented Gradient, and Support Vector Machines Method *Journal of Physics: : Conf. Ser.* **953** 012062

[8] Hati S and Sajeevan G 2013 Vietnam License Plate Recognition System Based On Edge Detection and Neural Networks *Plant Recognition from Leaf Image Through Artificial Neural Network* vol. 62 (17)

[9] Gonzales R C and Woods R E 1992 *Digital Image Processing. Reading, Ma, 2* (Wesley publishing company)

Coastal Land Use Change and the Need of Green Space Based Development Coastal Area Settlement of South Sulawesi

M. Husni Kotta[1], Arman Faslih[2], La Ode Amrul Hasan[3], Yudhi Dwi Hartono[4]

[1, 2, 3]Universitas Halu Oleo Kendari, Southeast Sulawesi, Indonesia
[4]Universitas Muhammadiyah Kendari, Southeast Sulawesi, Indonesia

E-mail: hkottahusni@yahoo.com

Abstract. Green space is a major concern due to imbalance and uncontrolled land use and the deterioration of the environment. This is caused by the increasing number of people that is reflected in the activity of open space and land-use change in the protected area of urban coastal are. Development of natural resources currently leads to the front view based on the needs with less concern on the environment. It caused environmental disturbance such as erosion and flood. Thus it would require land use management scenarios that can be the basis for the planning of development of sustainable coastal areas as a coastal settlement.

1. Introduction

One of the main issues of sustainability was the potential of adverse impact due to coastal area the land-use change that leads to the formation of the settlement area, thus causing imbalances soil water storage within the ecosystem [1]. The land use change in the coastal area would have an impact on global climate change, resulted in the percentage decrease in precipitation and increase potential evapotranspiration of green space [2]. The impact of land use change that results in imbalances for the development sustainable of the resources potential could be attenuated by the provision of green space that can be used as a component of an urban coastal ecosystem [3]. The initial idea of coastal area development sustainable was based on water resources planning as part of the transformation of the environment with the potential for coastal area urban green space [4]. Development of water resources coastal area edges becomes important to get attention because it was the first part integrating water resources with other physical aspects such as the beach among the land, and water, the minted and coastal waters [5]. Also, a riverside and beach were the most attractive area for human settlement [6].

Balance conditions need potential evapotranspiration on a land area of green space coastal, around the coastal areas of the urban coastal area is an important indicator in showing the quality of the environment [7]. This is related to population pressure, as reflected in the pattern of land use that is not controlled in the urban coastal area, the negative impact of the uncontrolled land use is the growing surface runoff, which can lead to erosion and landslides in areas with slope > 30 %. Besides, with the limited ability of the soil to infiltrate as it is covered by a waterproof layer that prevents the entry of water, there will be a decrease in the volume of groundwater and increases the potential for flooding [8]. This reduction is based on the ability of coastal beaches in the Cape region receives pressure and had fairly high sedimentation [9]. Furthermore, the environmental management measures on the coastal area and watershed areas should be systematically by using an integrated approach, thus the necessary analysis exact model and can answer environmental problems area and the coastal area [10-15].

The problems of environmental degradation, especially in potential coastal urban green space in the coastal area due to uncontrolled land use change occurs essentially throughout the urban coastal areas, especially the coastal and upstream systems that systematically affect land use for green space integrated coastal area.

Thus, this study compared the impact of coastal area development sustainable area caused by the transformation of land use change, which occurs in coastal urban areas and the methods used in analyzing the change induced land. The study area was Tanjung Bunga coastal area, Makassar, South Sulawesi.

2. Coastal Area and Transformation

Development of water resources is important and part of the firs to integrated the use of the potential of urban green space to coastal area development such as beach, river, and land [6]. The something is seen also on the side of the property benefit is easy to see [16], with regard to the concentration of coastal areas on the shore based in many cities in the world, coastal areas began to be used as a commercial center, and a transportation center of manufacturing, as the main focus of supporting regional coastal urban areas. Because of this problem is very complex and diverse, and this issue due to change in the coastal area [13, 17]. This condition was no longer a positive impact as a result of the economic supply and imaging of a region [18]. There are also negative impacts begin degradation in slump environmental value and social inequality as a result of uncontrolled land use [19]. On the other hand, the deterioration caused sedimentation impacts are quite high [20,21] using statistical and spatial methods are then modeled using basins, which states that the quality of the hydrology of a watershed depends on the type of land use on environmental management are taken around the other coastal area.

The study in the Amazon Basin of Peru [22] is a pre-plan to cause most of the deforestation in the area where cases with population pressure and net migration. Furthermore, the need for land use change in conditions causes the impact of a change in policy and excessive land [23]and it has been proved through the analysis of the state emperor in Peruvian Selva, Peru, the relationship between income and the cleanliness of the land is a positive identity but its Non monotonic [24]. When we look at history, especially the development of the coastal area in Indonesia in the past were more prevalent in the region -the coastal area [25]. It is also emphasized in detail again[5], which is most evident in the city's history-cities, such as Jakarta, Makassar, Semarang, and the cities along the Straits of Malacca and several other cities, especially areas with the purpose of transit cruise in trade center [19] but in the development of social change with the transformation of conflict society [26].This, then also makes an important impact on the development of the coastal area with the change of the social system [27,28], which leads to the use of land and its use for the benefit of society that sometimes leads to a major conflict in an area of urban coastal area green space.

3. Coastal Area Impact Based on Coastal Area Management

Territory coastal areas of coastal is a venue for a variety of environmental processes, economic and social activities of the rest of the container activity of waste. So that it becomes a waterfront coastal region as a whole have a variety of complex environmental problems [29, 30]. Makassar city coastal areas degraded environmental carrying capacity quite significantly which caused the uncontrolled use of space reclamation construction activities [31], as well as exacerbated with the condition of coastal geomorphology coastal areas prone to disaster risk and mitigated will result in damage to the coastal environment also affect the surrounding area [32].

Coastal area required integrated planning for the coastal area that dealt with how to develop the concept of disaster mitigation through appropriate zoning taking into account the level of risk and coastal area study area [33]. Other changes are also happening in Mindanao Region, Philippines, where the impact of coastal zoning against inland green infrastructures [34, 35]. Thus the need for a solution in describing, assessing, and evaluating each of these issues, so that the quality and utilization coastal area and open space land use around the region can be sustained and controlled [36].

4. Sustainable Coastal Area Land Use Based on Green Space Management

Development sustainability is determined by the potential availability of open space lands, this triggers the rapid development of the coastal area in line with the development of a physical environment that harms by the coastal area of Makassar [37, 38]. This is a linkage to climate and land-use patterns of land-use intensity, which in turn affects the quality of the green space quality and water infiltration. Consequently, the function occurs green space limitations in meeting the needs of society. Imbalances and the need for irrigation management towards the integration of sustainable development between the factors of green of quality, quantity, and distribution of open space availability of green space is increasingly limited, that was very worrying Makassar for reducing green space by about 20%. Therefore, there is a need for comprehensive regulation in addressing problems in the area of green space in coastal area, among others determining four factors that must be developed, namely: (1) Resource Potential Green space itself (2) Development of the Green Line (3) Recovery of Water Soil and (4) Utilization of Vacant land as Catchment area.

The present need to balance land use for human activities and open space is 70 : 30% [19]. The impact of land to potential users an open space on urban coastal areas can be measured based on population and need for open space functions that can be used as a component of urban ecosystem in the community. Use of open space land by Government regulation number: 63 of 2002 is to maintain the balance of the ecosystem which includes elements of the urban environment, plan social and cultural area that must be planned for approximately 25% of the total area [36], then it analyzed through special value of green open space [37].

The study analysis was residing in the coastal areas of urban coastal green space area of Tanjung Bunga, Makassar Mariso District, where degradation occurs and was influenced by two main factors, namely the pull-up and driven factors [38] as described in Figure 1.

By minimizing the factors driving through the realignment of open space land use patterns by the measurement criteria of land suitability using Geographical Information System (GIS), on each coast coastal areas green space, it can increase the potential availability of urban green space, It was confirmed by using Analytical methods for Benefit-Cost Ratio [36] that there is a need for expanding green space area in the coastal areas in the coming years.

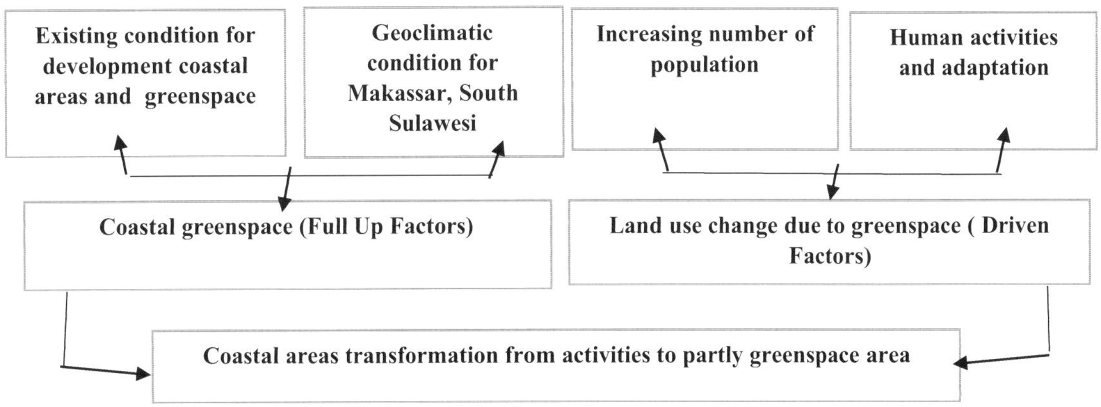

Figure 1 Driving pull up factors and driving factors in land use activities

5. Conclusion

There is a need for green space in coastal areas, especially coastal settlement is present. green space can reduce negative impacts coming from the environment such as sea pressure and activities of the settlement.

References

[1] H. Aalders. M.J. Aitkenhead, 2004, Agricultural census data landuse modeling, Computer, Environment and Urban Systems 30: 799-814.

[2] Ge Sun, Steven G, McNuly, Jennifer A. Moore and Erika C. Cohen, 2008. Impact of Climate Change, Population Growth, Landuse Change, and Groundwater Availability on Water Supply and Demand across the Conterminous U.S. AWRA Hydrology & Watershed Management Technical Committee Vol. 6 No.2.

[3] Daud Fery, 2008. Dampak Pembangunan Jalan Metro Tanjung Bunga Terhadap Komunitas Nelayan di Kota Makassar, Magister PascaSarjana UNDIP, Semarang.

[4] Singleton, Carleson and Berkers, 2005. Co-management Consepts and Methodological Implications, Journal of Environment Management 75,paper 05, pdf. Accesssed January 2012.

[5] Triyono dan Ahmad Arief, 2003. Modal Sosial Sebagai Mainstream Pengembangan Masyarakat Pesisir, Prosiding Lokakarya Nasional Menuju Pengelolaan Sumberdaya Wilayah Berbasis Ekosistem Untuk Mereduksi Potensi Konflik Antar Daerah, Fakultas Geografi UGM, Agustus 2003.

[6] Zhang L. 2002. An Evaluation on Urban Rifer front Park Sizekane, Washington Experincer and Lessons to Designer, Unpublished thesis, Washington State University, USA

[7] Warms, S, 2003.Konservasi Kawasan Pariwisata Kota Pantai (Obyek Kajian Wisata Wilayah Pesisir Galesong), Prosiding Lokakarya, Seminar Manajement Kota Pantai, UNHAS, Makassar, juli 2003.

[8] Nakayama Tadanobu, Shizuka Hashimoto. 2011. Analysis of the ability of water resources to reduce the urban heat island in the Tokyo megalopolis. Environmental Pollution 159 (2): 2164-2173.
[9] Benedit Notter, Lindsay MacMillan Daniel Vivoroli, Rolf Weingaetner, Hans Peter Uniger,2007. Impacts of environmental change on water resources in the Mt, Kenya region, Journal of Hydrology 343: 266-278.
[10] Angelo Doglioni, Francesca Primativo, Daniele Laucelli, Valeria Monno, Soon-ThainKhu, OrazioGiustolisi, 2009, An integrated modeling approach for the assessment of landuse change effect wastewater infrastructure Environmental, Modeling & Software 24: 1522 -1528.
[11] Daniel Anyalew Mangistu an Ayobami,T. Salami. 2007. Application of remote sensing and GIS in landuse/land cover mapping and change detection in a part of southwest in Nigeria. African Journal of Environmental Science and Technology 1 (5): 099-109.
[12] N. Banadda, I. Napi, and U.G. Wali, 2011. Determining and modeling the dispertion of non source pollutants in Lake Victoria: A Case Study of Gaba Landing site in Uganda. African Journal of Enviromental Science and Technology,5(3): 178-185.
[13] Ajibade, W. A, Ayodele, I.A. and Agbede, S.A. 2008. Water quality parameters in the major rivers of Kainji Lake National Park, Nigeria African Journal of Environmental Science and Technology 2 (7): 185-196.
[14] A. Sadoddin, V. Sheikh, R. Mostafa zadeh, M.Gh. Halili. 2010. Analysis of vegetation based management scenario using MCDM in the Ramian watershed, Golestan, Iran, International Journal of Plant Production 4(1): 51-62.
[15] Goemindari, Betty, Johan Silas, dan Rimadewi Supriharjo, 2010. Konsep Penataan Permukiman bantaran Sungai di Kota Banjarmasin, berdasarkan Masalah Budaya Setempat, Prosiding Seminar Nasional Perumahan/Permukiman dan Pembangunan Kota. JurusanArsitektur ITS, Maret 2010.
[16] Susanna T.Y. Young, Yu Sun, Thushara Ranatunga, Jie He, Y. Jeffrey Yang.2011. Predicting plausible impacts of sets of climate and landuse change scenario on water resources, Applied Geography 32: 477 – 489.
[17] T.Ilyas, Cf Leung, YK Chow, SS, Budi, 2004. Centrifugal Model Studi of Laterally Loaded File Groups in Clay, Journal of Geotechnical March, 2004, Vol.130, Nov.3,ISSN 1090-0241.
[18] Perz S. G, Aramburu C, and Bremner J. 2005. Population Land Use and Deforestation in the Pan Amazon Basin : Bolivia, A Comparison of Brazil, Bolivia, Columbia, Ecuador, Peru and Venezuela, Environment, Development and Sustainability, Vol.7,23-49.
[19] Barbier E.B, 2005, Natural resources and economic development, New York: Cambridge University Press.
[20] Zhane A.P.2007, Does Poverty Constrain Deforestation ? Econometric Evidence from Peru. Development Economic, Volume 84, pp.330-349.
[21] Madlazim, Santosa, BJ, and Utama, W,2008. The March 6, 2007, Earth quakes at Sianok Segment of the Sumatra Fault, Zone (Mw.max.63) Source Parameter and Identification of the Activated Fault Planes, Journal GEOFISIKA, Himpunan Ahli Geofisika Indonesia (HAGI),8(2).
[22] Veri Julianto, dan Dian Festianto, 2012. Pengetahuan Lingkungan dan Pengaruhnya Terhadap Interaksi Sosial, Politik Warga Kampung Code, Yogyakarta, Indonesia, Jurnal Suara Politika, Vol.12 No.13.

[23] Ferguson E. 2010 Adier's Innovative Contribution Regarding the Need to Belong, Journal of Individual Psychology, Vol.66 No.1

[24] Ali Mirchi, David Watkins, Jr, and Kaven, Madani, 2009, Modeling For Watershed Planning, Management, And Decision Making. Watersheds : Management, Restoration and Environmental, ISBN, 978-1-668-667.

[25] Inavonna, 2006, Thermal Confort, Waterfront and The Outdoor Utility Pattern, Case Study of Waterfront Leis me Area in Singapura, The 7thInternational Sustainable Environment and Architecture, November 20-21, (2006), Hasanuddin University, Makassar, Indonesia.,

[26] Langkoke, R. 2010. Morfo dinamika Pantai dan Prospek Sebaran Vegetasi Berdasarkan Sedimen Backshore Eustari Jeneberang, Makassar, Disertasi Program PascaSarjana, Universitas Hasanuddin, Makassar.

[27] Sasikumar K, and Mujundar P. 2000.Aplication of Fuzzy Probability in Water Quality Management of a River System. International Science Journal of System Science 31 (5),pp.575-592.

[28] Ferly S.P. 2001, Strategy Abroad towards Post Tsunami Rehabilitation and Rehabilitation and Conservation of The Srilangka Coast line (41), Proceeding The International Conference on Coastal Management, Institution of Civil and Health in Thomas telford, London on October 31, 2001.

[29] Buana G.M. and Mirabueno, 2001. The May it 1992, Event: Tsunami And Coastal Effect Eastern Mindanao, Philippines, Science of Tsunami Hazard The center Journal Tsunami 50 C,22(2): 61-68.

[30] Benedict M.A and McMahon E.T. 2001: Green Infrastructure Smart Conservation for The 21th Century, Sprawl Watch Clearing house Monograph Series.

[31] Saraswati, Isfa, 2003. Prinsip Perancangan Kawasan Tepi Air, Dalam Jurnal, Perencanaan Wilayah dan Kota, hal.95-118, PWK-ITB, Bandung.

[32] Buy Heart W, Rolando Calleri, Bert De Bievre, Felipe Cisneros, Guido Wyseure, Jozef Deckers, Robert Hofstede, 2006. Human Impact on the Hydrology of the Andreanparamos, Earth Science, Reviews 79: 53-72.

[33] Dodha Pradeep, and Ashvin K, Gosain, 2006. Impact on Watershed Management on livelyhoods: Quantization and Assessment, Landuse and Water Resources Research 8 (2008) 8,1-8.7.

[34] Omer Abden Mustafa, 2010, Water Resources Management and Sustainable Development in Sudan, International Journal of Water Resources and Environmental Engineering 2(8): 190-207.

[35] Francis P, 2001, Integrated Impact Assessment for Sustainable Development: A Case Study Approachch, Word Development Journal 29 (6): 1011-1024,

[36] Lyon J. 2000. An Analysis of Vegetation Communities the Lower Columbia River basin, Proceedings of the Pecora Symposium on Applications of Remote Sensing to Wildlife Management, Sioux Falla, pp. 321-327.

[37] Poernomohadi N, 2006. Tropical Urban Open Space, A key Factor for Sustainable Environmental, Presentation Proposal for The International, Seminar on Tropical Eco Settlement, Bali, Indonesia, November 14-16,2006 ISBN 918-9 74-849-344 age II, 142-147.

[38]Cho S, Pondyal N.C, and Roberts R.K, 2006. Spatial Analysis of The Equity Value of Green Open Space, International Journal of Ecological Economic, 66: 403-416.

AUTHOR INDEX

Institute of Physics
Dirac House, Temple Back
Bristol BS1 6BE UK

ISSN: 1757-8981
ISBN 978-1-7138-2672-9

CURRAN ASSOCIATES INC.
proceedings.com